AF781167

1

Biochemical Synthesis of Ag/AgCl Nanoparticles for Visible-Light-Driven Photocatalytic Removal of Colored Dyes

Xiyun Zhao [1,†], Jie Zhang [1,†,*], Binsong Wang [2,*], Amir Zada [3] and Muhammad Humayun [3]

[1] College of Life Science, Northeast Forestry University, Harbin 150040, China; E-Mail: zhaoxiyun1972@gmail.com

[2] Key Laboratory of Chemical Engineering Process & Technology for High-Efficiency Conversion, School of Chemistry and Material Sciences, Heilongjiang University, Harbin 150080, China

[3] Department of Chemistry, Abdul Wali Khan University Mardan, KPK, Mardan 23200, Pakistan; E-Mails: amirzada@awkum.edu.pk (A.Z.); humayunchem087@gmail.com (M.H.)

† These authors contributed equally to this work.

* Authors to whom correspondence should be addressed; E-Mails: zhangjie1972@nefu.edu.cn (J.Z.); wangbinsong@hlju.edu.cn (B.W.);

Academic Editor: Klara Hernadi

Abstract: Photocatalytic removal of organic pollution such as waste colored dyes was a promising technique for environment technique. However, effective photocatalysts were needed to enhance the photocatalytic efficiency. Ag/AgCl was regarded as high performance catalyst for photocatalytic degradation. Ag/AgCl nanoparticles were biochemically prepared with metabolin of living fungi which was used as reductant and characterized by X-Ray diffraction (XRD), UV-visible spectroscopy and transmission electron microscopy (TEM). The Ag/AgCl nanoparticle composites showed spherical aggregation shape with an average size of about 3–5 nm which is well inside the quantum regime. The UV-visible study showed that Ag/AgCl nanoparticles had strong visible light absorption and exhibited excellent visible-light-driven photocatalytic performance. Photocatalytic results indicated that the obtained Ag/AgCl nanoparticles were suitable for photocatalytic removal of RhB dye under visible light irradiation. The excellent photocatalytic activities could be attributed to the quantum size nanoparticles and the Plasmon resonance of Ag/AgCl composites.

Keywords: silver; silver chloride; plasmon resonance; photocatalysis; dye wastewater

1. Introduction

As the backbone for the economy of developed countries, dye industry plays significant role in spinning and other light industry, but brings some pollutions at the same time due to the improper discharge and other reasons [1]. The industrial effluents which contain some dyes cause much environment issues for their adverse effect on the aquatic and terrestrial life because of the water eutrophication by dyes and the presence of colored substances prevents the penetration of sunlight necessary for photosynthesis [2–4]. To solve this, a number of physical, chemical and biological methods have been used to remove pollutants from the industrial effluents but none of methods can be used to remove all the pollutants at the same time. Effective ways should be developed to solve this environment problem.

Color substances absorb visible radiation and can be effectively used as photosensitizers in photocatalytic reactions. And also, many photochemical reactions results in the degradation of organic compounds into CO_2 and water in the presence of a suitable conductive substrate. The use of an effective photocatalyst and the selection of optimum working parameters result in the complete removal of color substances. Photocatalytic technique is believed to be the most effective way to remove dye pollutants in water. A number of researchers have used TiO_2 and ZnO because their compounds have large band gaps, low cost, non-toxic and lower extensive recombination [5–7]. Moreover, exhaust TiO_2 could be reused for further degradation without any decrease in its efficiency because the pore structure of these compounds remains unaffected when the adsorbed dye is desorbed [8,9]. For example Mittal *et al.* [10] studied the UV degradation of Amaranth dye in the presence of H_2O_2 and TiO_2. Their experimental data revealed that the degradation efficiencies of UV, UV + H_2O_2, UV + TiO_2 and (UV + TiO_2 + H_2O_2) were 17%, 26%, 38% and 64%, respectively. Chung and Chen [11] investigated the photocatalytic decolourization of reactive violet 5 dye in the presence of TiO_2 at pH 4 and the efficiency of degradation was excellent at room temperature. However, these UV-light induced photocatalysts still have such shortcomings as the use of only 5% of the total visible light. Although the visible light driven photocatalysts such as CdS, Fe_2O_3, and WO_3 are also used to remove hazardous substances from drinking water [12,13]. However, the photocorrosion and the insufficient of reduction of them limited the extensive of utilization. Therefore, it is vital to develop effective visible-light photocatalytic materials to effectively improve the sunlight utilization [14].

The surface plasmon response of such noble metals as Au and Ag has been found to be excellent under visible light and exhibited high photocatalytic activity in a wide range of electromagnetic radiations [15]. Plasmon-induced Ag/AgX (Cl, Br, I) photocatalysts synthesized by ion-exchange, chemical precipitation and photo-assistant reduction are expensive and possess large particles. Prepared micrometer sized Ag/AgCl particles cause plasmon-induced electron-hole pairs to recombine before they reach the photocatalyst surface. The high recombination rate of charge carrier results in the decreased efficiency of a plasmonic photocatalytic system [16]. Owing to the rapid growth of bacterium, yeast, fungi and plants, biochemical synthesis is considered as a novel method for the bio-reduction of metal salts into nanomaterials [17–20]. Metal ions can easily be assimilated by micro-organisms and plants and converted into their respective metallic elements in a nano-regime size. The prepared nanoparticles can be collected using simple physical methods. Few works has been done on the preparation of nanomaterials using this method so far. However, Durán *et al.* [21–23] did lots of works about using biochemistry method to prepare nanomaterials. This inspires us to prepare Ag/AgCl nanoparticles by using our cultured *fungi*.

Based on the above consideration, we prepared Ag/AgCl nanoparticles aggregation composites with small and spherical size in the presence of *fungi* and used them effectively for the photocatalytic degradation of Rhodamine B. The synthesis of Ag/AgCl nanoparticles by this novel biochemical method are believed as a new and facile method to prepare nanoparticles.

2. Experimental Section

2.1. Preparation of Rose Bengal Medium

Rose Bengal medium was prepared by mixing of 5 g peptone, 10 g of glucose, 0.5 g of $MgSO_4 \cdot 7H_2O$, 20 g of agar, 0.03 g of Bengal rose, 0.1 g of chloramphenicol and 1.0 g of NaH_2PO_4 in 1 L of pure deionized water. The prepared Bengal medium was boiled for 1 min with constant stirring and then sterilized by autoclaving for 20 min.

2.2. Preparation of Potato Dextrose Agar

The Potato Dextrose Agar (PDA) medium was prepared by thoroughly washing 200 g of sliced potato with distilled water and then boiling it in 1 L of deionized water for 30 min. The prepared soup was filtered with cheesecloth and transferred into a volumetric flask with a volume of 1 L. Deionized water was added into the flask up to make a total volume of 1 L of suspension. 20 g each of sucrose and agar powder were then added into the soup solution. The prepared medium was then sterilized by autoclaving for 20 min.

The culture of fungi was grown in the soil taken from Mountain Maoer in Heilongjiang Province.

2.3. Screening of Fungal Strains and Identification of Fungi

10 g of soil was mixed with 90 mL of deionized water for preparation of fungal suspension. 1 mL of this suspension was taken and spread over the Rose Bengal medium. The fungal strain was obtained by incubating the rose Bengal medium doped with bacteria for 5 days at a constant temperature of 28 °C. The obtained strains were thoroughly purified with distilled water and then vaccinated to PDA medium with 0.5 mM of $AgNO_3$ solution and incubated for another 5 days at the same temperature of 28 °C. Finally, the fungal strains which could live in $AgNO_3$ solution were obtained and used to prepare Ag/AgCl nanoparticles. NaCl was used as an obligatory material in the nutrient solution of fungi because it provided Cl^- for the synthesis of AgCl.

The totals DNA of fungi were extracted using Fungal DNA Kit supplied by Omega Bio-Tek Company in China. The collected DNA was used as the template for Polymerase Chain Reaction (PCR) expansion. The primer used for PCR expansion of ITS1:5'-TCCGTAGGTGAACCTGCGG-3':ITS4:5'-TCCTCCGCTTATTGATATGC-3' was supplied by Shangon Biotech (Shanghai, China). The purified PCR product was sent to Huada Gene Company (Shenzhen, China) for Gene sequencing and the results were submitted to Genbank database. Blast searching software was used for the most similar Gene sequence through comparison. Software MEGA5.2 [24] was used to through phylogenetic analysis to construct the phylogenetic tree; HZQ-F160 orbital shaking incubator was supplied by Shanghai Yiheng Science and Technology Ltd. (Shanghai, China) and Gene Amp 9700 PCR master cycler was supplied by American Applied Bio-systems (Foster City, CA, USA).

2.4. Preparation of Ag/AgCl Nanoparticles

Freshly prepared active fungal strains were vaccinated to 300 mL of PDA medium with 7.5 mL of 2 mM NaCl solutions with a vaccination needle. The culture of fungi exposed to $AgNO_3$ and NaCl solutions was incubated for 5 days at 28 °C on an orbital shaking incubator at 120 rpm. The zymotic fluid was filtered with the help of filter paper. 200 mL of this fluid was mixed with 2 mL of 0.1 M $AgNO_3$ solution and then incubated for 48 h at 28 °C on an orbital shaking incubator at 120 rpm. Finally, the reaction solution was centrifuged at 14,000 rpm for 5 min to obtain Ag/AgCl nanoparticles. The prepared Ag/AgCl nanoparticles were washed four times with distilled water to remove the impurities absorbed on the surface of Ag/AgCl nanoparticles.

2.5. Characterization

The isolated Ag/AgCl nanoparticles were characterized by UV-Vis spectroscopy scanning (TU-1901, Beijing Puxi Inc., Beijing, China). The spectra were recorded in the wavelength range of 200 to 800 nm with an interval of 1 nm. The ultrathin sections of fungi were cut on LEICA ULTRACUT UCT instrument (Solms, Germany) and the TEM analysis of ultrathin sections was made on TECNAIG2 instrument (The Netherlands) operated at a voltage of 80 kV. The crystallographic structure of Ag/AgCl nanoparticles was investigated with X-ray diffractometer (Bruker D8, Karlsruhe, Germany) operated at a voltage of 40 kV in a scan range of 20°–80°.

2.6. Photocatalytic Test

The photocatalytic activity of the prepared Ag/AgCl nanoparticles for the photodegradation of RhB in an aqueous solution was evaluated by in a home-built reactor at ambient temperature (25 °C) under visible light irradiation. A 150 W Xe lamp with an IR filter was used as a solar light source. The visible light obtained by an optical filter ($\lambda > 420$ nm) was used to filter the UV light. The focused intensity on the flask was both ca. 10 mW/cm^{-2}. The reactor was open to the ambient air in order to reach the air-equilibrated condition. The concentration of RhB is 5 mg/L. For each run, 0.1 g of photocatalyst powders was dispersed in 20 mL of 5 mg/L RhB aqueous solution. Prior to the beginning of irradiation, the mixture was allowed to equilibrate in the dark with stirring for 30 min. Absorbance of irradiated samples was determined immediately after irradiation and removal of the powders by centrifugation and filtration. The photocatalytic performance of the obtained composite was also compared with N-TiO_2 sample. The N-TiO_2 was prepared by calcined commercial anatase TiO_2 particles (Aladdin) in NH_3 at 400 °C for 2 h. The photodegradation of RhB was determined by UV-vis absorbance at $\lambda_{max} = 552$ nm.

3. Results and Discussion

3.1. Identification of Fungi

The as-prepared *fungi* used for our present work were separated and cultured by ourselves in ways which were not used by any other researcher so far. In order to identify the fungi of us, the cultured pure fungal strains were accredited against the commonly used standard [25]. The fungal strains were PCR expanded by using the DNA of fungal strain genome as the template. The ITS serial number of our fungal strains was 558 bp in contrast to all the serial numbers in GenBank as determined using BLAST program to find the nucleotide homology and the similar base sequence. The ITS number of known strain with high homology was chosen and the ITS number of our strain's was used to construct the phylogenetic tree with the help of software MEGA5.2. 1000 times bootstrap was done to inspect the reliability of the phylogenetic tree. The strains of NYZJ03 formed the same ethnic groups with *Trichoderma hamatum* and the homology went up to 99.5% as shown in Figure 1. Our obtained fungus was considered to be *Trichoderma hamatum* from the result of morphological characteristics, homology of serials and the phylogenetic tree of fungal strains

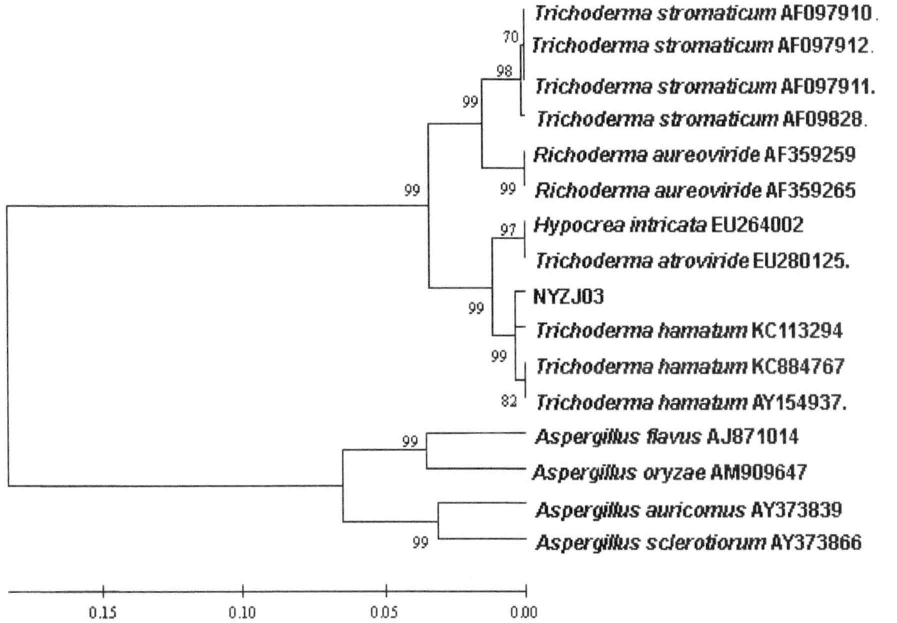

Figure 1. Phylogenetic tree of obtained fungal strains.

3.2. XRD Analysis

Ag/AgCl nanoparticles were characterized using X-ray diffractometer (Bruker D8, Karlsruhe, Germany) with Cu-Kα radiation operated at a voltage of 40 kV in the scan range of 20 to 80 degree. As shown in Figure 2. The characteristic XRD peaks at 38.2°, 44.6° and 75.2° were attributed to (111), (200), and (311) reflection planes for cubic Ag (JCPDS No. 65-2871). The clear and dominant eight peaks at 27.8°, 32.2°, 46.2°, 54.9°, 57.6°, 67.4°, 74.5° and 76.7° were attributed to planes (111), (200), (220), (311), (222), (400), (331), and (420) of cubic phase of AgCl crystal (JCPDS No. 31-1238). No other characteristic peaks could be attributed to impurities which indicated the high purity of Ag/AgCl. The strong and narrow diffraction peaks reveal the highly crystalline [26]. The particle size of

Ag and AgCl is calculated by Scherrer equation d = Kλ/βcosθ, which d is the mean size of the ordered (crystalline) domains; K is a dimensionless shape factor, with a value close to unity. The shape factor has a typical value of about 0.9, but varies with the actual shape of the crystallite; λ is the X-ray wavelength; β is the line broadening at half the maximum intensity (FWHM), after subtracting the instrumental line broadening, in radians. This quantity is also sometimes denoted as Δ(2θ); θ is the Bragg angle of the main crystal plane. We chose (111) of Ag and (200) of AgCl to calculate the particle size. The calculated particle size is 5.27 and 4.53 nm for Ag and AgCl, respectively.

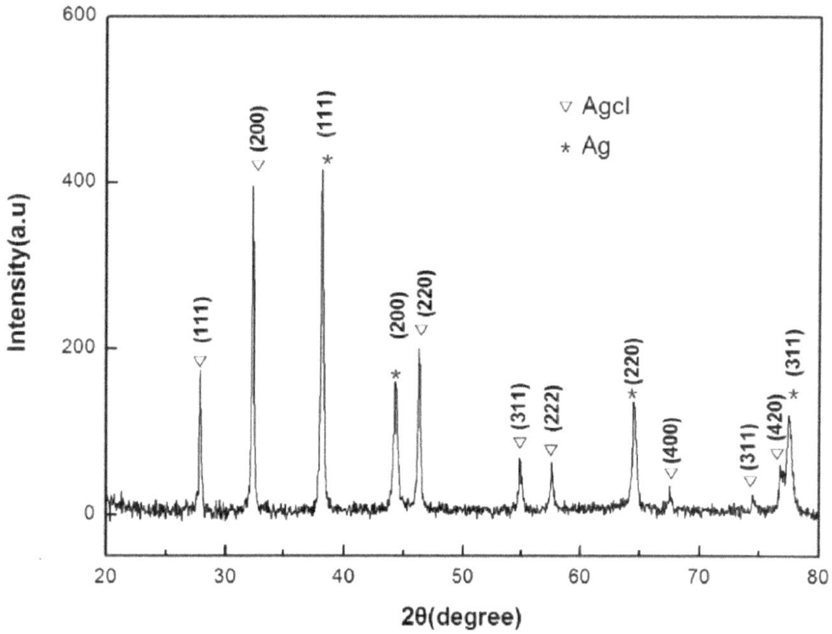

Figure 2. X-ray diffraction (XRD) pattern of Ag/AgCl nanoparticles.

3.3. TEM and HRTEM of Ag/AgCl Nanoparticles

The living fungi could produce some substances which have the property to reduce Ag^+ to Ag^0 nanoparticles. Durán *et al.* [22,23] has been proven that the free *cysteine* produced by *laccase* could be as a reducing group to reduce Ag^+. Thus, in this work, we believe that the reductant of Ag nanoparticles formation are *cysteine* and other reductive amino acids. Moreover, NaCl and chloramphenicol in the form of electrolytes are used as source of chlorine for the preparation of AgCl. Fresh AgCl and Ag nanoparticles were mixed to form Ag/AgCl composites. The Ag/AgCl composites were regarded as very good plasmon response materials for photocatalysis. The external morphology of Ag/AgCl nanoparticles was studied with TECNAIG2 instrument (Philips, Eindhoven, The Netherlands) at 80 kV. TEM and HRTEM images of Ag/AgCl composites were shown in Figure 3. TEM image in Figure 3A shows that the nanoparticles are almost spherical with approximately 3–5 nm in size and these nanoparticles aggregate with each other. This is consisting with the result by Scherrer equation from XRD peaks (5.27 and 4.53 nm for Ag and AgCl, respectively). HRTEM image in Figure 3B further investigated the Ag/AgCl composite material. The lattice fringes corresponding to (111) (d111 = 0.24 nm) were the clear crystallographic planes of Ag. The lattice fringes corresponding to (200) (d200 = 0.28 nm) were attributed to crystallographic planes of AgCl. The closed and compact interface interaction of Ag and AgCl could also be observed from HRTEM image. TEM-EDS further indicated the formation of

Ag/AgCl junction composite, as shown in Figure 4, that Cl dispersed uniformly in the composite. The content of AgCl in the composite is about 12 mol% from EDS. This result clearly shows that the Ag/AgCl heterogeneous junctions are formed in the composite material which favor the charge transport and separation and improve the photocatalytic activity of Ag/AgCl nanomaterials [27].

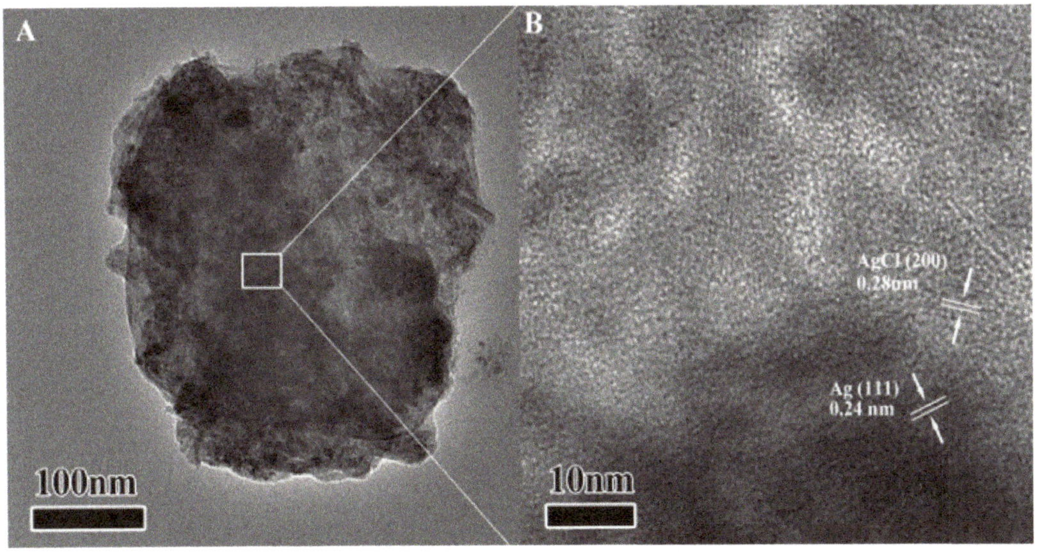

Figure 3. (**A**) Transmission electron microscopy (TEM) and (**B**) high-resolution transmission electron microscopy (HRTEM) images of Ag/AgCl nanoparticles.

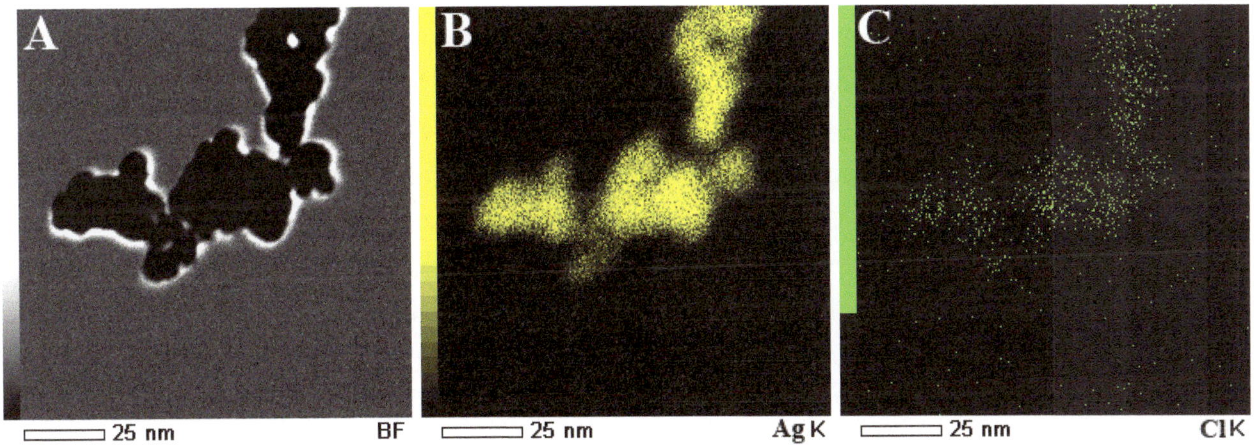

Figure 4. TEM-Energy Dispersive Spectroscopy (EDS) mapping of Ag/AgCl composites. (**A**) TEM image; (**B**) Mapping of Ag; (**C**) Mapping of Cl.

3.4. UV–vis. Diffuse Reflectance Spectra

It could be seen from Figure 5, there were strong absorption in both ultraviolet and visible-light regions of electromagnetic radiations. The absorption band between 200 and 350 nm could be attributed to the characteristic absorption of AgCl composite which possesses a direct band gap of 5.6 eV and an indirect band gap of 3.25 eV. The strong absorption band in the visible-light region was attributed to the surface plasmon resonance of Ag nanoparticles. Due to the small size of silver nanoparticles in the quantum regime, the sample exhibited a stronger absorption at 400–700 nm. The strong absorption of visible light was responsible for visible-light-driven photocatalytic activity of Ag/AgCl.

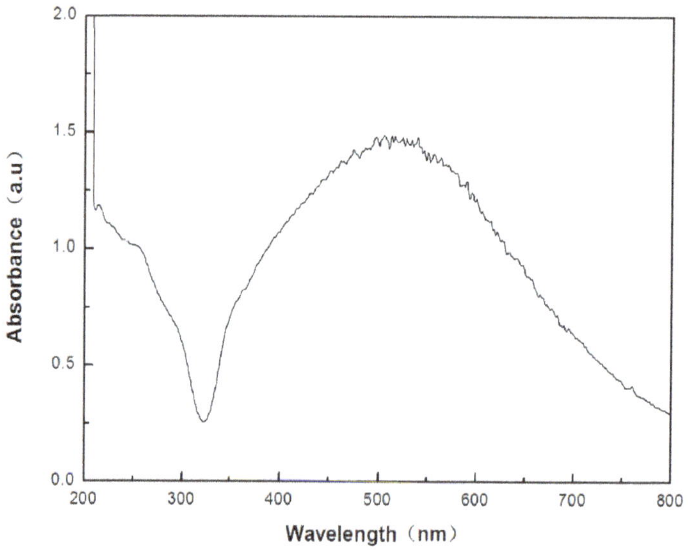

Figure 5. UV-visible absorption spectra of Ag/AgCl nanoparticles.

3.5. Photocatalytic Activity

Rhodamine B (RhB) was a carcinogenic chemical dye and it was often used as a tracer to determine the rate and direction of water flow [28–30]. RhB was used as a common modal to evaluate the performance of photocatalysts. The photocatalytic activities of Ag/AgCl nanoparticles for degradation of RhB were shown in Figure 6, in which C_0 is the equilibrium concentration of RhB at the equilibrium adsorption state and C was the concentration of RhB after visible light irradiation. Prior to visible-light irradiation, RhB and Ag/AgCl solutions were kept in the dark for 30 min to obtain the equilibrium adsorption state. When RhB solution was exposed to the visible light for 30 min, there was no apparent change could be observed in the concentration of RhB. The concentration of the RhB solution was slightly decreased. However, when Ag/AgCl nanoparticles was mixed and then exposed to visible light under the same conditions, excellent photocatlaytic performance was investigated 96% of RhB dye was decomposed, which was much higher than the reported work [31] and that of N-TiO_2 (in comparison to what is reported in reference [8]). The rate constant of N-TiO_2 and Ag/AgCl shown in Figure 6B is −0.01222 and −0.10519 min^{-1}, respectively. The excellent photocatalytic performance was mainly attributed to the closed junctions and the synergistic plasmon response of Ag and AgCl.

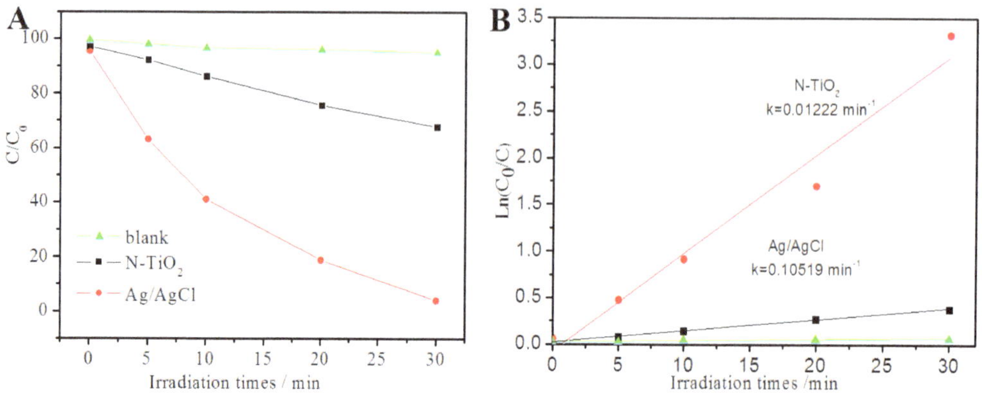

Figure 6. (**A**) Photodecomposition and rate of photodecomposition of RhB dyes over Ag/AgCl composites, N-TiO_2 and (**B**) the blank experiment without photocatalysts.

4. Conclusions

Summarily, it could be seen from the presentation above that Ag/AgCl nanoparticle composites could be biochemically prepared with the help of living fungi. The obtained Ag/AgCl nanoparticle composites showed excellent morphology and spherical shape with an average size of about 3–5 nm. The UV-visible study showed that Ag/AgCl nanoparticles have strong visible light absorption and exhibit excellent visible-light-driven photocatalytic performance. This property was utilized in the degradation of carcinogenic of RhB that after 30 f exposure to sun light the degradation rate was almost 96% indicating the good property. This biochemical method of synthesizing nanomaterials opens a new gate for the strategies and applications of nanoparticle composite materials.

Acknowledgments

This work was funded by National Science Foundation for Fostering Talents in Basic Research of National Natural Science Foundation of China (Grant No. J1210053), Natural Science Foundation of Heilongjiang Province of China (B201415), Project of Education Department of Heilongjiang Province of China (12541619) and Heilongjiang Postdoctoral Funds for Scientific Research Initiation (LBH-Q09184, LBH-Q11011).

Author Contributions

Experimental design was planned by Jie Zhang and Binsong Wang. Jie Zhang, Xiyun Zhao, Amir Zada and Muhammad Humayun did the experimental work and analysis of results. All the authors participated in the writing of the manuscript.

Conflicts of Interest

The authors declared no conflict of interest.

References

1. Rafatullah, M.; Sulaiman, O.; Hashim, R.; Ahmad, A. Adsorption of methylene blue on low-cost adsorbents: A review. *J. Hazard. Mater.* **2010**, *177*, 70–80.
2. Forgacs, E.; Cserhati, T.; Oros, G. Removal of synthetic dyes from wastewaters: A review. *Environ. Int.* **2004**, *30*, 953–971.
3. Robinson, T.; McMullan, G.; Marchant, R.; Nigam, P. Remediation of dyes in textile effluent: A critical review on current treatment technologies with a proposed alternative. *Bioresour. Technol.* **2001**, *77*, 247–255.
4. Alinsafi, A.; Evenou, F.; Abdulkarim, E.M.; Pons, M.N.; Zahraa, O.; Benhammou, A.; Yaacoubi, A.; Nejmeddine, A. Treatment of textile industry wastewater by supported photocatalysis. *Dyes Pigments* **2007**, *74*, 439–445.
5. Qu, Y.; Zhou, W.; Pan, K.; Tian, C.; Ren, Z.; Dong, Y.; Fu, H.G. Hierarchical anatase TiO_2 porous nanopillars with high crystallinity and controlled length: An effective candidate for dye-sensitized solar-cells. *Phys. Chem. Chem. Phys.* **2010**, *12*, 9205–9212.

6. Tian, C.; Li, W.; Zhang, Q.; Pan, K.; Fu, H.G. Controllable fabrication of various ZnO micro/nanostructures from a wire-like Zn–EG–AC precursor via a facile solution-based route. *Mater. Res. Bull.* **2011**, *46*, 1283–1289.
7. Tian, G.; Chen, Y.; Zhou, W.; Pan, K.; Dong, Y.; Tian, C.; Fu, H. Facile solvothermal synthesis of hierarchical flower-like Bi_2MoO_6 hollow spheres as high performance visible-light driven photocatalysts. *J. Mater. Chem.* **2011**, *21*, 887–892.
8. Chen, X.; Mao, S.S. Semiconductor-based photocatalytic hydrogen generation. *Chem. Rev.* **2007**, *107*, 2891–2959.
9. Asahi, R.; Morikawa, T.; Ohwaki, T.; Aoki, K.; Taga, Y. Visible-light photocatalysis in nitrogen-doped titanium oxides. *Science* **2001**, *293*, 269–271.
10. Gupta, V.K.; Jain, R.; Mittal, A.; Saleh, T.A.; Nayak, A.; Agawal, S.; Sikarwar, S. Photo-catalytic degradation of toxic dye amaranth on TiO_2/UV in aqueous suspensions. *Mater. Sci. Eng. C* **2012**, *32*, 12–17.
11. Chung, Y.C.; Chen, C.Y. Degradation of azo dye reactive violet 5 by TiO_2 photocatalysis. *Envir. Chem. Lett.* **2009**, *7*, 347–352.
12. Hwang, D.W.; Kim, J.; Park, T.J.; Lee, J.S. Mg-doped WO_3 as a novel photocatalyst for visible light-induced water splitting. *Catal. Lett.* **2002**, *80*, 53–57.
13. Hu, C.; Hu, X.X.; Wang, L.S.; Qu, J.H.; Wang, A.M. Visible-light-induced photocatalytic degradation of azodyes in aqueous AgI/TiO_2 Dispersion. *Environ. Sci. Technol.* **2006**, *40*, 7903–7907.
14. Chen, X.; Liu, L.; Yu, P.Y.; Mao, S.S. Increasing solar absorption for photocatalysis with black hydrogenated titanium dioxide nanocrystals. *Science* **2011**, *331*, 746–750.
15. Xin, B.; Jing, L.; Ren, Z.; Wang, B.; Fu, H. Effects of simultaneously doped and deposited Ag on the photocatalytic activity and surface states of TiO_2. *J. Phys. Chem. B* **2005**, *109*, 2805–2809.
16. Bi, Y.; Ye, J. *In situ* oxidation synthesis of Ag/AgCl core–shell nanowires and their photocatalytic properties. *Chem. Commun.* **2009**, *43*, 6551–6553.
17. Ahmad, A.; Mukherjee, P.; Senapati, S.; Mandal, M.; Khan, M.I.; Kumar, R.; Sastry, M. Extracellular biosynthesis of silver nanoparticles using the fungus Fusarium oxysporum. *Colloids Surf. B Biointer.* **2003**, *28*, 313–318.
18. Jain, N.; Bhargava, A.; Majumdar, S.; Tarafdar, J.C.; Panwar, J. Extracellular biosynthesis and characterization of silver nanoparticles using Aspergillus flavus NJP08: A mechanism perspective. *Nanoscale* **2011**, *3*, 635–641.
19. Acevedo, F.; Pizzul, L.; Castillo, M.P.; González, M.E.; Cea, M.; Gianfreda, L.; Diez, M.C. Degradation of polycyclic aromatic hydrocarbons by free and nanoclay-immobilized manganese peroxidase from Anthracophyllum discolor. *Chemosphere* **2012**, *80*, 271–278.
20. Liu, X.L.; Zhu, P.X.; Gao, Y.F.; Jin, R.H. Polyamine-promoted growth of one-dimensional nanostructure-based silica and its feature in catalyst design. *Materials* **2012**, *5*, 1787–1799.
21. Durán, N.; Marcato, P.D.; Souza, G.; Alves, O.L.; Esposito, E. Antibacterial effect of silver nanoparticles produced by fungal process on textile fabrics and their effluent treatment. *J. Biomed. Nanotechnol.* **2007**, *3*, 203–208.

22. Durán, N.; Marcato, P.D.; Durán, M.; Yadav, A.; Gade, A.; Rai, M. Mechanistic aspects in the biogenic synthesis of extracellular metal nanoparticles by peptides, bacteria, fungi, and plants. *Appl. Microbiol. Biotechnol.* **2011**, *90*, 1609–1624.
23. Durán, N.; Cuevas, R.; Cordi, L.; Rubilar, O.; Diez, M.C. Biogenic silver nanoparticles associated with silver chloride nanoparticles (Ag@AgCl) produced by laccase from Trametes versicolor. *Springer Plus* **2014**, *3*, 645.
24. Tamura, K.; Peterson, D.; Peterson, N.; Stecher, G.; Nei, M.; Kumar, S. MEGA5: Molecular evolutionary genetics analysis using maximum likelihood, evolutionary distance, and maximum parsimony methods. *Mol. Biol. Evol.* **2011**, *28*, 2731–2739.
25. Cheung, Y.C.; Liu, X.X.; Wang, W.Q.; Wu, J.Y. Ultrasonic disruption of fungal mycelia for efficient recovery of polysaccharide–protein complexes from viscous fermentation broth of a medicinal fungus. *Ultrason Sonochem.* **2015**, *22*, 243–248.
26. Wang, P.; Huang, B.; Lou, Z.; Zhang, X.; Qin, X.; Dai, Y.; Zheng, Z.; Wang, X. Synthesis of highly efficient Ag@AgCl plasmonic photocatalysts with various structures. *Chem. Eur. J.* **2010**, *16*, 538–544.
27. Wang, P.; Huang, B.; Qin, X.; Zhang, X.; Dai, Y.; Wei, J.; Whangbo, M.H. Ag@AgCl: A highly efficient and stable photocatalyst active under visible light. *Angew. Chem. Int. Ed.* **2008**, *47*, 7931–7933.
28. Ferrari, C.; Chen, H.; Lavezza, R.; Santinelli, C.; Longo, I.; Bramanti, E. Photodegradation of Rhodamine B Using the Microwave/UV/H_2O_2: Effect of Temperature. *Int. J. Photoenergy* **2013**, *2013*, 1–12.
29. Yang, S.; Huang, Y.; Wang, Y.; Yang, Y.; Xu, M.; Wang, G. Photocatalytic degradation of Rhodamine B with $H_3PW_{12}O_{40}/SiO_2$ sensitized by H_2O_2. *Int. J. Photoenergy* **2012**, *2012*, 1–6.
30. Luan, J.; Xu, Y. Photophysical property and photocatalytic activity of new Gd_2InSbO_7 and Gd_2FeSbO_7 compounds under visible light irradiation. *Int. J. Mol. Sci.* **2013**, *14*, 999–1021.
31. Xu, H.; Li, H.; Xia, J.; Yin, S.; Luo, Z.; Liu, L.; Xu, L. One-pot synthesis of visible-light-driven plasmonic photocatalyst Ag/AgCl in ionic liquid. *ACS Appl. Mater. Int.* **2011**, *3*, 22–29.

2

Medium-Term Function of a 3D Printed TCP/HA Structure as a New Osteoconductive Scaffold for Vertical Bone Augmentation: A Simulation by BMP-2 Activation

Mira Moussa [1], Jean-Pierre Carrel [2], Susanne Scherrer [1], Maria Cattani-Lorente [1], Anselm Wiskott [1] and Stéphane Durual [1,*]

[1] Division of Fixed Prosthodontics and Biomaterials, University of Geneva, University Clinics of Dental Medicine, 19, rue Barthélemy-Menn, Geneva 1205, Switzerland; E-Mails: mira.moussa@unige.ch (M.M.); susanne.scherrer@unige.ch (S.S.); maria.cattani-lorente@unige.ch (M.C.-L.); anselm.wiskott@unige.ch (A.W.)

[2] Department of Maxillofacial and Oral Surgery, Division of Oral and Maxillofacial Pathology (HUG), University Clinics of Dental Medicine, 19, rue Barthélemy-Menn, Geneva 1205, Switzerland; E-Mail: jean-pierre.carrel@unige.ch

* Author to whom correspondence should be addressed; E-Mail: stephane.durual@unige.ch;

Academic Editor: Naozumi Teramoto

Abstract: Introduction: A 3D-printed construct made of orthogonally layered strands of tricalcium phosphate (TCP) and hydroxyapatite has recently become available. The material provides excellent osteoconductivity. We simulated a medium-term experiment in a sheep calvarial model by priming the blocks with BMP-2. Vertical bone growth/maturation and material resorption were evaluated. **Materials and methods:** Titanium hemispherical caps were filled with either bare- or BMP-2 primed constructs and placed onto the calvaria of adult sheep (n = 8). Histomorphometry was performed after 8 and 16 weeks. **Results:** After 8 weeks, relative to bare constructs, BMP-2 stimulation led to a two-fold increase in bone volume (Bare: 22% ± 2.1%; BMP-2 primed: 50% ± 3%) and a 3-fold decrease in substitute volume (Bare: 47% ± 5%; BMP-2 primed: 18% ± 2%). These rates were still observed at 16 weeks. The new bone grew and matured to a haversian-like structure while the substitute material resorbed via cell- and chemical-mediation. **Conclusion:** By priming the 3D construct with BMP-2, bone metabolism was physiologically accelerated, that is, enhancing vertical bone growth and maturation as well as material bioresorption. The scaffolding function of the block was maintained, leaving time for the bone to grow and mature to a haversian-like structure. In parallel, the material resorbed via cell-mediated and chemical processes. These promising results must be confirmed in clinical tests.

Keywords: bone substitute block; 3D-printing; BMP-2; animal experiments; guided tissue regeneration; bone regeneration

---

1. Introduction

The contemporary approach to regeneration is using resorbable matrices as scaffolds to guide the growth of new tissue. In the initial stages, the scaffold induces and guides cell growth and development. Later it supports the newly formed tissue while progressively resorbing. Ideally, the products of the resorptive process would be incorporated as natural components into the neosynthesized tissue.

Within the specific domain of bone reconstruction, the optimal matrix is "osteopromotive", that is, it functions as a pattern for osteoconduction, releases growth factors and provides scaffolding to maintain the volume in which osteogenesis will take place [1]. At the cellular level, the scaffold is perceived as a form of extracellular matrix which promotes both cell migration and stabilization in a controlled and physiologically effective manner. To this effect, the scaffold should present a moderate (2–3 μm) surface roughness to permit cell adhesion and guidance [2,3]. At the macroscale level (0.1–0.4 mm range), the scaffold must be porous (about 60 vol%) so that it can be penetrated by cells and new vessels. Further, the pores must be interconnected for the osteoconductive process to spread into the depth of the scaffold [4,5]. Finally, the scaffold should resorb at a rate that harmonizes with the formation and maturation of new bone. In its superlative form, the scaffold would also support physiological loads in the early stages of the process. These loads would then be transferred to the newly formed bone as it matures while the scaffold disappears.

Calcium orthophosphates bioceramics, e.g. hydroxyapatite (HA), α- or β-tricalciumphosphates (α- or β-TCP) are chemically and structurally similar to the natural components of bone and teeth [6]—a resemblance which provides them with excellent osteoconductive properties [7]. As their rate of resorption can be adjusted by adjusting the Ca/P and HA/TCP ratios [8], calcium phosphates enter into the composition of many synthetic bone substitutes. A most prominent representative in this category is biphasic calcium phosphate (BCP)—a combination of HA and TCP formulated as an optimum balance between the most stable form of HA and most soluble form of TCP [9]. Recently, a 3D-printed construct (OsteoFlux®) made of BCP has been developed to form a regular arrangement of orthogonally layered cylindrical filaments with inner channels of 250 μm. In a sheep calvarial model, the osteoconductivity of this construct was 2–4 times greater 8 weeks after implantation than particulate bone substitutes of natural- (bovine bone) or synthetic (β-TCP) origin [10]. At 16 weeks, bone formation had largely evened out for all three materials. Still, four month was a duration that was sufficient to evaluate the potential for osteoconductivity, but insufficient for the bone to fully mature and the material to resorb.

Bone morphogenetic protein 2 (BMP-2) is a potent activator of bone metabolism [11]. Therefore, our working hypothesis was that adding BMP-2 to the constructs would accelerate bone metabolism and thus permit an investigation of bone maturation and remodeling as well as material resorption within the 16 weeks of the study. Our objective, therefore, was to histologically and histomorphometrically evaluate bone maturation on OsteoFlux® constructs and their resorption after an initial stimulation with BMP-2 in a model of guided bone regeneration in sheep.

2. Materials and Methods

2.1. Experimental Design

This experiment was a subproject of a larger study conducted on 12 sheep in which the osteoconductive performance of 3D printed constructs was compared to two other bone substitutes using titanium caps placed on sheep skulls [12]. Six locations were available (occipital left and right, median left and right, frontal left and right). Six randomly allotted titanium hemispheres filled with (i) 3D printed constructs; (ii) 3D printed constructs + BMP-2; (iii) particulate bone substitute of bovine origin; (iv) particulate bone substitute of synthetic origin or (v) plain blood coagulum (negative control) were placed on the calvarium of each animal. Randomization was obtained by drawing lots with the following constraints. (i) Each sheep was to receive at least one of the three substitutes but (ii) no sheep could be fitted with more than two samples of the same substitute; (iii) each of the three substitutes was to be placed at least once in one of the six positions; (iv) the four controls (no substitute) had to be placed on different animals and (v) distributed over the occipital, the median and the frontal sites. Six animals were euthanized after 8 weeks and six after 16 weeks. The skulls were block-sectioned and subjected to histomorphometric analysis. Outcomes were analyzed in terms of the relative volumes of bone substitute and new bone tissue.

The data on the synthetic construct and the particulate bone substitutes have been published elsewhere [10]. The following report specifically addresses the response of the synthetic construct with and without BMP-2.

Samples

The following conditions were evaluated (Table 1):

1. **3D-printed constructs** (OsteoFlux®, Vivos-Dental, CH), **no BMP added**, serving as baseline, $n = 8$. These constructs are made of orthogonally layered cylindrical filaments, 350 μm in diameter separated by spaces 250 μm in width. Upon printing, the filaments are deposited as a high viscosity paste which progressively sets to a hard consistency. The paste is synthetic calcium phosphate with a calcium-to-phosphate ratio of 1.43 (range: 1.35 to 1.5). Calcium phosphates are present in the form of α-TCP and microcrystalline, calcium deficient HA. The constructs' macroporosity ranges between 40% and 50% and the total porosity between 50% and 65% [10].

2. **3D-printed block, BMP-2 added**, $n = 8$. Recombinant human BMP-2 (Inductos®, Medtronic, USA) was prepared under sterile conditions according to the manufacturer's specifications (1.5 mg/mL). At surgery it was spread as a single dose of 100 μg onto the 3D-printed block. After complete absorption, autologous blood was added prior to placement of the titanium hemisphere onto the skull.

For the sake of simplicity, plain OsteoFlux® specimens and OsteoFlux® + BMP-2 specimens will be referred to as OF-plain and OF-BMP-2 hereafter.

Table 1. Specimens analyzed.

Healing period	Samples	Number of specimen analyzed (n)
	Coagulated blood	4
8 weeks	3D-printed block (OF-plain)	8
	3D-printed block + BMP-2 (OF-BMP-2)	8
	Coagulated blood	4
16 weeks	3D-printed block (OF-plain)	8
	3D-printed block + BMP-2 (OF-BMP-2)	8
Total	Total placed: 40	Total analyzed: 40

2.2. Hemispheres and Placement of Bone Substitutes

The hemispheres were machined out of titanium (cpTi-gr2) with an inner diameter of 10 mm, an outer diameter of 11 mm and a height of 5 mm. OF constructs were printed to a perfect fit inside the hemispheres (Figure 1A) and were gamma sterilized before use.

At the time of surgery, the OF constructs were placed into the hemispheres and impregnated with autologous blood. BMP-2 was added to the test specimens. The negative control spheres were merely filled with blood (no construct). The hemispheres were placed onto the skull immediately after being filled.

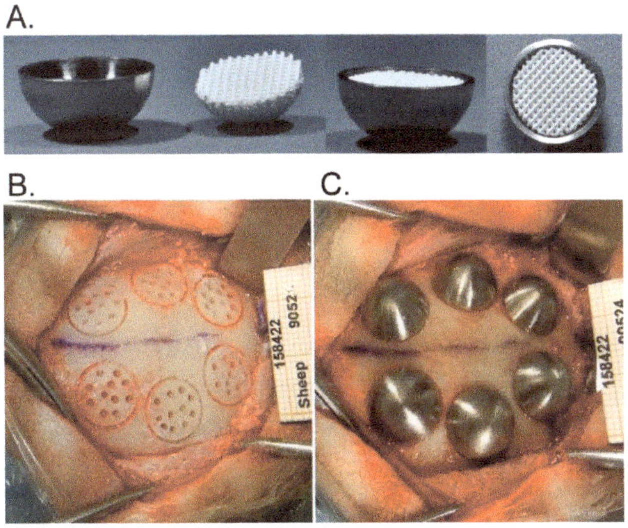

Figure 1. (**A**) Titanium hemispheres filled with 3D printed construct; (**B**) circular grooves with transcortical perforations prior to the fixation of the hemispheres filled with bone substitute (**C**).

2.3. Animals

Twelve adult BMC sheep (female, 2.5–4 years, 69–87 kg) were included in the study (Eymin Breeding, Seyssuel, France). The animals underwent an acclimation period of one week prior to surgery. The experiment was conducted in a dedicated facility (NAMSA, Chasse sur Rhône,

F (ISO/CEI 17025) [13]). In line with European requirements (European Directive 2010/63/EU [14]), all animal experiments were approved by veterinary committees (NAMSA Ethical Committee and French Ministry of Agriculture).

2.4. Surgical Procedure

Placement of the titanium caps was performed under general anesthesia. The animals were deprived of food (24 h) and water (12 h) prior to surgery to prevent vomiting. A prophylactic antibiotic coverage was dispensed 1 day before and up to 2 weeks after the surgery (amoxicillin-Duphamox, Pfizer, Fort Dodge, IA, USA, i.m. every 2 days (15 mg/kg) and enrofloxacin-Baytril 5%, Bayer Pharma, Leverkusen, Germany, s.c. daily (5 mg/kg)). The animals first received an analgesic treatment (flunixine-Meflosyl i.m. Pfizer/Fort Dodge, NY, USA (2 mg/kg), and buprenorphine-Buprecare s.c. (0.01 mg/kg), Animalcare, York, England). Anesthesia was induced by an intravenous injection of a mixture of thiopental-Nesdonal (750 mg, Merial, Lyon, France)—Pentobarbital sodium (273 mg, CEVA, Libourne, France)—Atropine sulfate (1 mg, Aguettant, Lyon, France) before intubation. Deep anesthesia was obtained using 2% isoflurane-Aerrane (Baxter, Deerfield, IL, USA) in pure oxygen. A rectal temperature probe and a rumen tube were placed. Heart function, temperature and oxygen saturation were monitored. Iso-volumetric conditions were maintained by infusion of a ringer lactate solution during the entire procedure.

The surgical areas were shaved and the skin was scrubbed with povidone iodine (Vetoquinol, Magny-Vernois, France), wiped with isopropyl alcohol, painted with povidone iodine and draped. First a midline incision was made through the skin, from the orbits to the external occipital protuberance. The temporalis muscles were elevated subperiosteally from the frontal and parietal bone and bilaterally retracted. On each side of the median suture, 3 circumferential grooves, 10 mm in diameter and approximately 0.5 mm in depth, were trephined under saline irrigation. In each of the resulting circles 11 holes that penetrated the medulla (1 mm in diameter, *ca.* 2 mm in depth) were drilled with a round bur to allow bone cell migration from the marrow to the surface. Thus in each animal, 6 grafting sites made of a cortical bone plate penetrated by 11 transcortical perforations were produced (Figure 1B). Primary stability of the hemispheres was ensured by the clipping effect of the titanium caps onto the bone (Figure 1C). Each animal received at least one of each bone substitute. Wound closure was carried out in 3 planes. The deeper tissues were closed using a discontinuous resorbable suture (Vicryl 3-0, Ethicon, Somerville, NJ, USA). A povidone iodine solution was applied before closing the subcutaneous tissues with the resorbable suture material. The skin was closed with a continuous non-resorbable suture (Prolene 3-0, Ethicon, Somerville, NJ, USA) and disinfected with a spray of oxytetracyclin (Oxytetrin, Summit, NJ, USA).

Post-operative pain was minimized by subcutaneous injections of buprenorphine (Buprecare, Animalcare, UK) 0.01 mg/kg, twice daily for 2 days. Inflammation was controlled for 7 days by daily intramuscular injections of flunixine (Meflosyl, Zoetis, Florham Park, NJ, USA) 2 mg/kg. The sutures were removed after complete healing of the skin. The wounds were disinfected until 2 days thereafter.

After sacrifice, the skulls were dissected and immersed in neutral buffer and 10% formalin.

2.5. Histological Preparation, Histopathologic and Histomorphometric Analysis

After complete fixation, the skulls were block-sectioned to isolate each hemisphere. The blocks were (i) rinsed for 3 h with tap water, (ii) dehydrated in alcohol solutions of increasing concentration; (iii) cleared in xylene and (iv) embedded in polymethyl methacrylate resin (Merck, Kenilworth, NJ, USA). Using a precision band saw (EXAKT system, Norderstedt, Germany), undecalcified specimens were obtained by sectioning the titanium caps perpendicular to the bone surface mid-way through the hemispheres. Then the specimens were thinned out to a thickness of 30–40 μm by microgrinding and polished (Mecapol P320, Presi, F, Grenoble, France). Finally they were stained with modified paragon. The sections were digitized with a light microscope (Eclipse 80i, Nikon, Tokyo, Japan) coupled to a digitizing camera (Allied Vision Technologies, D, Stadtroda, Germany). Single pictures of high power fields at 10x magnification were assembled to generate large overview images.

The outlines of bone substitute remnants and new bone tissue were scribed using a software for image analysis (Calopix viewer, Tribun, Châtillon, France). The structures of interest were delimited manually using a graphical pad on the entire surface inside the titanium caps, that is, from the external cortical limit of the bony bed to the internal limit of the caps. The amounts of new bone and bone substitute were expressed as percentages of the total volume under the hemispheres. Artifacts and/or broken tissues were excluded from the computations. The histopathologic evaluation was performed using the scoring system described in Table 2 (ISO 10993-6 standard [15]).

2.6. Statistical Analysis

The histomorphometrical data were checked for normal distribution and equivalence of variances. Unpaired t-tests were used to compare OF-plain (n = 8) to OF-BMP-2 (n = 8), at 8 and 16 weeks. For the analyses on horizontal segmentation, only groups within a slice were compared. The null hypothesis was rejected at $p < 0.05$.

3. Results

At necroscopy, macroscopically none of the 40 sites presented signs of inflammation. All hemispheres remained as placed both at 8 and 16 weeks.

3.1. Histopathologic Evaluation

Prior to histomorphometric analysis, a histopathologic quantification of inflammatory and bone cells was conducted. The scoring system and the resulting data are presented in Tables 2 and 3. No inflammatory cell infiltrate was detected at any site. Regarding phagocytic cells however, that is those in charge of degradation and remodeling processes, the application of BMP-2 led to heavy infiltrates of macrophages after 8 weeks while only a few cells were observed in control specimens. After 16 weeks, the situation had evened out with a slight presence of macrophages both in experimental and control specimens. BMP-2 had no effect on osteoclasts and giant cells, which were rarely seen both at 8 or 16 weeks. Conversely there was a marked presence of osteoblasts on both surface types at 8 weeks, which slightly decreased after 16 weeks.

Table 2. Histopathologic scoring system (semi-quantitative).

Cell type/response	Score				
	0	1	2	3	4
Polymorphonuclear cells (PMN)	0	Rare, 1–5/phf	5–10/phf	Heavy infiltrate	Dense
Lymphocytes (Lc)	0	Rare, 1–5/phf	5–10/phf	Heavy infiltrate	Dense
Plasma cells (Plc)	0	Rare, 1–5/phf	5–10/phf	Heavy infiltrate	Dense
Macrophages (Ma)	0	Rare, 1–5/phf	5–10/phf	Heavy infiltrate	Dense
Giant cells/osteoclastic cells (Gc/Oc)	0	Rare, 1–2/phf	3–5/phf	Heavy infiltrate	Sheets
Osteoblasts (Ob)	0	Slight, equivalent to normal bone	Moderate, >normal bone	Marked, >normal bone	Highly marked

phf = per high powered (400×) field

Table 3. Semi quantitative histopathologic evaluation (Scoring Table 2, $n = 8$, mean ± SD).

Time	Samples $n = 8$	Polymorphomuclear cells	Lymphocytes	Plasma cells	Macrophages	Giant Cells/ osteoclastic cells	Osteoblastic cells
8	OF-plain	0	0	0	1.1 ± 0.3	0.9 ± 0.3	2.6 ± 0.5
weeks	OF-BMP-2	0	0	0	2.6 ± 0.5	1	3
16	OF-plain	0	0	0	1.3 ± 0.4	1	2 ± 0.5
weeks	OF-BMP-2	0	0	0	1.6 ± 0.5	1.1 ± 0.3	1.8 ± 0.7

3.2. Histologic Evaluation

Generally, from 8 weeks on, the hemispheres were covered by a fibrous layer. The circular grooves that stabilized the hemispheres were often filled with new bone (Figure 2C,D) which somewhat lifted the caps out of their bone bed. As a consequence, on the inside of the caps a layer of connective tissue developed between the metal and the substitute materials.

(Figure 2A–C). In the negative control sites (plain blood coagulum) the hemispheres were almost empty although some minor bone growth was noted [10].

At 8 weeks, OF-plain samples showed evident signs of vertical and horizontal osteoconduction along the canals of the construct (Figure 2A). The new bone primarily grew out of the transcortical holes but also developed from the cortical bone bed and often spread to the top of the structure. The osteoid and the fibrous connective tissue were largely vascularized and showed marked osteoblastic activity. Large parts of the bone substitute were well integrated into the new bone tissue while others were lined by a thin layer of osteoid tissue and rows of osteoblasts. Signs of material degradation were observed at the base of the scaffold.

Adding BMP-2 led to the growth of large amounts of new bone that filled a vast surface. At 8 weeks, signs of maturity and remodeling were observable (Figure 2B). The remaining space was colonized by an active osteoid tissue. It also contained remnants of bone substitute which demonstrated evident and substantial signs of degradation, primarily mediated by macrophages (Figure 3A,E,F) and occasionally by osteoclasts (Figure 3A–C). A form of dissolution of the material into the newly formed bone tissue but without apparent cell mediation was also often observed (Figure 3A,D).

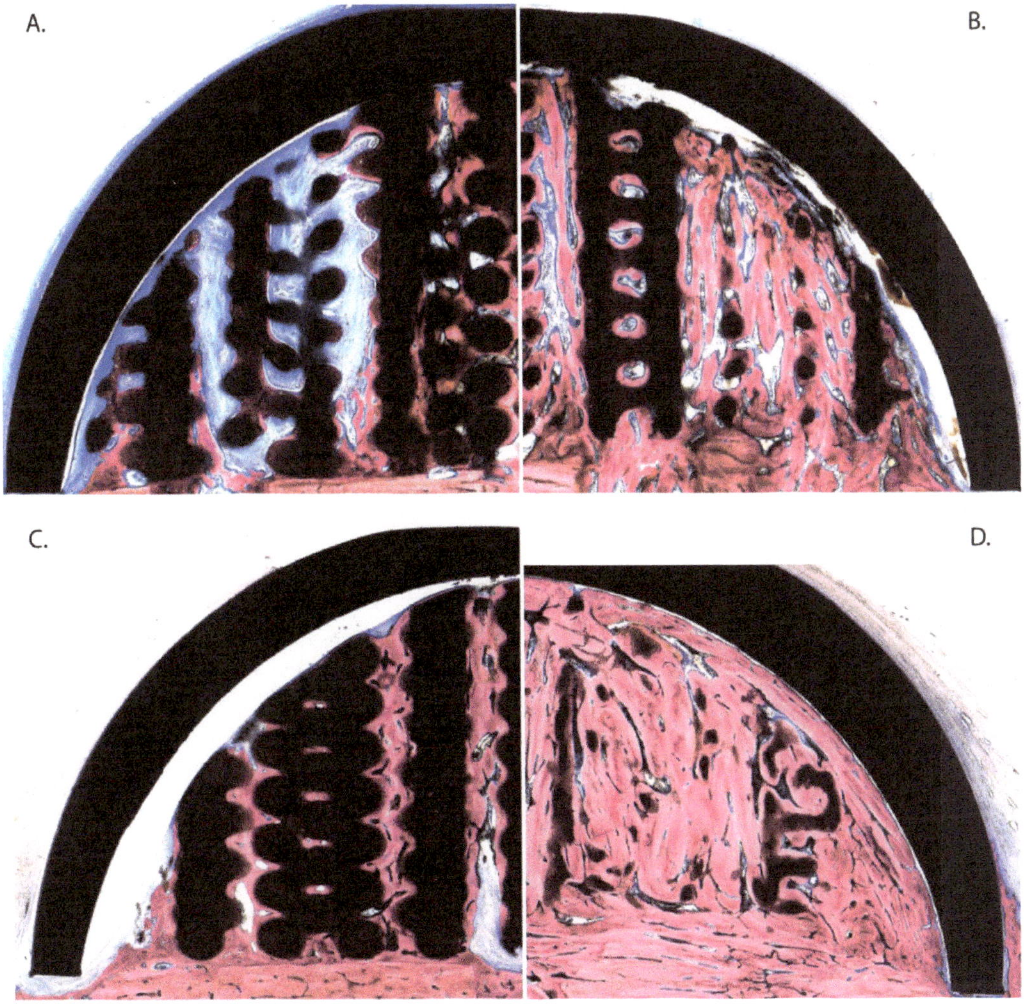

Figure 2. Hemi-slide showing the bone substitutes' engraftment under the titanium caps at 8 and 16 weeks. (**A**) 3D construct (8 weeks); the construct appears as brown circles or rounded polygons. The newly formed bone (pink) sprouts from the transcortical perforations and the bony bed. It then develops into vertical columns up to the titanium caps. The remainder of the volume is filled with fibroid and osteoid tissue (blue); (**B**) BMP-2 primed construct (8 weeks); large portions are filled with new bone already entering the remodeling process. Some particles of bone substitute appear as largely degraded and intimately nested into the new bone. Other portions of bone substitute remain undegraded; (**C**) 3D construct (16 weeks); fibroid and osteoid tissue is replaced by maturing new bone. The bone substitute did not resorb; (**D**) BMP-2 primed construct (16 weeks); traces of bone substitute (largely imbricated into a mature bone) are still observable.

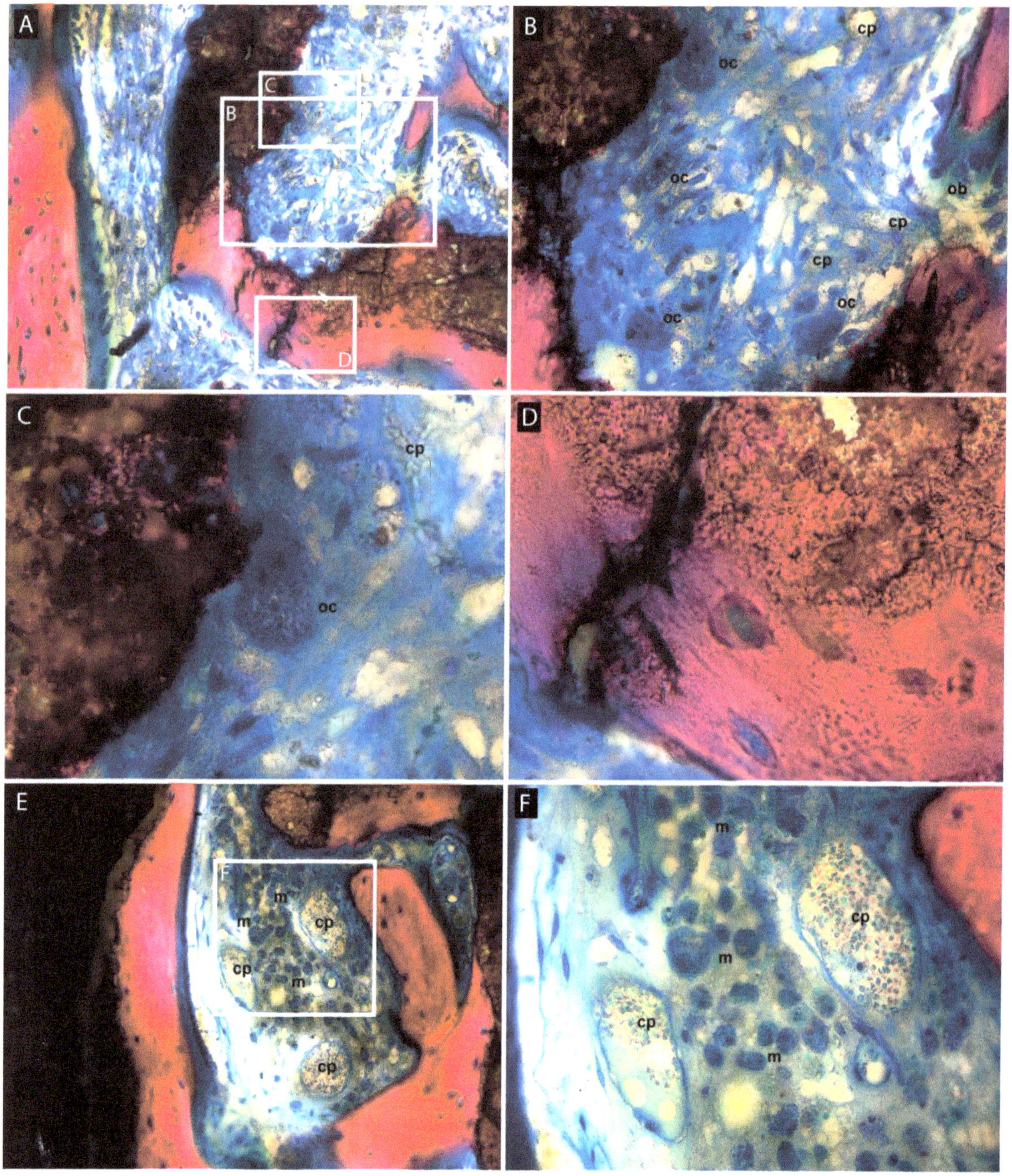

Figure 3. Higher magnifications of active zones of resorption in a BMP-2 primed construct (**A**,**E**) showing various processes of resorption of the scaffold at 8 weeks. The fibroid and osteoid tissue (blue) is highly vascularized (**B**,**C**,**F**), as shown by the broad presence of capillaries (cp), and is rich in plump macrophages (m) harboring intracytoplasmic debris; some osteoclasts (oc) are present and active (**B**,**C**). The bone substitute is not only degraded via cellular activity, but is also biochemically dissolved into the newly synthesized bone (**D**).

At 16 weeks, the new bone had matured and had replaced the osteoid and connective tissue in the OF-plain samples (Figure 2C). The bone substitute was tightly surrounded by the new bone tissue but degradation was light. In OF-BMP-2 samples, the cap was entirely filled by new mature bone (Figure 2D). The remodeling and maturation process led to a homogenous tissue which merged with the native cortical bone. Haversian-like structures were occasionally observed (Figure 4A) and newly formed osteons developed close to substitute remnants (Figure 4D). A differentiation into fatty bone marrow was also noted in some lacunae (Figure 4A–C). Traces of bone substitute remained which progressively dissolved into the new bone tissue (Figure 4E.)

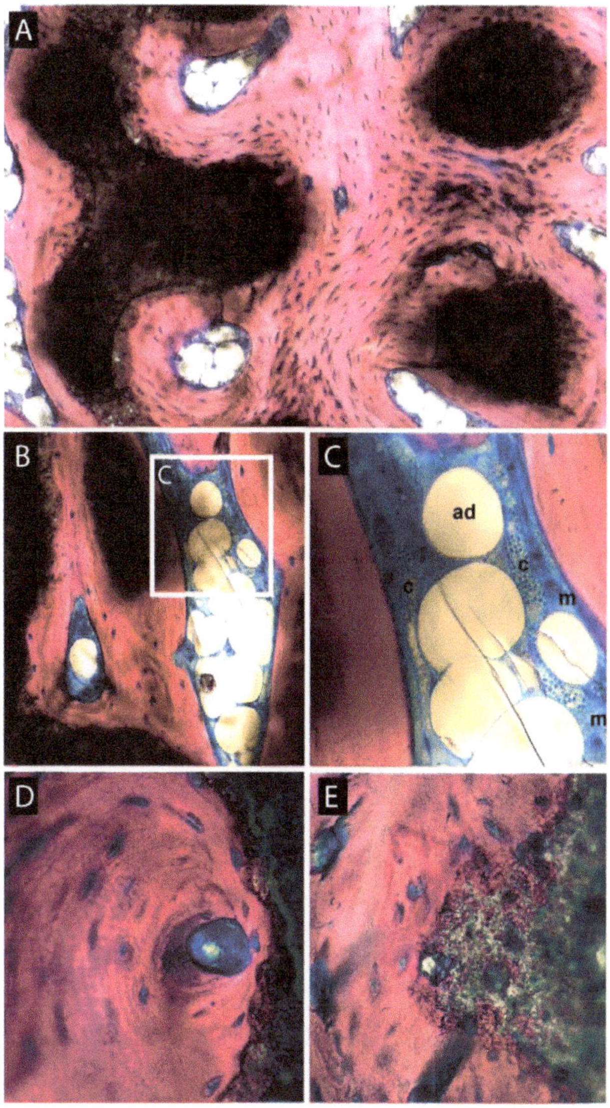

Figure 4. Higher magnifications showing mature bone at 16 weeks in BMP-2 primed constructs. (**A**) Presence of "haversian like" structures with adipocytic inclusions. Note the tight integration of the bone substitute remnants; (**B**) adipocytic inclusion within mature bone and higher magnification (**C**) showing adipocytes (ad), numerous capillaries (cp) and plump macrophages (m); (**D**) presence of newly formed osteons with central vasculature in close contact to the bone substitute that "dissolves" into the mature bone (**E**) in many locations.

3.3. Histomorphometric Evaluation

The histomorphometric analyses addressed the bone substitute and the new bone tissue that had formed under the hemispheres. At 8 weeks, the new bone volume (NBV) in OF-plain specimens was 22% ± 2.1%. The addition of BMP-2 significantly stimulated bone growth by 2.2, NBV reaching 50% ± 3% ($p = 1.5 \times 10^{-5}$) (Figure 5A). This augmentation was observed from the base to the top of the scaffold (Figure 6A). At 16 weeks, a rate of growth of 1.5 was observed in both conditions, whatever the compartment analyzed (Figure 6A), NBV amounting to 35% ± 2% on OF-plain and 68% ± 4% on OF-BMP-2 samples ($p = 2 \times 10^{-6}$) (Figure 5A).

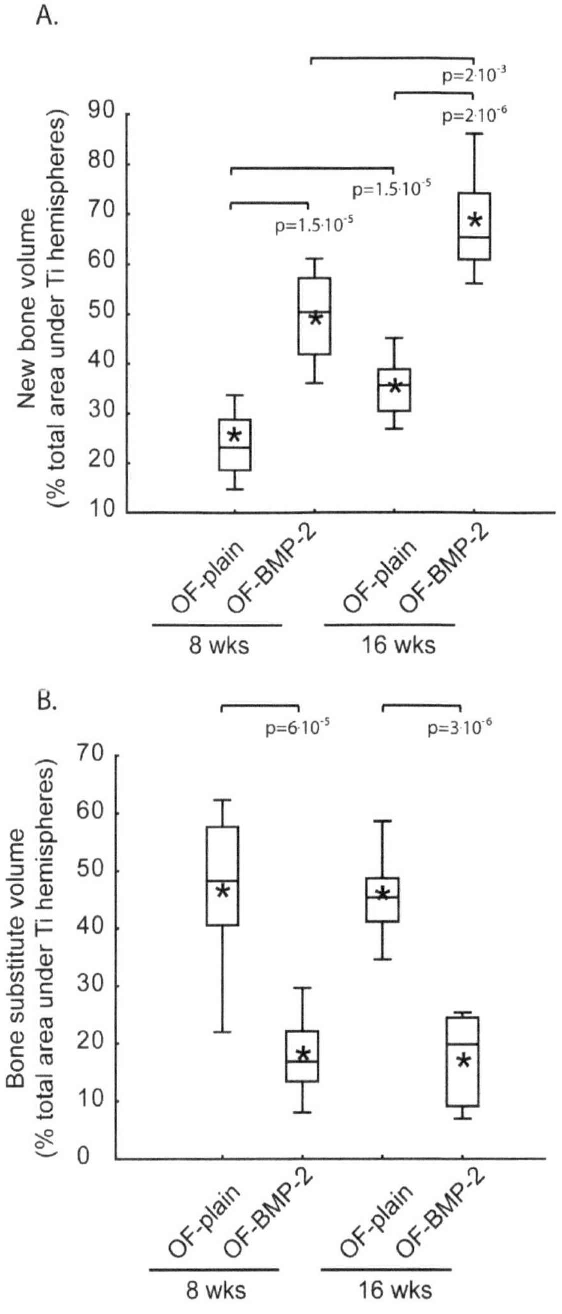

Figure 5. Histomorphometrical assessment of new bone volume (**A**) and volume of bone substitute (**B**) under the hemispheres at 8 and 16 weeks ($n = 8$ for each column). Brackets and p values indicate significant differences. Horizontal line: median. Boxes: 25%–75%. *: mean. Bars: range of non-outliers. (OF-plain: bare construct; OF-BMP-2: BMP-2 primed construct).

Regarding the volume occupied by the bone substitute (BSV), an opposite effect was observed. After 8 weeks, the addition of BMP-2 led to a material degradation that was 2.7 higher than that of OF-plain specimens (OF-plain: 47% ± 5%, OF-BMP-2: 18% ± 2%, $p = 6 \times 10^{-5}$) (Figure 5B). These ratios remained nearly constant after 16 weeks (OF-plain: 45% ± 2%; OF-BMP-2: 17% ± 3%;

$p = 3 \times 10^{-6}$), the distribution of calculated values tending to be more constricted when compared to 8 weeks (Figure 5B). These differences were significant from 0 to 4 mm up to the bony bed, at 8 and 16 weeks (Figure 6B).

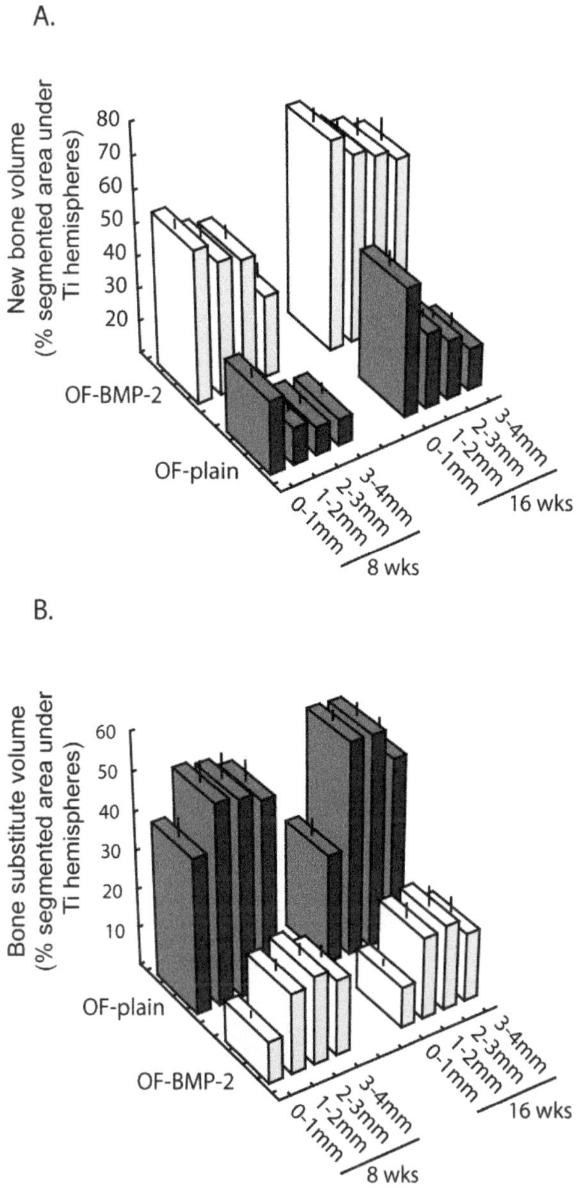

Figure 6. New bone volume (**A**) and bone substitute volume (**B**) under the hemispheres, segmented in 4 horizontal planes of 1 mm in height at 8 weeks and 16 weeks in OF-plain (grey) and OF-BMP-2 (white) samples. ($n = 8$ for each column). (Bars = SEM, OF-plain: bare construct; OF-BMP-2: BMP-2 primed construct).

4. Discussion

The present work aimed at simulating the performance of a 3D-printed construct of bone substitute (OsteoFlux®) over a period of time beyond that observed under standard conditions. Bone metabolism was considerably accelerated by an initial stimulation with BMP-2. Osteoconduction, bone maturation as well as material resorption were histologically and histomorphometrically evaluated 8 and 16 weeks after implantation. Unstimulated constructs served as controls.

The calvarial model of a one-wall defect in which bone substitutes are maintained and protected by titanium caps was first developed for rabbits [16,17] and pigs [12]. The model has recently been transferred to sheep and proved satisfactory when comparing several bone substitutes [10] (sheep have been widely used in cranioplasty, distraction and reconstruction studies [18–20]).

The main advantages of this model are as follows: (i) several conditions may be compared simultaneously (e.g., different bone substitutes, addition of cytokines); (ii) in comparison to bone defect models, the cortical bone bed is preserved, thereby testing the true potential of a material for promoting bone augmentation; (iii) medullar cells may migrate towards the experimental via transcortical perforations.

The study herein described was a part of a larger study aimed at comparing a synthetic 3D construct to two other bone substitutes [10]. When investigating the osteoconductive potential of a material, a timeframe of 16 weeks is generally considered adequate [12,16,17]. Problematically, the response of phosphoceramics regarding their ability to resorb must be approached in timeframes from months to years [21]. Yet the current trend in animal testing is to minimize the number of animals and the duration of the experiments. For this reason, it was felt that that accelerating bone metabolism (*i.e.*, bone synthesis and bone remodeling) was an elegant solution to investigate the process of bone maturation and material resorption at later stages of development.

The bone morphogenetic protein (BMP) family of cytokines and growth factors is known for their potential for osteoinduction, both orthotopically and ectopically. Among these, BMP-2 is a most prominent representative to the extent that it found clinical applications in bone repair [11]. Still, side effects such as heterotopic bone formation [22], bone resorption and inflammation [23], and cervical or gingival swelling [24] were noted. Due to the well-delineated zone of effect of BMP-2 in our setup, none of these signs were observed in our study (except for local bone overgrowth out of the caps).

BMP-2 is typically used at supraphysiological concentrations (in the tenths of a milligram range) to compensate for its proteolytic degradation. Therefore scaffolds that would release BMP-2 at a steady rate (instead of a one-time massive discharge) would be of great interest to (i) decrease dosages; (ii) maintain a sustained stimulatory effect on bone induction and (iii) decrease side effects [11]. In numerous clinical applications, collagen sponges were used. Still, due to their rapid degradation, they tend to release BMP-2 at high doses and over too short a time span. In contrast, incorporating BMP-2 into a phosphocalcic bioceramic may have dual advantages—first, a release of BMP-2 according to a natural flow, and second, a reinforcement of the device's mechanical resistance [25,26]. In the present experiment, loading the 3D construct with 100 μg BMP-2 led to a bone growth that was twice that of controls while still retaining its normal characteristics in terms of physiological processes, that is, abundant neovascularization, important osteoblastic activity and matrix mineralization.

After sufficient amounts of new mineralized tissue have been formed, the next step consists in remodeling to set up the bone's definitive architecture: structurally (haversian lamellae), morphologically (trabecular, cortical) and with regard to the overall anatomy. Eight weeks after implantation, the first signs of bone remodeling were observed in BMP-2 treated specimens. These became evident at 16 weeks, as shown by the presence of newly formed osteons, harversian like structures, adipocytic bone marrow inclusions and the merging of the new bone with the cortical bony bed.

In conclusion, the addition of 100 μg BMP-2 onto the scaffold succeeded in physiologically accelerating bone development, leading to the development of pre-mature haversian bone.

The ultimate goal of the study was to evaluate the 3D constructs regarding scaffolding for new bone as well as their ability to progressively resorb. The biodegradation process of phosphoceramics is a combination of chemical dissolution and cell-mediated resorption [6,27]. The chemical dissolution primarily depends on the nature and crystallinity of the ceramic [6,28]. On the biological end, the cells involved in the degradation process are mainly of the phagocytic lineage, that is, osteoclasts, giant cells, macrophages and monocytes [27].

After 16 weeks, the resorption of the material was 2–3 times higher in BMP-2 treated specimens but then nearly halted for the next two months. Hence, most material resorption took place during the first 8 weeks.

BMP-2 is known for its bifunctional stimulation, that is, it affects bone formation as well as bone resorption. The latter is mediated by osteoclasts which are activated via (i) the RANK-L/RANK/OPG regulation loop [29,30] or (ii) a direct interaction with BMP-2 receptors on osteoclasts [31,32]. High doses of BMP-2 activate both osteoblasts and osteoclasts at early healing stages (1 week to 1 month) (nonetheless there is a preponderance of osteoclasts in the very initial times [33]). Subsequently, resorption units, made of a pool of osteoclasts which are continuously renewed, are recruited and maintained for a period of several weeks to months [34].

In our study osteoclasts were occasionally observed in BMP-2 treated specimens but not in substantial amounts, neither were resorption units. Furthermore, the cortical bone bed, which normally would have been heavily remodeled by massive osteoclast infiltrations, remained largely intact. Therefore, it appears that no sizeable osteoclast-mediated bone/material resorption process was active in the first 8 weeks. In contrast, 2–3 times more macrophages harboring debris of material were found in close vicinity of BMP-2 treated scaffolds. Macrophages are known precursors of osteoclasts but may also degrade foreign materials on their own [27]. They are able to synthetize BMP-2 and thus activate bone synthesis [35] but their own response to BMP-2 is poorly documented. As observed here, the sequence of events takes place as if BMP-2 induced the recruitment and activation of macrophages to degrade the 3D construct instead of engaging the cells to differentiate into osteoclasts. Still, the process left sufficient time for the scaffold to support the migration of bone cells and the formation of new bone as demonstrated by the bone's morphology at 8 weeks (made of vertical columns reflecting the original scaffold structure, by now largely dissolved).

As mentioned above, phosphoceramics, and especially TCP, are known to dissolve in water [6,28]. Dissolution was observed both on controls and on BMP-2 specimens, but was not actually increased by BMP-2 treatment. During the last two months of the experiment, the material continued its resorption process, but at a very slow pace.

Finally, considering the ability of the 3D construct to support the formation of a mature bone, there was clear evidence that the construct had promoted a large amount of bone ingrowth during the first 8 weeks, followed by a remodeling and maturation process finally leading to a harversian-like bone. The construct’s remnants after 16 weeks were fully osteointegrated and showed signs of dissolution. Eventually the process should lead to the complete disappearance of the material. This, however, must be confirmed by further experiments. The anatomical shape of the newly synthesized bone was very regular, patterned by the scaffold and delimited by the titanium cap.

5. Conclusions

Priming a 3D printed construct with BMP-2 accelerated bone metabolism whereby this acceleration proceeded within physiological boundaries. The rate of vertical bone growth, bone maturation and material bioresorption nearly doubled after implantation in sheep calvaria. The scaffolding function of the constructs was maintained, thus leaving sufficient time for the bone to grow and mature until finally acquiring a haversian-like structure. In parallel, the biomaterial resorbed via cell mediated and chemical processes. These are promising results which must be confirmed in clinical tests.

Acknowledgments

The authors are indebted to NAMSA-Lyon, France. Particular thanks go to Sophie Chirat, Agnès Letessier and Antoine Alves.

Author Contributions

Mira Moussa: design, analysis and wrote the paper; Jean-Pierre Carrel: surgeries; Susanne Scherrer: surgeries and analysis; Maria Cattani-Lorente: analysis and statistics; Anselm Wiskott: design, funding and reviewed the paper; Stéphane Durual: design, funding, analysis and wrote the paper.

Conflicts of Interest

The authors declare no conflict of interest. No benefit of any kind has been or will be received either directly or indirectly by the authors.

References

1. Draenert, F.G.; Huetzen, D.; Neff, A.; Mueller, W.E. Vertical bone augmentation procedures: Basics and techniques in dental implantology. *J. Biomed. Mater. Res. A* **2014**, *102*, 1605–1613.
2. Sato, M.; Webster, T.J. Designing orthopedic implant surfaces: Harmonization of nanotopographical and chemical aspects. *Nanomedicine (Lond.)* **2006**, *1*, 351–354.
3. Fini, M.; Giardino, R.; Borsari, V.; Torricelli, P.; Rimondini, L.; Giavaresi, G.; Nicoli Aldini, N. *In vitro* behaviour of osteoblasts cultured on orthopaedic biomaterials with different surface roughness, uncoated and fluorohydroxyapatite-coated, relative to the *in vivo* osteointegration rate. *Int. J. Artif. Organs* **2003**, *26*, 520–528.
4. Seitz, H.; Rieder, W.; Irsen, S.; Leukers, B.; Tille, C. Three-dimensional printing of porous ceramic scaffolds for bone tissue engineering. *J. Biomed. Mater. Res. B Appl. Biomater.* **2005**, *74*, 782–788.
5. Hollister, S.J.; Lin, C.Y.; Saito, E.; Schek, R.D.; Taboas, J.M.; Williams, J.M.; Partee, B.; Flanagan, C.L.; Diggs, A.; Wilke, E.N.; *et al*. Engineering craniofacial scaffolds. *Orthod Craniofac Res.* **2005**, *8*, 162–173.
6. Dorozhkin, S.V. Bioceramics of calcium orthophosphates. *Biomaterials* **2010**, *31*, 1465–1485.
7. Barrere, F.; van Blitterswijk, C.A.; de Groot, K. Bone regeneration: Molecular and cellular interactions with calcium phosphate ceramics. *Int. J. Nanomed.* **2006**, *1*, 317–332.

8. Daculsi, G.; Laboux, O.; Malard, O.; Weiss, P. Current state of the art of biphasic calcium phosphate bioceramics. *J. Mater. Sci. Mater. Med.* **2003**, *14*, 195–200.
9. Daculsi, G. Biphasic calcium phosphate concept applied to artificial bone, implant coating and injectable bone substitute. *Biomaterials* **1998**, *19*, 1473–1478.
10. Carrel, J.P.; Wiskott, A.; Moussa, M.; Rieder, P.; Scherrer, S.; Durual, S. A 3D printed TCP/HA structure as a new osteoconductive scaffold for vertical bone augmentation. *Clin. Oral Implants Res.* **2014**, *2014*, doi: 10.1111/clr.12503.
11. Carreira, A.C.; Lojudice, F.H.; Halcsik, E.; Navarro, R.D.; Sogayar, M.C.; Granjeiro, J.M. Bone morphogenetic proteins: Facts, challenges, and future perspectives. *J. Dent Res.* **2014**, *93*, 335–345.
12. Busenlechner, D.; Tangl, S.; Mair, B.; Fugger, G.; Gruber, R.; Redl, H.; Watzek, G. Simultaneous *in vivo* comparison of bone substitutes in a guided bone regeneration model. *Biomaterials* **2008**, *29*, 3195–3200.
13. ISO/IEC 17025:2005. *General Requirements for the Competence of Testing and Calibration Laboratories*; International Organization for Standardization (ISO): Geneva, Switzerland, 2005.
14. 2010/63/EU. *European Horizontal Legislation for the Protection of Animals Used for Scientific Purposes*; European Commission: Brussels, Belgium, 2010.
15. ISO 10993-6:2007. *Biological Evaluation of Medical Devices—Part 6: Tests for Local Effects after Implantation*; International Organization for Standardization (ISO): Geneva, Switzerland, 2007.
16. Yamada, Y.; Tamura, T.; Hariu, K.; Asano, Y.; Sato, S.; Ito, K. Angiogenesis in newly augmented bone observed in rabbit calvarium using a titanium cap. *Clin. Oral Implants Res.* **2008**, *19*, 1003–1009.
17. Yamada, Y.; Nanba, K.; Ito, K. Effects of occlusiveness of a titanium cap on bone generation beyond the skeletal envelope in the rabbit calvarium. *Clin. Oral Implants Res.* **2003**, *14*, 455–463.
18. Kuemmerle, J.M.; Oberle, A.; Oechslin, C.; Bohner, M.; Frei, C.; Boecken, I.; Von Rechenberg, B. Assessment of the suitability of a new brushite calcium phosphate cement for cranioplasty—An experimental study in sheep. *J. Craniomaxillofac Surg.* **2005**, *33*, 37–44.
19. Gosain, A.K.; Riordan, P.A.; Song, L.; Amarante, M.T.; Kalantarian, B.; Nagy, P.G.; Wilson, C.R.; Toth, J.M.; McIntyre, B.L. A 1-year study of osteoinduction in hydroxyapatite-derived biomaterials in an adult sheep model: Part ii. Bioengineering implants to optimize bone replacement in reconstruction of cranial defects. *Plast Reconstr. Surg.* **2004**, *114*, 1155–1163.
20. Eufinger, H.; Rasche, C.; Lehmbrock, J.; Wehmoller, M.; Weihe, S.; Schmitz, I.; Schiller, C.; Epple, M. Performance of functionally graded implants of polylactides and calcium phosphate/calcium carbonate in an ovine model for computer assisted craniectomy and cranioplasty. *Biomaterials* **2007**, *28*, 475–485.
21. Artzi, Z.; Weinreb, M.; Givol, N.; Rohrer, M.D.; Nemcovsky, C.E.; Prasad, H.S.; Tal, H. Biomaterial resorption rate and healing site morphology of inorganic bovine bone and beta-tricalcium phosphate in the canine: A 24-month longitudinal histologic study and morphometric analysis. *Int. J. Oral Maxillofac Implants* **2004**, *19*, 357–368.
22. Walker, D.H.; Wright, N.M. Bone morphogenetic proteins and spinal fusion. *Neurosurg. Focus* **2002**, *13*, e3.

23. Ritting, A.W.; Weber, E.W.; Lee, M.C. Exaggerated inflammatory response and bony resorption from bmp-2 use in a pediatric forearm nonunion. *J. Hand. Surg. Am.* **2012**, *37*, 316–321.
24. Perri, B.; Cooper, M.; Lauryssen, C.; Anand, N. Adverse swelling associated with use of RH-BMP-2 in anterior cervical discectomy and fusion: A case study. *Spine J.* **2007**, *7*, 235–239.
25. Oh, S.H.; Kim, T.H.; Lee, J.H. Creating growth factor gradients in three dimensional porous matrix by centrifugation and surface immobilization. *Biomaterials* **2011**, *32*, 8254–8260.
26. Hagi, T.T.; Wu, G.; Liu, Y.; Hunziker, E.B. Cell-mediated BMP-2 liberation promotes bone formation in a mechanically unstable implant environment. *Bone* **2010**, *46*, 1322–1327.
27. Lu, J.; Descamps, M.; Dejou, J.; Koubi, G.; Hardouin, P.; Lemaitre, J.; Proust, J.P. The biodegradation mechanism of calcium phosphate biomaterials in bone. *J. Biomed. Mater. Res.* **2002**, *63*, 408–412.
28. Kamitakahara, M.; Ohtsuki, C.; Miyazaki, T. Review paper: Behavior of ceramic biomaterials derived from tricalcium phosphate in physiological condition. *J. Biomater. Appl.* **2008**, *23*, 197–212.
29. Kamiya, N.; Ye, L.; Kobayashi, T.; Mochida, Y.; Yamauchi, M.; Kronenberg, H.M.; Feng, J.Q.; Mishina, Y. Bmp signaling negatively regulates bone mass through sclerostin by inhibiting the canonical wnt pathway. *Development* **2008**, *135*, 3801–3811.
30. Glass, D.A., 2nd; Bialek, P.; Ahn, J.D.; Starbuck, M.; Patel, M.S.; Clevers, H.; Taketo, M.M.; Long, F.; McMahon, A.P.; Lang, R.A.; *et al*. Canonical wnt signaling in differentiated osteoblasts controls osteoclast differentiation. *Dev. Cell* **2005**, *8*, 751–764.
31. Kaneko, H.; Arakawa, T.; Mano, H.; Kaneda, T.; Ogasawara, A.; Nakagawa, M.; Toyama, Y.; Yabe, Y.; Kumegawa, M.; Hakeda, Y. Direct stimulation of osteoclastic bone resorption by bone morphogenetic protein (BMP)-2 and expression of bmp receptors in mature osteoclasts. *Bone* **2000**, *27*, 479–486.
32. Itoh, K.; Udagawa, N.; Katagiri, T.; Iemura, S.; Ueno, N.; Yasuda, H.; Higashio, K.; Quinn, J.M.; Gillespie, M.T.; Martin, T.J.; *et al*. Bone morphogenetic protein 2 stimulates osteoclast differentiation and survival supported by receptor activator of nuclear factor-kappab ligand. *Endocrinology* **2001**, *142*, 3656–3662.
33. Kim, R.Y.; Oh, J.H.; Lee, B.S.; Seo, Y.K.; Hwang, S.J.; Kim, I.S. The effect of dose on rhBMP-2 signaling, delivered via collagen sponge, on osteoclast activation and *in vivo* bone resorption. *Biomaterials* **2014**, *35*, 1869–1881.
34. Robert, L.; Jilka, T.B.; Almeid, M.; Plotkin, L.I.; O'Brien, C.A.; Manolagas, R.S.W.A.S.C. Chapter 13. Apoptosis of bone cells. In *Principles of Bone Biology*; Bilezikian, J.P., Raisz, L.G., Martin, T.J., Eds.; Academic Press: Elsevier, Amsterdam, The Netherlands, 2008.
35. Chen, Z.; Wu, C.; Gu, W.; Klein, T.; Crawford, R.; Xiao, Y. Osteogenic differentiation of bone marrow MSCs by beta-tricalcium phosphate stimulating macrophages via BMP2 signalling pathway. *Biomaterials* **2013**, *35*, 1507–1518.

3

A Novel Porcine Graft for Regeneration of Bone Defects

Eisner Salamanca [1,†], Wei-Fang Lee [2,†], Chin-Yi Lin [1], Haw-Ming Huang [3], Che-Tong Lin [1], Sheng-Wei Feng [1] and Wei-Jen Chang [1,4,*]

1 School of Dentistry, College of Oral Medicine, Taipei Medical University, 250 Wu-Hsing Street, Taipei 110, Taiwan; E-Mails: eisnergab@hotmail.com (E.S.); drjerrylin@gmail.com (C.-Y.L.); chetong@tmu.edu.tw (C.-T.L.); b8702070@tmu.edu.tw (S.-W.F.)
2 School of Dental Technology, College of Oral Medicine, Taipei Medical University, 250 Wu-Hsing Street, Taipei 110, Taiwan; E-Mail: weiwei@tmu.edu.tw
3 Graduate Institute of Biomedical Materials & Tissue Engineering, College of Oral Medicine, Taipei Medical University, 250 Wu-Hsing Street, Taipei 110, Taiwan; E-Mail: hhm@tmu.edu.tw
4 Dental Department of Taipei Medical University, Shuang-Ho Hospital, Taipei 110, Taiwan

† These authors contributed equally to this work.

* Author to whom correspondence should be addressed; E-Mail: cweijen1@ tmu.edu.tw;

Academic Editor: Franz E. Weber

Abstract: Bone regeneration procedures require alternative graft biomaterials to those for autogenous bone. Therefore, we developed a novel porcine graft using particle sizes of 250–500 μm and 500–1000 μm in rabbit calvarial bone defects and compared the graft properties with those of commercial hydroxyapatite (HA)/beta-tricalcium phosphate (β-TCP) over eight weeks. Surgery was performed in 20 adult male New Zealand white rabbits. During a standardized surgical procedure, four calvarial critical-size defects of 5 mm diameter and 3 mm depth were prepared. The defects were filled with HA/β-TCP, 250–500 μm or 500–1000 μm porcine graft, and control defects were not filled. The animals were grouped for sacrifice at 1, 2, 4, and 8 weeks post-surgery. Subsequently, sample blocks were prepared for micro-computed tomography (micro-CT) scanning and histological sectioning. Similar bone formations were observed in all three treatment groups, although the 250–500 μm porcine graft performed slightly better. Rabbit calvarial bone tissue positively responded to porcine grafts and commercial HA/β-TCP, structural analyses showed similar crystallinity and porosity of the porcine and HA/β-TCP grafts, which facilitated bone formation through osteoconduction. These porcine grafts can be considered as graft substitutes, although further development is required for clinical applications.

Keywords: bone regeneration; bone graft; porcine tissue; *in vivo* test; hydroxyapatite

1. Introduction

In past decades, multiple reconstructive procedures have been assessed as periodontal regenerative treatments for deep infrabony defects that are associated with periodontal pockets [1,2] and furcation defects [3] and to prevent bone crest resorption after tooth extraction [4]. Currently, autogenous bone grafts are considered to be the gold standard in bone regeneration procedures because they contain viable osteoblasts, organic and inorganic matrices, and biological modifiers [5]. However, disadvantages of autogenous bone grafts include limited availability, tendency toward partial resorption, requirement of additional surgery, and increased morbidity [5]. Thus, further studies of alternative biomaterials, such as allogenic and xenogenic grafts, and alloplastic materials, are required to determine their efficacy in osteoconduction, osteoinduction, and osteogenesis [3] and to assess their clinical suitability for bone regeneration [6–8].

Biphasic calcium phosphate (BCP) is a bioinert and bioactive alloplastic material that comprises mixtures of hydroxyapatite (HA) and beta-tricalcium phosphate (β-TCP) at varying HA/β-TCP ratios [9,10]. Previous studies showed that BCP comprising approximately 60% HA and 40% β-TCP allows control of resorption and maintenance of osteoconductive properties [11–16]. Several investigators found that the optimal macroporosity of this material for the ingrowth of bone tissue is between 150 and 500 μm [12]. Accordingly, the commercial biphasic ceramic product, Micro-Macro Biphasic Calcium Phosphate (MBCP, Biomatlante Co. Ltd., Vigneux-de-Bretagne, France), comprises 60% HA and 40% β-TCP and has a complete interconnected porosity of 70% with macropores (300–600 μm) and micropores (<10 μm) at a ratio of 2:1. Among xenogenic materials, osteoconductive and osteoinductive properties of porcine bone substitutes have been reported in numerous histological and morphological studies [17–20]. However, most of the xenogenic grafts were developed from the bovine bone. The Creutzfeldt–Jakob disease, fetal prion disease in humans, is a problem for the clinical consideration [21]. Therefore, given the similarity of porcine and human genomes, porcine xenograft materials have been widely considered as substitutes for osseous grafts and grafting material within dental implants treatments [18,22–24]. In this study, a new porcine graft was developed due to the clinical consideration. The material properties and the bony defect filling effect of the new porcine grafts were investigated in this study.

2. Results

2.1. Scanning Electron Microscopy

The scanning electron microscopy (SEM) analyses revealed similar structures to those of bovine grafts and HA. Specifically, the porcine grafts had rough surfaces within porous structures of particle surfaces, and induced osteoblastic cell attachment to a 3-dimensional microstructure, and facilitated bone formation. The surfaces also had a spherical morphology comprising a mixture of 250–500 μm and 500–1000 μm particles, which provided a favorable matrix for attachment and proliferation of osteoblasts cells (Figure 1).

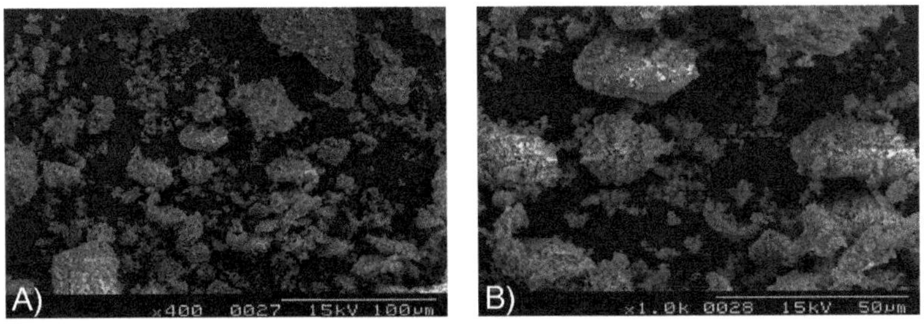

Figure 1. Scanning electron microscope (SEM) images of porcine bone particles: (**A**) Original magnification ×400; (**B**) Original magnification ×1000.

2.2. Energy Dispersive Spectrometry

The energy dispersive spectrometry (EDS) analyses showed very high accuracy between calcium (Ca) and phosphorus (P), which were present at a Ca/P ratio of 1.668 (28.16% and 16.88%, respectively), and the porcine graft profile resembled that of the HA/β-TCP structure (Table 1).

Table 1. Energy dispersive X-ray spectrometer: table of analyzed elements.

Element	Weight (%)	Atomic (%)
C	8.01	13.90
O	44.32	57.77
Na	0.86	0.78
Mg	1.78	1.53
P	16.88	11.37
Ca	28.16	14.65
Total	100.00	100.00

2.3. X-ray Diffraction

In the XRD analyses, the highest intensity (600 a.u.) of the porcine graft was observed at 27°, and the porcine grafts had peaks of 200 a.u. at 25° and 40°. However, the corresponding wide peak widths indicated low crystallinity (Figure 2).

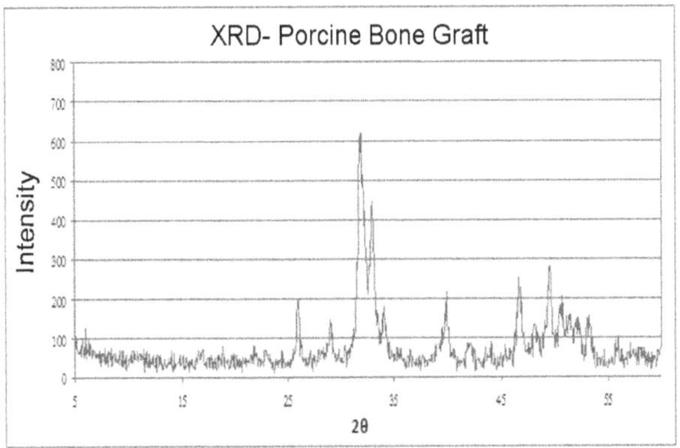

Figure 2. X-ray diffraction: Porcine graft intensities at scattering angles 2θ.

2.4. MTT Proliferation Assay

Absorbance values decreased by 4% in the positive control cells compared with those in the control media. However, osteoblast cell proliferation was 2% greater in the porcine grafts than in the control media with the MG-63 cells progressing from attachment to spreading and showing enhanced cell viability (Figure 3). No significant difference was noted among the different groups.

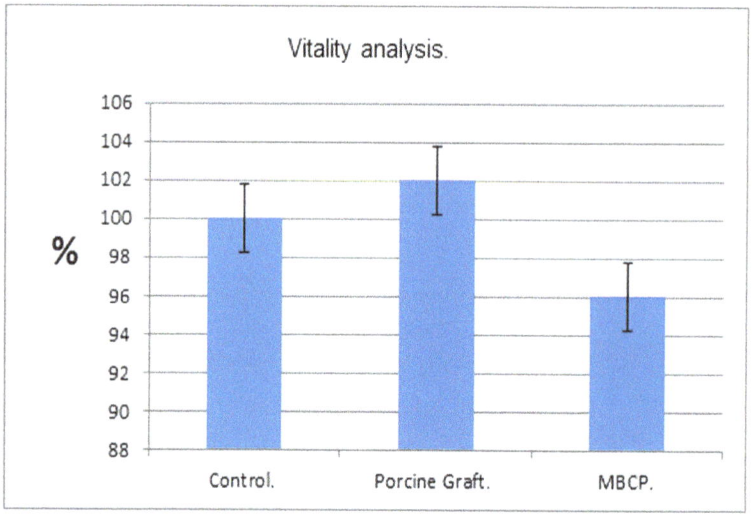

Figure 3. MTT proliferation assays: Lower absorbance values indicate reduced cell proliferation. Porcine grafts enhanced proliferation of osteoblast-like cells.

2.5. Micro-CT Scanning

Porcine graft bone volume densities were calculated using micro-CT and are presented for each time point in Table 2. During week 1, the 250–500 μm porcine graft produced 43% more new bone, whereas MBCP and the 500–1000 μm porcine graft both produced 40% growth and the control media produced 31% growth ($p < 0.05$). New bone growth was reduced in subsequent weeks under all conditions (Figure 4A).

Table 2. Micro-computed tomography (micro-CT) assessments of bone formation. Data are presented as means and standard deviations.

Micro CT Average-SD Values for Newly Formed Bone				
Time	**Control**	**MBCP**	**Porcine Graft 250–500 μm**	**Porcine Graft 500–1000 μm**
Week 1	30.57 ± 7.57	40.42 ± 2.64	43.29 ± 8.37	40 ± 2.33
Week 2	41.57 ± 8.71	32.59 ± 5.05	31.52 ± 3.51	35.24 ± 2.33
Week 4	30.38 ± 10.54	30.7 ± 6.5	32.47 ± 3.83	31.05 ± 1.3
Week 8	31.55 ± 7.25	27.81 ± 3.86	35.02 ± 2.92	36.19 ± 2.85

The rate of new bone production was the greatest in the control group during week 2, with a 42% increase compared with 33%, 32%, and 35% increases in the MBCP, 250–500 μm porcine graft, and 500–1000 μm porcine graft groups, respectively. New bone formation in all filled defects was less in week 2 than in week 1 (Figure 4B).

During week 4, the 250–500 µm porcine graft produced 33% new bone, which did not significantly differ with that of the MBCP (31%) or 500–1000 µm porcine graft (31%) groups. In contrast, new bone growth was only 12% in the control group (Figure 4C).

After eight weeks, the 250–500 µm porcine graft produced 35% new bone growth similar to that seen at weeks 1 and 4 and was superior to the other treatments. MBCP only produced 28% new bone growth, which was equal to that in the control group. The 500–1000 µm porcine graft showed improved performance from week 4, with 36% new bone formation at week 8 (Figure 4D).

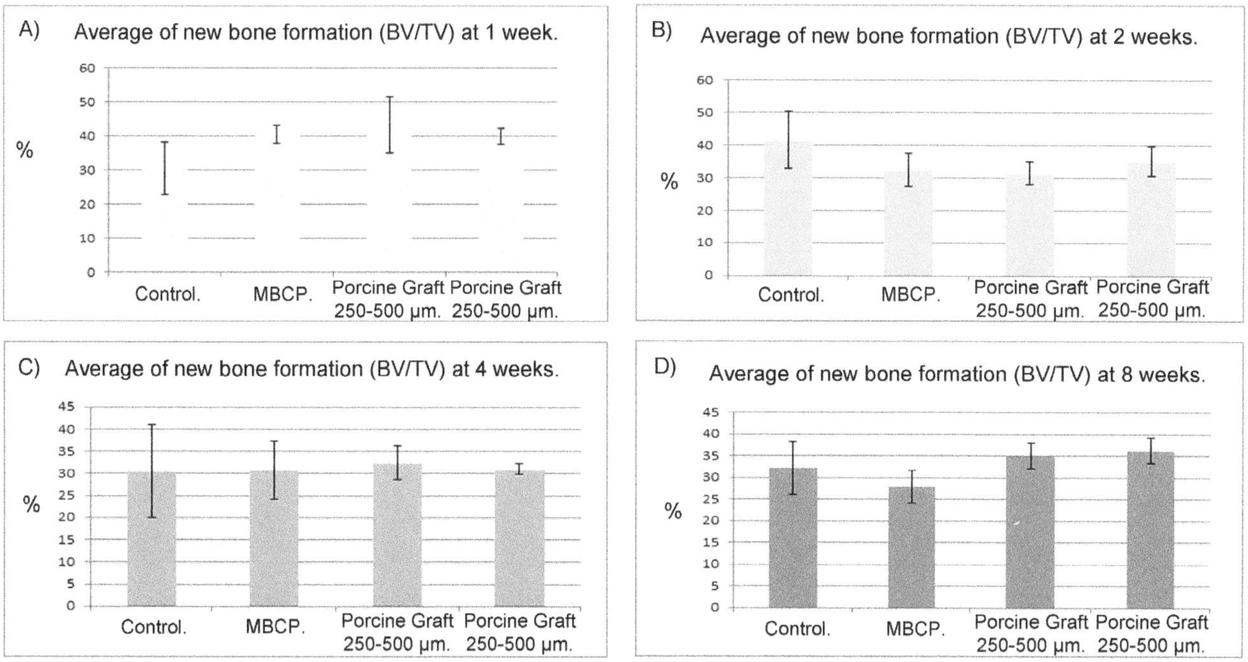

Figure 4. Percentage of bone formation in the total volume of the defect: The bars represent the percentage of bone formation calculated from the relation between bone and tissue volumes in the four calvarial defects at (**A**) one, (**B**) two, (**C**) four and (**D**) eight weeks, in all rabbits.

2.6. Histological Examination

Photomicrographs of H&E stained sections from the rabbit calvaria showed granulation tissue in the control defect and in the defects filled with MBCP, 250–500 µm, and 500–1000 µm porcine grafts. Specifically, the defects contained particles in direct contact with woven bone at the border and granulation tissue in the middle of the defect, and all filled defects presented acceptable bone formation (Figure 5A–D).

Light miscroscope photographs of the H&E stained sections from the week 2 showed similar patterns to those from the week 1. However, the granulation tissue was replaced with loose fibrous connective tissue, and although this connective tissue encapsulated some graft particles, other particles were surrounded by new bone (Figure 5E–H).

At the week 4, the analyses showed newly formed bone in contact with biomaterials, and whereas new bone was observed in the contours of the control defects, more new bone was observed around the particles in the MBCP-filled defects and around and across particles in defects filled with 250–500 µm and 500–1000 μm porcine grafts. All the defects had less connective tissue at this time point than at previous time points. (Figure 5I–L).

At the week 8, the control defects showed partial closures with bone around the defect and well organized connective tissue in the middle. Moreover, the bone areas increased with decreasing MBCP, 250–500 μm, and 500–1000 μm porcine graft areas. Finally, the bone was more mature at week 8, and borders between newly formed bone and graft materials were easily observed (Figure 5M–P).

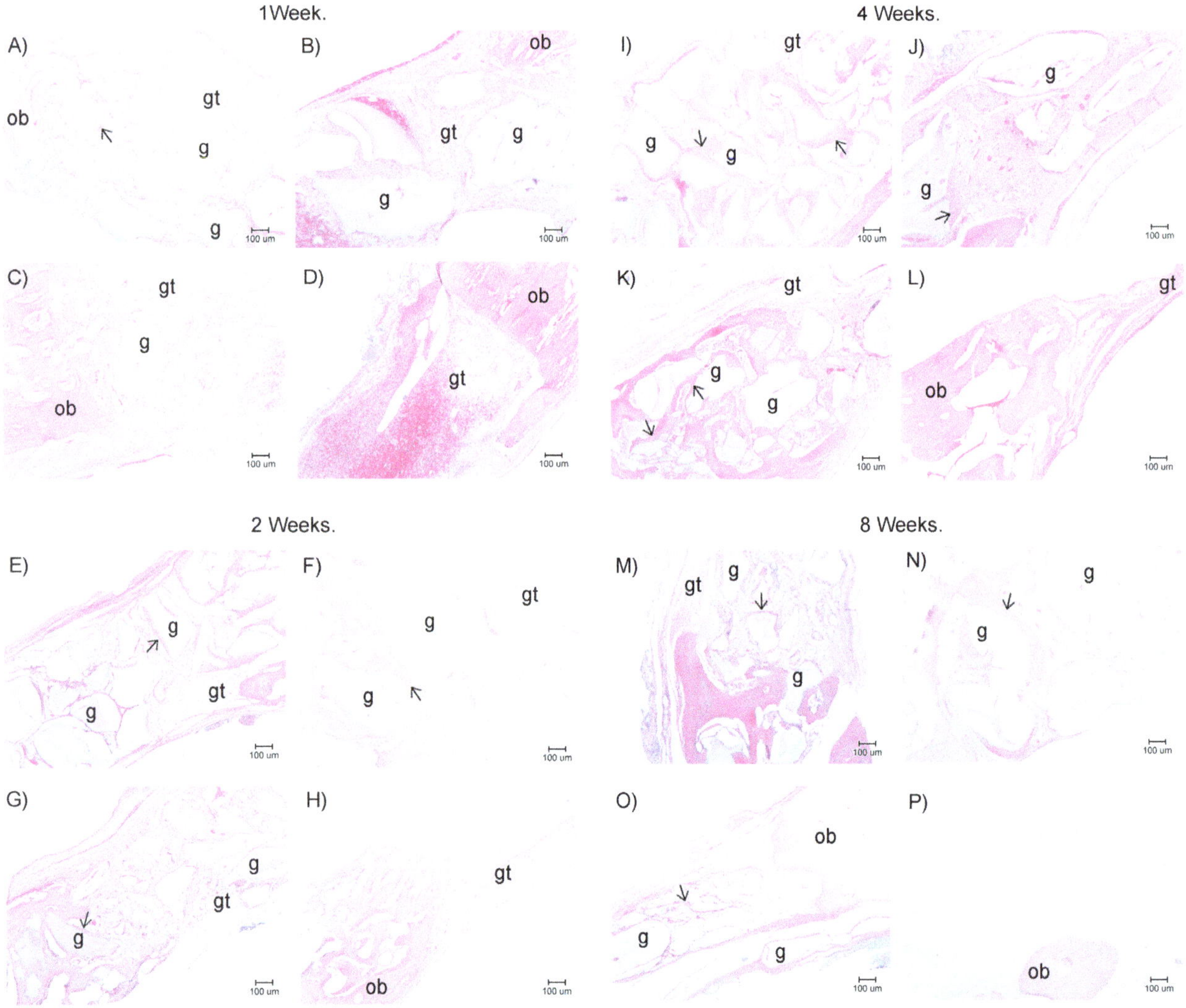

Figure 5. Histology of porcine graft in rabbit calvarial bone defect. Panoramic ×40 Hematoxylin–Eosin stain images of the different defects at one, two, four, eight weeks. (**D**, **H**, **L**, **P**) Control defects; (**A**, **E**, **I**, **M**) MBCP; (**C**, **G**, **K**, **O**) Porcine Graft 250–500 μm; (**B**, **F**, **J**, **N**) Porcine Graft 500–1000 μm. Arrows indicate new bone formation, g = graft, gt = granulation tissue, ob = old bone.

3. Discussion

In comparison with models using self-contained and/or dehiscence-type defects, the critical-size defect model represents a challenging situation for bone formation because bone proliferation can only occur from the borders of the calvarial defect [25]. Hence, this model has been used to study regeneration of defects with and without the use of barrier membranes. In the present study, bone formation was assessed in the rabbit calvarial critical-size defects without barrier membranes in the presence and absence of novel porcine graft materials with varying particle sizes.

Implantation of 250–500 μm and 500–1000 μm porcine grafts into the rabbit calvarial bone led to equal or better healing than implantation of MBCP. Moreover, micro-CT scanning demonstrated greater percentage increases in bone volume density in the presence of the three graft materials than in the control defects. Micro-CT scanning provides accurate determinations of both cancellous and cortical bone and of spatial distributions of scaffold mineralization, particularly in calvarial defects [26]. Nonetheless, histological examinations are considered the gold standard for assessments of healing after fracture wounds and surgical procedures [27]. Thus, bone formation were investigated using micro-CT scanning and light microscopy. In the microscopic observation, the new bone formations were found around the surface of the porcine graft, as Albrektsson *et al.* (2001) indicated osteoconduction means that bone grows on a surface [28]. The histological results demonstrate the osteoconductive potential of the porcine graft. The similar results also showed as qualitative description in the micro-CT observation and further validated the application of micro-CT scanning for measurements of new bone formation.

In agreement with the present data, Calvo *et al.* (2013) showed biocompability, bioabsorbability, and osteoconductive characteristics of porcine grafts that did not inhibit normal bone repair [18]. Moreover, previous studies showed promising clinical application of porcine bone [5,6,29], although sinus augmentation from subsequently placed dental implants revealed condensation properties in one-year histological analyses, indicating absorption of the material with time [19]. In agreement, Barone *et al.* (2012) performed six-month hystomorphometric evaluations of maxillary sinus augmentation following implantation of prehydrated corticocancellous porcine bone, and showed excellent osteoconductive properties that facilitated sinus augmentation. Moreover, porcine bone showed a high percentage of reabsorption after six months, potentially reflecting the collagen contents and porosity of the graft material [30].

In the present X-Ray Diffraction (XRD) analyses, the porcine grafts exhibited similar Ca:P ratios to those of human bone and showed a maximum intensity of 600 a.u. at 27°, similar to that in HA (21°). Moreover, peaks of 200 a.u. were observed at 25° and 40°, which correspond with those in adult human and bones from 26 day old fetuses, although peak widths indicated low crystallinity at these angles. Nonetheless, these similarities of porcine graft materials and human bones indicate that these graft materials facilitate the production of new bone (Figure 2).

Slotte *et al.* (2013) demonstrated considerable gain of new bone in clinical cases of maxillary sinus bone augmentation using porcine bone and showed reabsorption and replacement with new bone over time [31]. In the present study, formation of new bone was observed *in vivo* in bone defects that were filled with 250–500 μm and 500–1000 μm porcine graft materials, which were slowly reabsorbed. Similar efficacies of these materials may reflect interporotic connections, synthetic temperatures (800 °C), and particle sizes of 100 μm, which is regarded as the minimum pore size for guaranteed bone growth. Go *et al.* (2013) [32] investigated the efficacy of low crystallinity materials with high surfaces areas, large pore sizes, and large pore volumes, which are crucial properties for osteoblast-like cell affinity and for degradation of graft materials during bone regeneration. In that study, differences in chemical and physical properties of porcine-derived xenografts that were heat treated at 400 °C and 1200 °C were described. Heat treatment at 400 °C led to slightly greater increases in new bone volume (45% *vs.* 43%), and the resulting porous structures had larger surface areas and pore volumes than those following heat treatment at 1200 °C [32]. Regardless of particle sizes, the present xenogenic porcine grafts were made at 800 °C and demonstrated identical crystallinity on XRD analyses. However, concomitantly wider peaks were indicative of limited crystallinity of these grafts, which may facilitate direct decomposition by osteoclasts and remodeling of endogenous bone materials [32]. Although crystallinity did not differ between the porcine grafts of differing particle sizes, bone formation was slightly greater with the 250–500 μm porcine graft material. However, both 250–500 μm and 500–1000 μm graft materials were synthesized at temperatures that produce mesopores and macropores and are, therefore, both viable substitutes for autogenous grafts in bone regenerating treatments.

The use of porcine xenogenic grafts in humans carries the risk retroviral transmission from pigs to humans. However, numerous studies showed no infrequent pig endogenous retrovirus transmissions, indicating the safety of transplantation into humans. Nonetheless, these risks require examination in animals models of pig to human xenotransplantations and in clinical studies [17]. Kim *et al.* (2013) assessed the safety of bovine-derived graft biomaterials that may transmit bovine spongiform encephalopathy prions. However, this risk remains poorly characterized, and although bovine derived bone materials are effective for the treatment of human intrabony periodontal lesions, these materials are dangerous in humans [21]. Taken together, the present data indicate the efficacy and safety of porcine bonc grafts and warrant further investigation in human studies.

4. Materials and methods

4.1. Biomaterials

The biphasic ceramic material MBCP comprises 60% HA and 40% β-TCP and has a complete interconnected porosity of 70%, comprising macropores of >10 μm and micropores of <10 μm at a 2:1 ratio [33].

The present porcine grafts were produced with particle sizes of 250–500 μm and 500–1000 μm at Taipei Medical University. Porcine graft material with a high porosity was produced using cortical porcine bone. In this procedure, pig bones were collocated in water and were settled at 300–400 °C for 15–30 min to remove fat, residual meat, and other soft tissues. Subsequently, the bones were re-cut into 5–10 mm pieces and were immersed in 0.1–0.5 M hydrochloric acid (HCl) for 5–10 min to etch and remove proteins. The bones were then rinsed with deionized water and dried and were subjected to the following temperature gradients in crucibles with a heating process:

- 5 °C/min heated to 100 °C and then maintained for 30 min;
- 5 °C/min heated to 300 °C and then maintained for 60 min;
- 10 °C/minute heated to 800 °C and then maintained for 120 min.

The samples were then allowed to cool to room temperature, and the resulting products were sieved to the desired particle size in a mortar. Finally, bone aliquots were placed into glass bottles and were sterilized using γ-rays.

4.2. Porcine Graft Analysis

4.2.1. Scanning Electron Microscopy and Energy Dispersive Spectrometry

Elemental analyses of porcine graft samples were performed using a scanning electron microscope (SEM) equipped with an energy dispersive X-ray spectrometer (EDS).

4.2.2. X-ray Diffraction

Porcine graft diffraction patterns were analyzed using X-ray diffraction (XRD), and crystal structures and constituent materials were determined.

4.2.3. Cell Proliferation (MTT Assay)

Cell metabolic activity was evaluated according to succinic dehydrogenase (SDH) activity using the spectrophotometric methyl tetrazolium assay (MTT assay) without washing steps. The main advantages of the colorimetric assay are rapid, precision, and without any radioisotope [34]. In these experiments, Dulbecco's Modified Eagle's Medium (DMEM-S) solutions containing Phosphate Buffered Saline (BLS) + polysialyltransferases (PST) with or without (control) porcine or commercial graft materials were incubated in a humidified incubator at 37 °C. After 72 h, the media solutions were separated from porcine and commercial graft materials and were placed on MG-63 cells, which were then allowed to recover for 24 h. Cell metabolic activity was then analyzed in 12 cell cultures for each media solution.

The MG-63 cells were seeded (30,000 cells/cm^2) onto 24-well plates (Costar Corporation, Cambridge, MA, USA) and were maintained in a humidified incubator with 5% CO_2 and 95% air at 37 °C for 72 h. Thereafter, the culture medium was aspirated, and all media solutions were added to the cell culture wells. After exposure for 24 h, the media solutions were aspirated and replaced with 500 μL of culture medium (α-MEM) and 500 μL of MTT solution (5 mg/mL phosphate buffered saline-PBS) and were incubated at 37 °C for 24 h. Thereafter, the solution was replaced with 500 μL of acidified 0.5% Triton X-100, and absorbance was measured at 570 nm using a spectrophotometer (Plate Chameleon Multilabel Detection Platform; HIDEX Co. Ltd., Turku, Finland).

Absorbance values were standardized to solutions of tetrazolium bromide, and toxicity was determined according to cell survival. The data were analysis with a Student *t*-test.

4.2.4. Statistical Analysis

The non parametric Kruskal–Wallis test was used to analyze the histomorphometric from the Micro CT data among experimental times within each tested material. MTT assay were analyzed by Student *t*-test. The significant level was set at $p < 0.05$.

5. Surgical Procedure

Surgical procedures were performed in 20 adult male New Zealand white rabbits with a mean age of 10 weeks and a mean weight of 2.1 kg. The animals were housed in cages at 19 °C and 55% humidity in the Taipei Medical University Laboratory Animal Center and were fed standard rabbit chow and water *ad libitum*. Anesthesia was administered using an intramuscular injection of Zoletil 50 (50 mg/mL) at a dosage of 15 mg/kg into the gluteal region, and surgery was performed in the sedated animals after 10 min. The calvarial region was then shaved, draped, and sterilized using iodine, and 1.8 mL of 2% lidocaine with epinephrine 1/100,000 was then injected as a hemostatic. Subsequently, 2 cm longitudinal vertical skin midline incisions were made, periosteums were retracted to expose calvarial bones, and four critical calvarial defects of 5 mm diameter and 3 mm depth were prepared using a trephine bur (3I implant innovation, Palm Beach Gardens, FL, USA) bilaterally in the parietal and frontal bones of each rabbit. The upper left, upper right, and lower left defects were filled with 250–500 μm porcine graft, 500–1000 μm porcine graft, and HA/β-TCP (MBCP; Biomatlante biologics solutions), respectively, and the lower right defects were used as unfilled controls. The animals were kept in cages under surveillance for the first 24 h and were then examined every 3 days for 2 weeks, and weekly thereafter [25,35] (Figure 6).

The animals were grouped for sacrifice at 1, 2, 4, and 8 weeks after surgery. Euthanasia was performed by CO_2 asphyxiation 10 min after the intramuscular injection of Zoletil 50 (50 mg/mL) at 15 mg/kg into the gluteal region. Subsequently, a section was made through the middle of the frontal plane between the 4 filled critical defects. Samples blocks were prepared in formalin and micro-computed tomography (micro-CT) scanning analyses were performed within 2 weeks using Skyscan 1076 (Skyscan, Antwerp, Belgium).

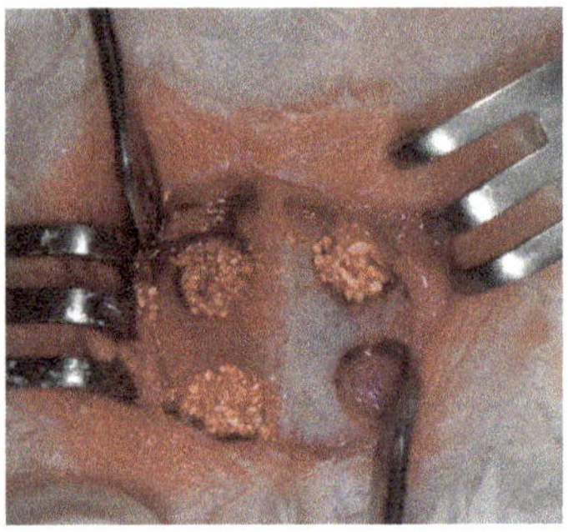

Figure 6. Graft materials and calvarial critical-size defects. The four calvarial critical-size defects were filled with 250–500 μm porcine graft (upper left), 500–1000 μm porcine graft (upper right), or MBCP (lower left), and the lower right defect was used as an unfilled control.

After setting the micro-CT images, coronal images of the center and peripheral areas of the defect were saved in the database and 2-dimensional (2D) morphological analyses were performed for 64 samples. Among these, 8 were not analyzed because of premature deaths of two animals.

Morphometric parameters were calculated on individual binarized cross-section images, and 2D morphometric parameters were determined slice-by-slice and were then integrated over multiple slices. After integration of whole volumes of interest (VOI), percentages of VOI occupied by binarized solid objects were calculated. This parameter comprises percentages of the total volume (TV) of VOI and the total binarized volume (BV) of objects within VOI. However, this relationship is only relevant if the studied volume is fully contained within a well-defined biphasic region of solid and space, such as the trabecular bone region, and does not extend into or beyond the boundary of the object, such as the cortical boundary of the bone sample. Moreover, measured percent volumes are dependent on the criteria for selection of VOI. Thus, binary selections of samples from the morphometric analyses were made according to grayscale density between units of 20 and 80. The morphometric analyses were performed using Skyscan 1076 Skyscan CTan software according to the manufacturer's instructions (Figure 7).

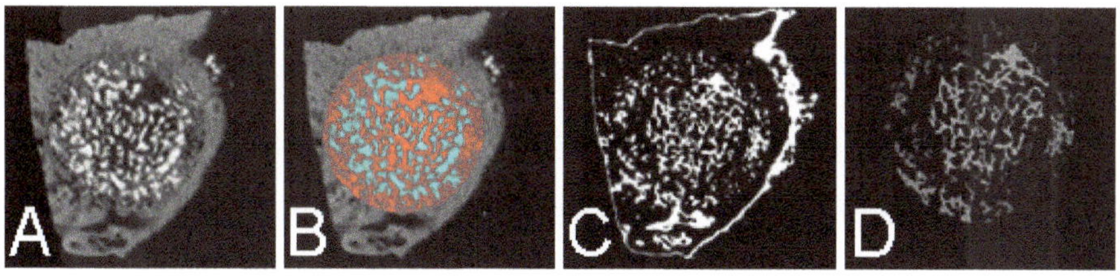

Figure 7. Morphometric analyses in grayscale density using Skyscan 1076 Skyscan CTan software. Morphometric analyses were done in grayscale density using Skyscan 1076 Skyscan CTan software (**A**) Raw image, coronal view; (**B**) Determine region of interest; (**C**) Binari selection of the sample; (**D**) Morphometric analysis.

Tissue sections were taken from all micro-CT sample blocks using the following procedures:

- Decalcification: The bone specimens were trimmed to a thickness of less than 0.5 cm (to accelerate decalcification) and were decalcified in fresh fluid. Decalcification times, specimen thicknesses, temperature, decalcification solution freshness, and block samples' decalcification conditions were recorded.
- Washing with water: After decalcification, the specimens were washed in running water for several hours to neutralize the strongly acidic decalcifying solution.
- Dehydration and embedding: The bone tissues were dehydrated in alcohol and were then embedded in liquid paraffin.
- Sectioning: Slice thicknesses were set at 5–7 μm.
- Staining: Hematoxylin and Eosin (H&E) staining was performed on all paraffin embedded tissues.
- Visualization: The histological slides were viewed using an optical microscope (Olympus BH-2, Tokyo, Japan), and the images were captured at 40× magnification using a camera SPOT idea tm Camera that was connected to the microscope and were analyzed using the corresponding software (SPOT imaging software, Sterling Heights, MI, USA).

6. Conclusions

The present porcine grafts showed great promise and resembled human bone in the micro-CT, XRD, and histological analyses, and promoted cell survival and growth of new bone. The rabbit calvarial bone tissue responded well to both the 250–500 μm and 500–1000 μm porcine grafts, which had similar efficacy to that of commercial HA/β-TCP (MBCP). The present observations also indicate that porcine grafts do not interfere with wound healing and promote bone formation through osteoconduction, warranting further investigation as graft substitutes. Although minimal graft reabsorption was observed in this study, longer experiments may show sufficient resorption in humans.

Acknowledgments

The authors would like to thank Ministry of Science and Technology, Taiwan for grant support (NSC 102-2314-B-038-020), and Enago (Kent, DE, USA) for the English language review.

Author Contributions

All authors contributed to this work, discussed the results and implications and commented on the manuscript at all stages. Wei-Jen Chang, Chin-Yi Lin, Haw-Ming Huang and Eisner Salamanca conceived the strategies. Che-Tong Lin supervised the project. Wei-Jen Chang, Eisner Salamanca and Wei-Fang Lee designed and performed experiments, analyzed data and wrote the paper. Wei-Jen Chang also funded in this study.

Conflicts of Interest

All authors state no conflicts of interest.

References

1. Trombelli, L. Which reconstructive procedures are effective for treating the periodontal intraosseous defect? *Periodontology 2000* **2005**, *37*, 88–105.
2. Susin, C.; Wikesjö, U.M.E. Regenerative periodontal therapy: 30 years of lessons learned and unlearned. *Periodontology 2000* **2013**, *62*, 232–242.
3. Ramseier, C.A.; Rasperini, G.; Batia, S.; Giannobile, W.V. Advanced reconstructive technologies for periodontal tissue repair. *Periodontology 2000* **2012**, *59*, 185–202.
4. Weiss, P.; Layrolle, P.; Clergeau, L.P.; Enckel, B.; Pilet, P.; Amouriq, Y.; Daculsi, G.; Giumelli, B. The safety and efficacy of an injectable bone substitute in dental sockets demonstrated in a human clinical trial. *Biomaterials* **2007**, *28*, 3295–3305.
5. Scarano, A.; Piattelli, A.; Perrotti, V.; Manzon, L.; Iezzi, G. Maxillary sinus augmentation in humans using cortical porcine bone: A histological and histomorphometrical evaluation after 4 and 6 months. *Clin. Implant Dent. Relat. Res.* **2011**, *13*, 13–18.
6. Orsini, G.; Scarano, A.; Piattelli, M.; Piccirilli, M.; Caputi, S.; Piattelli, A. Histologic and ultrastructural analysis of regenerated bone in maxillary sinus augmentation using a porcine bone—Derived biomaterial. *J. Periodontol.* **2006**, *77*, 1984–1990.
7. Trombelli, L.; Heitz-Mayfield, L.J.A.; Needleman, I.; Moles, D.; Scabbia, A. A systematic review of graft materials and biological agents for periodontal intraosseous defects. *J. Clin. Periodontol.* **2002**, *29*, 117–135.
8. Ramírez-Fernández, M.; Calvo-Guirado, J.L.; Delgado-Ruiz, R.A.; Maté-Sánchez del Val, J.E.; Vicente-Ortega, V.; Meseguer-Olmos, L. Bone response to hydroxyapatites with open porosity of animal origin (porcine [osteobiol®mp3] and bovine [endobon®]): A radiological and histomorphometric study. *Clin. Oral Implants Res.* **2011**, *22*, 767–773.
9. Tamimi, F.M.; Torres, J.; Tresguerres, I.; Clemente, C.; López-Cabarcos, E.; Blanco, L.J. Bone augmentation in rabbit calvariae: Comparative study between Bio-Oss® and a novel β-TCP/DCPD granulate. *J. Clin. Periodontol.* **2006**, *33*, 922–928.
10. Nery, E.B.; LeGeros, R.Z.; Lynch, K.L.; Lee, K. Tissue response to biphasic calcium phosphate ceramic with different ratios of HA/βTCP in periodontal osseous defects. *J. Periodontol.* **1992**, *63*, 729–735.

11. Garrido, C.A.; Lobo, S.E.; Turíbio, F.M.; LeGeros, R.Z. Biphasic calcium phosphate bioceramics for orthopaedic reconstructions: Clinical outcomes. *Int. J. Biomater.* **2011**, *2011*, doi:10.1155/2011/129727.
12. Blokhuis, T.J.; Termaat, M.F.; den Boer, F.C.; Patka, P.; Bakker, F.C.; Haarman, H.J. Properties of calcium phosphate ceramics in relation to their *in vivo* behavior. *J. Trauma Inj. Infect. Crit. Care* **2000**, *48*, 179–186.
13. Huffer, W.E.; Benedict, J.J.; Turner, A.S.; Briest, A.; Rettenmaier, R.; Springer, M.; Walboomers, X.F. Repair of sheep long bone cortical defects filled with COLLOSS®, COLLOSS® E, OSSAPLAST®, and fresh iliac crest autograft. *J. Biomed. Mater. Res. B Appl. Biomater.* **2007**, *82B*, 460–470.
14. Habibovic, P.; Kruyt, M.C.; Juhl, M.V.; Clyens, S.; Martinetti, R.; Dolcini, L.; Theilgaard, N.; van Blitterswijk, C.A. Comparative *in vivo* study of six hydroxyapatite-based bone graft substitutes. *J. Orthop. Res.* **2008**, *26*, 1363–1370.
15. Maté-Sánchez de Val, J.E.; Calvo-Guirado, J.L.; Delgado-Ruiz, R.A.; Ramírez-Fernández, M.P.; Negri, B.; Abboud, M.; Martínez, I.M.; de Aza, P.N. Physical properties, mechanical behavior, and electron microscopy study of a new α-TCP block graft with silicon in an animal model. *J. Biomed. Mater. Res. A* **2012**, *100*, 3446–3454.
16. Calvo-Guirado, J.L.; Ramírez-Fernández, M.P.; Delgado-Ruíz, R.A.; Maté-Sánchez, J.E.; Velasquez, P.; de Aza, P.N. Influence of biphasic β-TCP with and without the use of collagen membranes on bone healing of surgically critical size defects. A radiological, histological, and histomorphometric study. *Clin. Oral Implants Res.* **2014**, *25*, 1228–1238.
17. Magre, S.; Takeuchi, Y.; Bartosch, B. Xenotransplantation and pig endogenous retroviruses. *Rev. Med. Virol.* **2003**, *13*, 311–329.
18. Calvo Guirado, J.L.; Ramírez Fernández, M.P.; Negri, B.; Delgado Ruiz, R.A.; Maté Sánchez de-Val, J.E.; Gómez-Moreno, G. Experimental model of bone response to collagenized xenografts of porcine origin (OsteoBiol® mp3): A radiological and histomorphometric study. *Clin. Implant Dent. Relat. Res.* **2013**, *15*, 143–151.
19. Pagliani, L.; Andersson, P.; Lanza, M.; Nappo, A.; Verrocchi, D.; Volpe, S.; Sennerby, L. A collagenated porcine bone substitute for augmentation at neoss implant sites: A prospective 1-year multicenter case series study with histology. *Clin. Implant Dent. Relat. Res.* **2012**, *14*, 746–758.
20. Barone, A.; Ricci, M.; Calvo-Guirado, J.L.; Covani, U. Bone remodelling after regenerative procedures around implants placed in fresh extraction sockets: An experimental study in beagle dogs. *Clin. Oral Implants Res.* **2011**, *22*, 1131–1137.
21. Kim, Y.; Nowzari, H.; Rich, S.K. Risk of prion disease transmission through bovine-derived bone substitutes: A systematic review. *Clin. Implant Dent. Relat. Res.* **2013**, *15*, 645–653.
22. Calvo-Guirado, J.L.; Gómez-Moreno, G.; López-Marí, L.; Guardia, J.; Marínez-González, J.M.; Barone, A.; Tresguerres, I.F.; Paredes, S.D.; Fuentes-Breto, L. Actions of melatonin mixed with collagenized porcine bone versus porcine bone only on osteointegration of dental implants. *J. Pineal Res.* **2010**, *48*, 194–203.
23. Calvo-Guirado, J.L.; Gómez-Moreno, G.; Guardia, J.; Ortiz-Ruiz, A.; Piatelli, A.; Barone, A.; Martínez-González, J.M.; Meseguer-Olmo, L.; López-Marí, L.; Dorado, C.B. Biological response to porcine xenograft implants: An experimental study in rabbits. *Implant Dent.* **2012**, *21*, 112–117.

24. Zhu, H.; Li, J.; Wang, S.; Liu, K.; Wang, L.; Huang, L. Hmgb1-TLR4-IL-23-IL-17A axis promote ischemia-reperfusion injury in a cardiac transplantation model. *Transplantation* **2013**, *95*, 1448–1454.
25. Donos, N.; Lang, N.P.; Karoussis, I.K.; Bosshardt, D.; Tonetti, M.; Kostopoulos, L. Effect of GBR in combination with deproteinized bovine bone mineral and/or enamel matrix proteins on the healing of critical-size defects. *Clin. Oral Implants Res.* **2004**, *15*, 101–111.
26. Sawyer, A.A.; Song, S.J.; Susanto, E.; Chuan, P.; Lam, C.X.F.; Woodruff, M.A.; Hutmacher, D.W.; Cool, S.M. The stimulation of healing within a rat calvarial defect by mPCL–TCP/collagen scaffolds loaded with rhBMP-2. *Biomaterials* **2009**, *30*, 2479–2488.
27. Yeom, H.; Blanchard, S.; Kim, S.; Zunt, S.; Chu, T.-M.G. Correlation between micro-computed tomography and histomorphometry for assessment of new bone formation in a calvarial experimental model. *J. Craniofac. Surg.* **2008**, *19*, 446–452.
28. Albrektsson, T.; Johansson, C. Osteoinduction, osteoconduction and osseointegration. *Eur. Spine J.* **2001**, *10*, S96–S101.
29. Traini, T.; Piattelli, A.; Caputi, S.; Degidi, M.; Mangano, C.; Scarano, A.; Perrotti, V.; Iezzi, G. Regeneration of human bone using different bone substitute biomaterials. *Clin. Implant Dent. Relat. Res.* **2013**, *17*, 150–162.
30. Barone, A.; Ricci, M.; Covani, U.; Nannmark, U.; Azarmehr, I.; Calvo-Guirado, J.L. Maxillary sinus augmentation using prehydrated corticocancellous porcine bone: Hystomorphometric evaluation after 6 months. *Clin. Implant Dent. Relat. Res.* **2012**, *14*, 373–379.
31. Slotte, C.; Lindfors, N.; Nannmark, U. Surgical reconstruction of peri-implant bone defects with prehydrated and collagenated porcine bone and collagen barriers: Case presentations. *Clin. Implant Dent. Relat. Res.* **2013**, *15*, 714–723.
32. Go, A.; Kim, S.E.; Shim, K.M.; Lee, S.-M.; Choi, S.H.; Son, J.S.; Kang, S.S. Osteogenic effect of low-temperature-heated porcine bone particles in a rat calvarial defect model. *J. Biomed. Mater. Res. A* **2014**, *102*, 3609–3617.
33. Le Guehennec, L.; Goyenvalle, E.; Aguado, E.; Pilet, P.; Bagot D'Arc, M.; Bilban, M.; Spaethe, R.; Daculsi, G. MBCP® biphasic calcium phosphate granules and Tissucol® fibrin sealant in rabbit femoral defects: The effect of fibrin on bone ingrowth. *J. Mater. Sci. Mater. Med.* **2005**, *16*, 29–35.
34. Mosmann, T. Rapid colorimetric assay for cellular growth and survival: Application to proliferation and cytotoxicity assays. *J. Immunol. Methods* **1983**, *65*, 55–63.
35. Messora, M.R.; Nagata, M.J.; Mariano, R.C.; Dornelles, R.C.; Bomfim, S.R.; Fucini, S.E.; Garcia, V.G.; Bosco, A.F. Bone healing in critical-size defects treated with platelet-rich plasma: A histologic and histometric study in rat calvaria. *J. Periodontal Res.* **2008**, *43*, 217–223.

4

Characterization of Dentine to Assess Bond Strength of Dental Composites

Saad Liaqat [1], Anas Aljabo [1], Muhammad Adnan Khan [1], Hesham Ben Nuba [1], Laurent Bozec [1], Paul Ashley [2] and Anne Young [1,*]

[1] Division of Biomaterials and Tissue Engineering, UCL Eastman Dental Institute, 256 Gray's Inn Road, London WC1X 8LD, UK; E-Mails: saad.liaqat.11@ucl.ac.uk (S.L.); anas.aljabo.10@ucl.ac.uk (A.A.); adnan.khan.11@ucl.ac.uk (M.A.K.); hesham.nuba.09@ucl.ac.uk (H.B.N.); l.bozec@ucl.ac.uk (L.B.)

[2] Department of Paediatric Dentistry, UCL Eastman Dental Institute, 256 Gray's Inn Road, London WC1X 8LD, UK; E-Mail: p.ashley@ucl.ac.uk

* Author to whom correspondence should be addressed; E-Mail: anne.young@ucl.ac.uk;

Academic Editor: Jung Kwon (John) Oh

Abstract: This study was performed to develop alternating dentine adhesion models that could help in the evaluation of a self-bonding dental composite. For this purpose dentine from human and ivory was characterized chemically and microscopically before and after acid etching using Raman and SEM. Mechanical properties of dentine were determined using 3 point bend test. Composite bonding to dentine, with and without use of acid pre-treatment and/or the adhesive, were assessed using a shear bond test. Furthermore, micro gap formation after restoration of 3 mm diameter cavities in dentine was assessed by SEM. Initial hydroxyapatite level in ivory was half that in human dentine. Surface hydroxyapatites decreased by approximately half with every 23 s of acid etch. The human dentine strength (56 MPa) was approximately double that of ivory, while the modulus was almost comparable to that of ivory. With adhesive use, average shear bond strengths were 30 and 26 MPa with and without acid etching. With no adhesive, average bond strength was 6 MPa for conventional composites. This, however, increased to 14 MPa with a commercial flowable "self–bonding" composite or upon addition of low levels of an acidic monomer to the experimental composite. The acidic monomer additionally reduced micro-gap formation with the experimental composite. Improved bonding and mechanical properties should reduce composite failures due to recurrent caries or fracture respectively.

Keywords: dental composite; dentine; 4-META; self-adhesive; hydroxyapatite; collagen; micro-gap

1. Introduction

Dental caries, otherwise known as tooth decay, is one of the most prevalent chronic diseases of people worldwide [1]. It is caused by bacteria producing acid that enhances the solubility of hydroxyapatite in enamel and underlying dentine. Over time this process will lead to demineralization and cavitation [2]. Once the cavity is through to the dentine the infected areas must be removed and replaced by a filling material to restore tooth function and prevent continuing decay. Following the Minamata convention on mercury, phasing out of amalgam fillings will become likely. Dental composites will then be the main direct tooth restorative material. Composites have good esthetics, adhesion, and requires minimal cavity preparation. But on the other hand they are technically more difficult to place, and have higher failure rates due primarily to secondary caries beneath the restoration [3–5].

Conventionally, the most reliable composite bond has been achieved by first acid etching the dentine [6]. This provides a porous hydroxyapatite depleted surface collagen mesh and opens up aqueous fluid filled dentine tubules. Flowable hydrophilic dentine primers and adhesives can penetrate and upon polymerization physically interlock with both collagen and tubules [7]. The adhesives can additionally contain acidic monomers that can form ionic bonds with the dentine. Furthermore, the adhesive is able to chemically bond with the viscous hydrophobic composites. In an attempt to increase simplicity and reduce complexity, in the last decade, a major drive has been towards "single step" adhesives that may bond without etching. Their clinical success, however, has been variable. More recently, "flowable" composites, that have the potential to bond without etching or adhesive, have also been produced [8]. These flowable composites, however, can have high shrinkage enhancing micro-gap formation.

Predicting clinical success and optimization of new dental adhesives has been difficult. Many studies have used extracted human and bovine teeth to evaluate the adhesive strength of restorative materials. Human teeth, however, are difficult to obtain. Other problems include their small size, variability with age and disease level [9–11], infection hazard [12], and ethical issues [13]. Bovine teeth on the other hand are easy to obtain, but have same drawbacks like human teeth, which includes small size, variability, and cross infection. Alternative dentine models to quantify and provide better understanding of factors affecting bonding are therefore required. Confiscated ivory destined for destruction is available from customs but must be for academic purposes only. The relatively massive size of a single tusk, however, provides a potentially more reproducible substrate for academic research. The aim of this study was therefore to assess if ivory is a good replacement model for human dentine. The study additionally compares the dentine bonding of various composites with and without a single step adhesive and/or acid etching. The composites include conventional and flowable materials in addition to a new viscous composite with potential for self-bonding to dentine.

2. Results and Discussion

2.1. Results

2.1.1. Raman and SEM of Dentine

The normalized average spectra of both dentine were practically identical between 1200 and 1800 cm^{-1}. This range included the 1670 cm^{-1} amide I, 1453 cm^{-1} amide II, and 1260 cm^{-1} amide III peaks due to collagen (Figure 1a,b). Below 1200 cm^{-1} both the hydroxyapatite (961 cm^{-1}) and β- CO_3^{-2} (1073 cm^{-1}) peaks were more intense for the human dentine. After acid etching with 37% phosphoric acid for 20, 60 and 120 s both dentine hydroxyapatite and carbonate peaks declined (Figure 1c). The 20 s etched human dentine was almost identical to that of ivory except for having a much stronger carbonate peak.

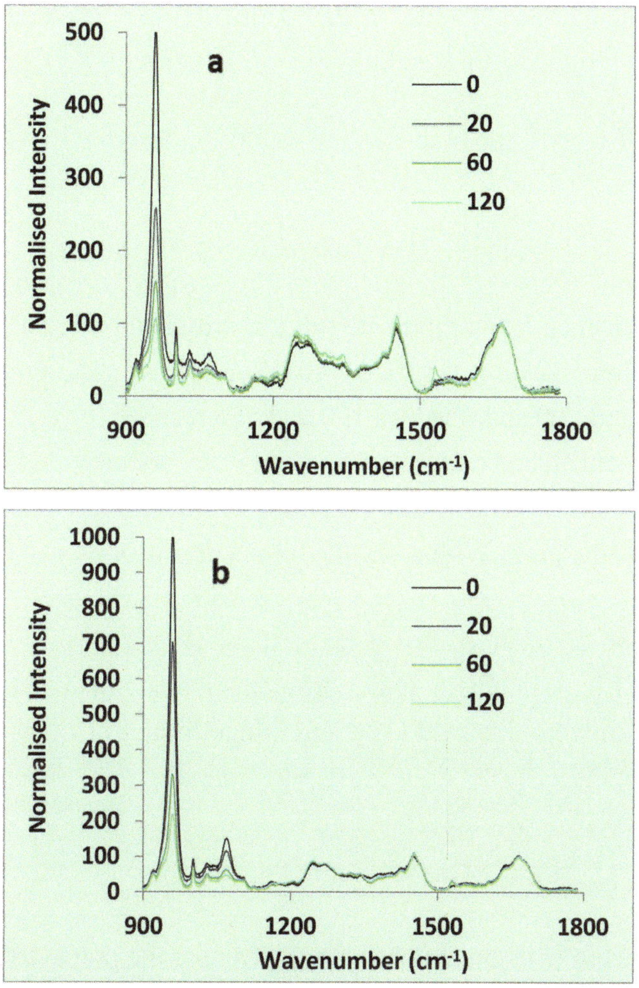

Figure 1. *Cont.*

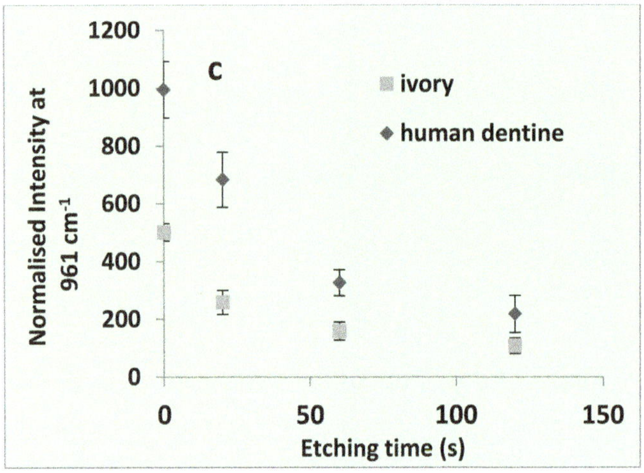

Figure 1. Chemical analysis of dentine using Raman spectroscopy. Raman spectra of (**a**) ivory; and (**b**) human dentine both after 0, 20, 60, or 120 s of acid etching; (**c**) Normalized intensity at 961 cm^{-1} is plotted against etching time in seconds (n = 5).

Normalized intensity at 961 cm^{-1} for both dentine could be described well by an Equation (1) of the form:

$$[\ln\left[\frac{I_t - I_f}{I_0 - I_f}\right] = -0.03t]\ R^2 > 0.99 \quad (1)$$

where I is intensity and subscripts t, 0 and f indicate times t, initial and final. This indicates the time for half maximum surface reaction is 23 s (−(ln (0.5))/0.03) for both dentine type. For ivory and human dentine I_0 values were 1000 and 500 and I_f values 190 and 105 respectively.

Example SEM images of acid etched human dentine and ivory are provided in Figure 2. With human dentine the dentine tubule density (number/mm^2) increased progressively from approximately 15,000 near the crown to 30,000 near the pulp. With ivory the tubule density was ~10,000 mm^{-2}.

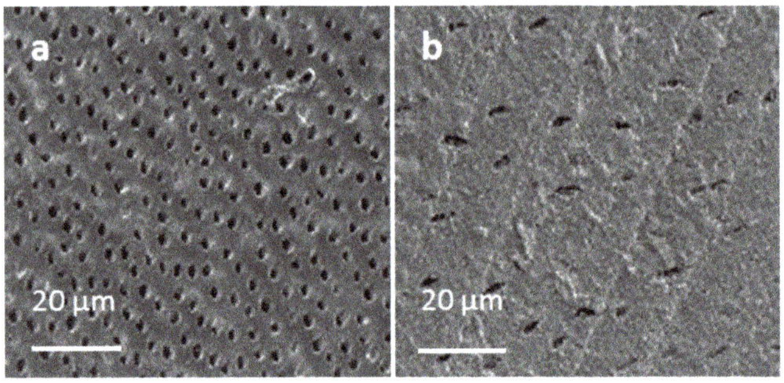

Figure 2. SEM images of (**a**) human dentinal tubules near the pulp; (**b**) tubules in ivory.

2.1.2. Flexural Properties

P_f (Equation (5)) was plotted *vs.* bending strength, modulus, and flexural strain of dentine in Figure 3.

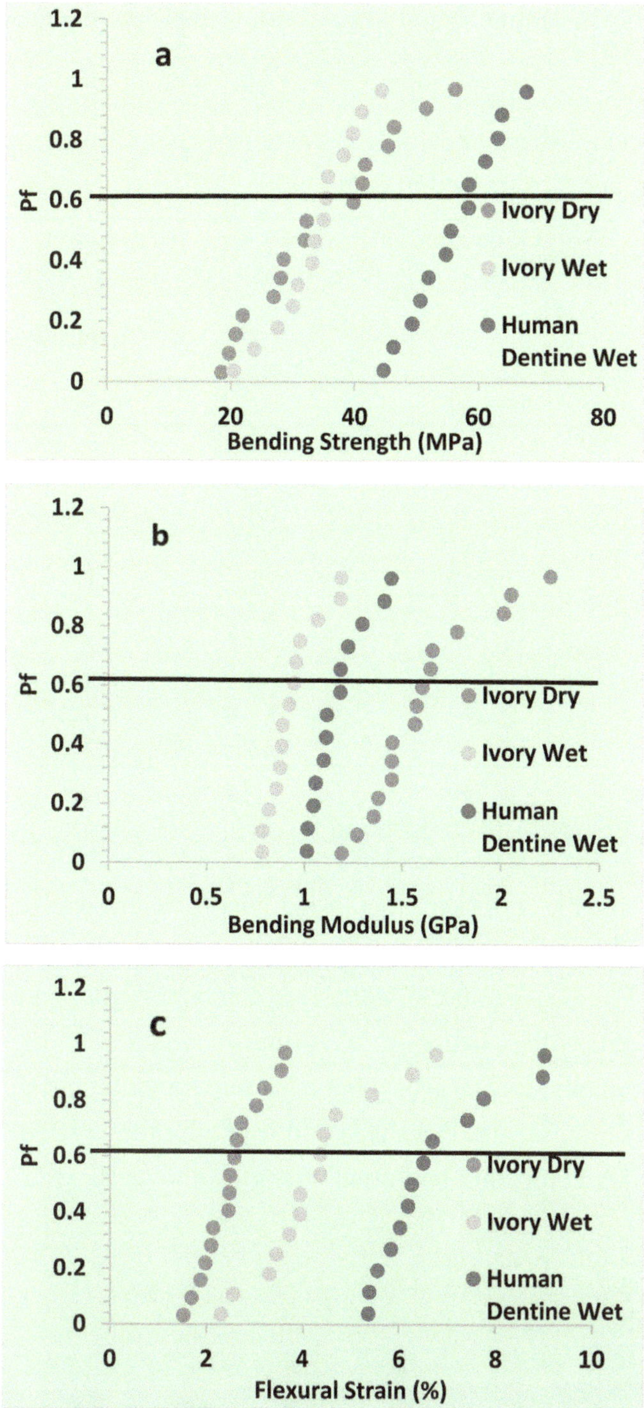

Figure 3. P_f was plotted against (**a**) three point bend strength; (**b**) modulus; and (**c**) flexural strain of ivory (dry & wet) and human dentine (wet).

Mean results and Weibull parameters obtained by fitting Equation (6) are provided in Table 1. The high R^2 values indicated the Weibull distribution described strength variation particularly well but could also fit modulus and strain data. The Weibull strength scale parameter, σ_θ, and mean strength were significantly higher for human dentine as compared to ivory dry and wet. With dry ivory, the reduced shape parameter, *m*, and greater 95% CI, indicated a broadened distribution of strengths compared with wet ivory or human dentine. Conversely, the Weibull scale parameter and mean modulus, increased in the order wet ivory < human wet dentine < dry ivory. Flexural strain increased in the order dry ivory < wet ivory < human wet dentine. Modulus and strain shape parameters, were not significantly affected by dentine type.

Table 1. Weibull scale σ_0 and shape *m* parameters obtained upon fitting Equation (6) to three point strength, modulus, and flexural strain data.

Three point strength	Weibull parameter	Ivory dry	Ivory wet	Human dentine wet
Bending Strength (MPa)	σ_0	38	36	59
	m	3	6	8
	R^2	0.97	0.99	0.99
	Mean	34	34	56
	95% Cl	6	4	4
Bending Modulus (GPa)	σ_0	1.7	1.0	1.2
	m	6	7	8
	R^2	0.90	0.90	0.91
	Mean	1.6	0.9	1.2
	95% Cl	0.1	0.1	0.1
Flexural Strain (%)	σ_0	2.7	4.7	7.2
	m	4	4	5
	R^2	0.96	0.96	0.90
	Mean	2.5	4.3	6.7
	95% Cl	0.3	0.7	0.7

2.1.3. Shear Bond Strengths of Composites to Dentine

From the combined results the biggest factor increasing bond strength was use of the adhesive ibond (average of 28 with *vs.* 9 MPa without). Acid etching increased average bond strengths both with (26 to 32 MPa) and without (7 to 12 MPa) use of adhesive (compare Figure 4a,b).

With ibond use, bond strengths with both dentine were almost comparable (Figure 4a). Without ibond, Gradia and Z250 bonded best to acid etched ivory whist the experimental C-HEMA composite bonded better to acid etched human dentine. Vertise flow had bond strengths above 10 MPa to all surfaces except un-etched ivory. The C-4META composite had bond strengths above 10 MPa for all dentine surfaces (Figure 4b).

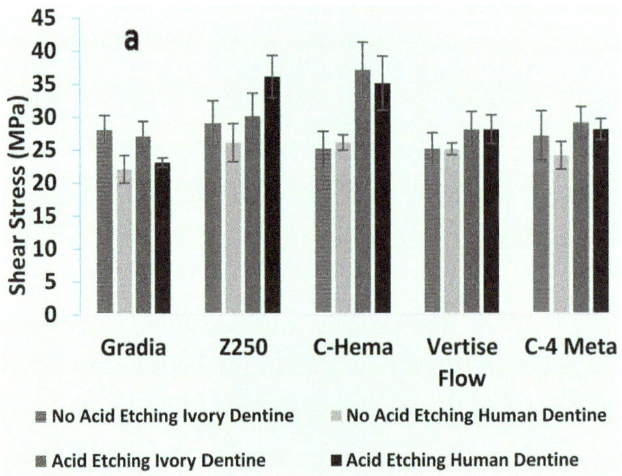

Figure 4. *Cont.*

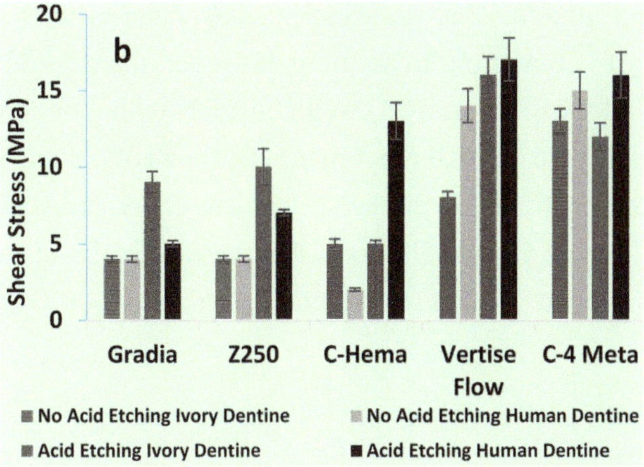

Figure 4. Shear bond strength of experimental and commercial composites, using dentine. (**a**) Shows the effect of ibond along with or without acid etching (20 s); (**b**) shows the effect of No ibond with or without acid etching (20 s) (n = 5).

2.1.4. SEM of Composite/Dentine Interfaces

The SEM images of interfaces were comparable with ivory and human dentine. With the use of ibond adhesive and no acid etching a thin intact interface layer of <2 micron was typically observed between the dentine and most composites (e.g., Figure 5a). The only exception was with Z250. In this case, on one side of the restoration, some cracking of the dentine and/or interface was observed (Figure 5b). If the dentine was acid etched prior to ibond use this problem was overcome. Additionally a thicker adhesive/dentine hybrid layer of ~20 micron thickness was detected irrespective of composite or dentine type (Figure 5c).

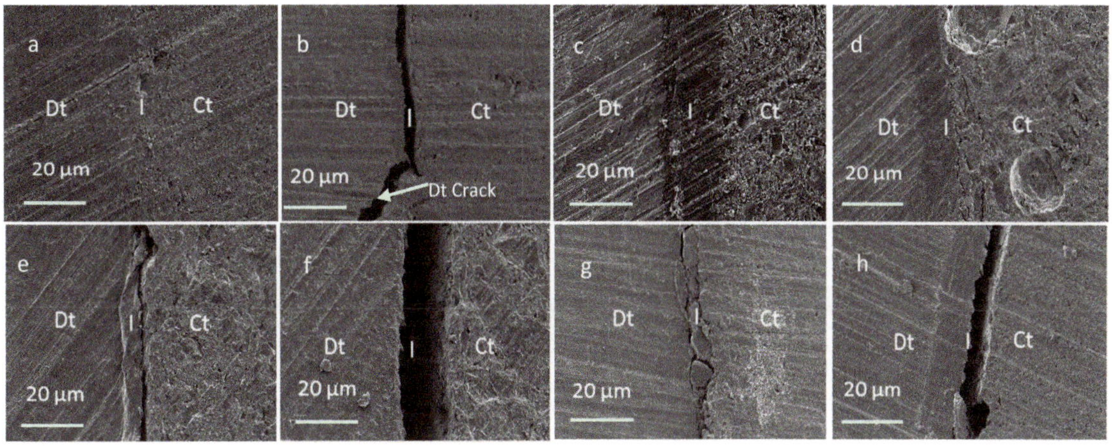

Figure 5. SEM images of interfaces between composites and dentine (**a**) ibond treated dentine with 4-META composite shows 3 different zones (dentine (Dt), interface (I), & composite (Ct)); (**b**) with ibond treated dentine and Z250 some interface and dentine cracking is observed; (**c**) acid and ibond treated dentine with 4-META composite has an ~20 micron thick interface layer; (**d**) Acid etched dentine with 4-META composite shows only minor cracking within the composite; (**e**) Un-treated dentine with 4-META composite has gaps of <2 micron whilst (**f**) un-treated dentine with HEMA composite has gaps >10 micron; (**g**) ibond treated dentine and Z250, failure at dentine interface can be seen; (**h**) Is acid and ibond treated dentine and Z250, the failure can be seen at dentine-ibond interface.

With acid etched dentine and no adhesive, the experimental material with 4-META could form a dentine/composite hybrid layer of ~10 micron. In addition, however, there could be some cracks visible within the compos[ite near this interface (Figure 5d). With no acid etching or adhesive and the 4-META composite gaps were <2 micron (Figure 5e). All other materials showed variable gaps of up to 20 micron between the composite and dentine when no adhesive was employed irrespective of acid treatment (Figure 5f). The SEM images in Figure 5g, and 5h shows that the weak point in composite-ibond-dentine complex is most of the time due to weakness at dentine interface, or with in the dentine itself (Figure 5b).

2.2. Discussion

2.2.1. Chemical Properties of Dentine

The Raman spectra observed above for dentine and acid etched dentine were comparable with those observed previously for intact and carious dentine. Earlier studies additionally showed that etch depth was ~8 micron with 15 s etch of non-carious dentine but that this could be doubled if the dentine was carious [14]. The Raman microscope used in the above study provides the chemistry of the top 10–20 micron thick surface layer and is therefore ideal for assessing changes in chemistry upon acid etching. The relative levels of surface collagen and hydroxyapatite in the above new study suggests that chemically both 20 s etched human dentine and un-etched ivory are a good model for carious dentine. 20 s etched ivory would then be a model for etched carious dentine. The above new results also show acid clearly dissolves the carbonate associated with hydroxyapatite. Carbonate is formed by dissolution of carbon dioxide (e.g., from carbonated drinks) in saliva and is able to replace phosphate ions in hydroxyapatite [15]. This increases the solubility of the hydroxyapatite [16]. The lack of carbonate in ivory is consistent with lack of contact with food and saliva.

2.2.2. Microscopic Properties of Dentine

Microscopic analysis showed that ivory tubules were of comparable size and shape to those in human dentine but of lower density. Sizes observed were comparable with those observed previously with human dentine [17–19]. Tubule density increasing in human dentine closer to the pulp has been previously reported [20].

2.2.3. Mechanical Properties

The above bending strength of the human dentine is comparable with that observed in the literature (between 15 [21], and 200 MPa [22–25]). The broad range of values observed was a consequence of the large number of factors that affect this property. Strength has been shown to increase if the tubules are aligned parallel to the direction of applied force [25] or if the dentine is heated [26]. A reduction in tubule density would also enhance strength. Changes in chemistry, including hydroxyapatite and water content or collagen structure, might also modify strength. Collagen consists of rigid rod like triple helices with crosslinks that bind them into fibrils. Water can penetrate between the helices and expand the fibrils [27]. Hydroxyapatite precipitates around the fibrils [28] stabilizing the collagen against excessive water sorption induced expansion and attack by enzymes. The above new studies suggest that enhancing water content of the ivory has little effect on strength. This might be a consequence of the plasticizing effect of excessive water sorption being compensated by ionization of the collagen enhancing forces of attraction between the collagen fibrils. The higher strength of the human dentine compared with wet ivory could be a consequence of the higher hydroxyapatite content limiting water sorption and thereby compensating the higher density of tubules. On average the maximum amount of water absorbed by human and dry ivory dentine was ~10 w/w %, and 15 w/w % respectively [29].

Increasing water content of collagen can reduce its dry modulus from ~5 GPa down to a few MPa [30]. Increasing volume density of the tubules will decrease modulus but addition of rigid hydroxyapatite (modulus ~100 GPa) enhances modulus [31]. The above observed reduction in modulus of the ivory upon hydration will be due to plasticization of the collagen. The higher modulus of human dentine compared with wet ivory could be primarily due to the increased hydroxyapatite.

From the above study it is clear that water sorption enhances the flexural strain for ivory but that it is even higher for the more mineralized human dentine. A possible explanation could be that in water, positive and negative charges may form (on e.g., lysine and aspartic acid amino acids respectively) in collagen providing inter-fibril interactions. The precipitation of hydroxyapatite could further enhance these ionic interactions. Stronger ionic attractions between the collagen fibrils would enable them to be pulled further apart before breaking [32].

2.2.4. Adhesion Properties

Although there has been much criticism of the reproducibility of the shear bond test, it is still commonly used as a first choice for dentine adhesion studies [33]. From the above, the biggest factor increasing bond strength was use of the adhesive ibond. The values obtained for human dentine using ibond are in agreement with previous studies using un-etched human dentine [34]. In the presence of water the anhydride group in the 4-META within ibond is hydrolyzed to provide two carboxylic acid groups. It is proposed that these may partially demineralize the dentine to allow some micro-mechanical interlocking, but in addition enable a chemical bond with calcium in remaining hydroxyapatite. Furthermore, it may bond with basic amino acid groups in the collagen. The low viscosity as a result of the solvent in ibond also enables increased penetration into demineralized dentine. After solvent evaporation upon air drying, adhesive polymerization additionally provides strong chemical bonds with the monomers in the composite [35]. The ibond UDMA hydrophobicity aids intermixing with the composites. The above SEM images suggest that most often, bond failure occurs at the dentine interface or in the dentine itself rather than at any composite-ibond interface.

Upon acid etching of the human dentine, the above results showed the adhesive forms a much thicker interface layer. This, however, caused only a small increase in the average bond strength consistent with previous studies [36]. The interface layer thickness observed above was comparable to the depth of acid etching previously observed with phosphoric acid use [14]. This layer will consist largely of a mixture of residual hydroxyapatite crystallites and collagen fibrils with adhesive monomers replacing displaced unbound water [37]. The solvents, hydrophilic monomers and low viscosity, aid adhesive penetration into water filled collagen and tubules. Acid etching also enables greater penetration of adhesives into tubules which may potentially further enhance interlocking between the adhesive and dentine [38]. The lack of any significant difference in the bond strengths for ivory with lower density of tubules instead of human dentine, suggests this mechanism of bonding had limited additional benefit when ibond was employed.

The very low bond strength of the conventional composites (Z250, Gradia and C-HEMA) to un-etched human dentine will be due to lack of any mechanism for chemical or micromechanical bonding. The greater bond strengths obtained with Vertise flow is in agreement with previous studies [39]. This has been attributed to both lower viscosity as a result of lower filler content and the addition of glycerol phosphate dimethacrylate (GPDM) [40,41] which can form ionic bonds to calcium. The comparable bond strength for the higher viscosity C-4META composite suggests the ionic interactions may be more important in this case than filler loading.

The increased bond strength after etching of human dentine for the C-HEMA composite but lack of improvement upon using ivory could suggest that this composite may require penetration into tubules to improve bonding. This may have been improved by the addition of the low viscosity and hydrophilic HEMA [42]. With Gradia and Z250 improved bonding to etched ivory may be a consequence of these materials having some weak micro-mechanical interaction with highly demineralized collagen. Vertise flow may be able to bind un-etched ivory less well due to lack of hydroxyapatite. This problem may have been reduced with the C-4META composite by enhanced acidity [41] and therefore increased etching and chemical interaction.

The advantages of higher filler content in the 4-META composites compared with Vertise Flow include increased strength and lower shrinkage of the composite upon polymerization [43–45]. The SEM images show that when the bond strength is low the material can upon polymerization generate a gap between the dentine and composite. The size of the gaps observed were comparable with what might be expected from the known sample dimensions and composite shrinkages upon polymerization. These are typically ~3% for conventional composites with around 80 wt % filler [46] but 4.5% for Vertise flow with 70 wt % filler [47]. The higher filler content, in combination with moderate bond strengths would explain the limited gaps when C-4META was employed.

3. Experimental Section

3.1. Dentine Source

Elephant ivory dentine was used as a model substrate and was provided by the U.K Border Agency, Heathrow airport (CITES Reference 08/2012) for research purpose only. It should be noted that the authors do not advocate the use of ivory for any purposes and are thankful for CITES to allow the use of confiscated ivory that would be otherwise destroyed for academic research.

Human dentine was collected from non-carious adult human teeth. The teeth were collected through the UCL Eastman Bio bank after ethical approval and patient consent (Bio bank ID number 1304). The uncut teeth were stored in a 0.2% thymol solution at 4 °C for up to 4 weeks prior to use. The dentine was taken directly beneath the occlusal part of the enamel.

3.2. Composite Preparation/Source

The commercial dentine adhesive and etchant employed were ibond total etch (batch 010037) and ibond 35% phosphoric acid gel (batch 395074) (Heraeus Kulzer, Germany). ibond total etch consists primarily of the high molecular weight dental monomer urethane dimethacrylate (UDMA), adhesion promoting and demineralizing 4-methacryloxyethyl trimellitic anhydride (4-META), acetone and water [48]. The low viscosity of this adhesive enables better flow and penetration into rough dentine surfaces and tubules. Commercial composites examined included Z250 (3M™, Bracknell, UK, batch 202464, shade B3), Gradia Direct PA-2 (GC Corporation, Newport Pagnell, UK, batch 105936, shade P-A2) and Vertise flow (Kerr Dental Supply, Peterborough, UK, batch 1110402, shade A1). Z250 contains 82 wt % inorganic fillers. Gradia and Vertise flow contain 77 and 70 wt % fillers respectively some of which is pre-polymerized polymer. All these composites contain light curable high molecular weight hydrophobic dimethacrylate monomers. In addition, Vertise Flow contains glycerol phosphate dimethacrylate which has the potential to bond to hydroxyapatite. Their detailed composition has been provided elsewhere [49,50].

Experimental light curable dental composites were prepared with 80 wt % silanated barium alumino silicate glass with an average particle diameter of 7 μm (DMG, Hamburg, Germany). The monomer phase was formulated using UDMA and triethylene glycol dimethacrylate (TEGDMA) (DMG, Hamburg, Germany) in 3:1 weight ratio. 1 wt % camphorquinone (Polysciences, Warrington, PA, USA) and 1 wt % N, N-dimethyl-p-toluidine (Sigma Aldrich, UK) initiator and activator were added. Furthermore, 5 wt % of either the hydrophilic monomer 2-Hydroxy ethyl methacrylate (HEMA) (DMG, Hamburg, Germany) or 4-META (Polysciences, Warrington, PA, USA) were added. Formulations were mixed using a Speed Mixer™ DAC 150.1 FVZ (Synergy Devices Ltd, High Wycombe, UK) at room temperature for 20 s at 3000 rpm. The experimental composites had greater viscosity than Vertise flow but lower viscosity than Z250 and Gradia.

3.3. Raman Spectra of Dentine

Raman mapping spectroscopy (Horiba Jobin Yvon, Paris, France) was used to generate average Raman spectra of the surfaces of both dentine before and after treatment with 35% phosphoric acid gel for 20, 60, or 120 s. After etching, each sample was thoroughly washed with distilled water for about 30 s, and blotted dry.

Raman were obtained between 800 and 1800 cm^{-1} using a confocal hole of 150 μm, 632.8 nm He-Ne laser and 50x microscope objective. For each specimen, spectra were obtained by mapping areas of 40 × 40 μm. In each area 16 spectra were obtained for 20 s each. After baseline subtraction, spectra were normalized by the Amide I peak at 1670 cm^{-1} and then averaged. This was repeated 5 times for each treatment and dentine type using multiple sites. Average normalized intensities as a function of wavenumber and 95% confidence intervals were calculated. Intensity of the hydroxyapatite peaks at 961 cm^{-1} were plotted against the etchant time.

3.4. Flexural Properties of Dentine

For assessment of dentine strength, modulus, and strain, both dentine were cut into rectangular sections of 15 × 5 × 2 mm. The ivory samples were stored dry at room humidity/temperature or in distilled water for 24 h at room temperature. Cut human dentine specimens were all stored in 0.2% thymol at 4 °C before testing. Specimen thickness was checked using digital vernier calipers (Moore and Wright, West Yorkshire, UK). Strength was assessed using a “3-point bending” jig. This consisted of two support rollers 5.0 mm in diameter. The centers of support rollers were 10.0 mm apart. Load was applied at the midpoint, between the supports by means of a third roller 3 mm in diameter.

The load was applied with a 1 kN load cell, at a cross head speed of 1 mm/min using a computer-controlled universal testing machine (Instron 4502, Bucks, UK). The modulus E, 3-point bending strength σ, and strain ϵ were determined using Equations (2)–(4).

$$E = \frac{\mathrm{m}L^3}{4bh^3} \tag{2}$$

$$\sigma = \frac{3FL}{2bh^2} \tag{3}$$

$$\epsilon = \frac{6\mathrm{Dh}}{L^2} \tag{4}$$

F is the maximum load on the load deflection curve, L is the length of support span, b is the width of tested specimen, h is the thickness of tested specimens, m is the gradient (*i.e.*, slope) of the initial straight-line portion of the load deflection curve, and D is the maximum displacement of the tested specimen from its original position to the point of highest load. Sample repetition was 15.

Data were fitted to a Weibull type expression:

$$P_f = 1 - \exp(-\sigma/\sigma_\theta)^m \tag{5}$$

m–Weibull shape parameter, σ–bending strength, (which can be replaced by modulus, or flexural strain) of each specimen and σ_θ–Weibull scale parameter. When $\sigma = \sigma_\theta$, $P_f = 63.2\%$. P_f was defined as $(i - 0.5)/n$, where n is the number of specimens and i is the rank of a specimen in a list when strength, modulus or strain are ordered from lowest to highest values. Rearranging and taking double logs of Equation (5) gives:

$$\ln\ln[1/(1 - P_f)] = m \ln \sigma - m \ln \sigma_\theta \tag{6}$$

Weibull parameters were calculated from the slope and intercept of the left hand side of Equation (6) plotted *vs.* ln σ [51].

3.5. Shear Bond Strength of Composites to Dentine

Cut ivory dentine cubes (~1 × 1 × 1 cm) obtained from middle of a single tusk were placed in water for 24 h at 37 °C. Thereafter they were kept in small sealed containers to reduce water evaporation and used within 48 h. Fifteen human teeth were used in shear bond test. Both dentine cubes, and 0.2% thymol stored human teeth were cut vertically in half, and embedded in slow-setting viscous self-curing resin such that dentine tubules were perpendicular to the top resin surface. P120 paper was used to grind the resin surface until sufficient dentine was exposed. The dentine was then polished with P500 paper until the surface was even and smooth when visually inspected.

To assess bond strength, composite pastes were poured in 2 mm increments into a brass tube of 3 mm internal diameter, and 6 mm long placed on the surface of the dentine. The end of the tube in contact with dentine was chamfered at 45 degrees to reduce its contact area. Each 2 mm increment was cured for 40 s. Immediately before application of the composite, the dentine was pretreated by

(1) ibond application and light cure for 20 s as per manufacturer's instructions; or
(2) acid etch application for 20 s followed by water rinsing, gentle drying and ibond application and cure; or
(3) acid etch for 20 s, rinse and dry; or
(4) no acid or ibond treatment.

The shear bond test was done according to ISO standard [52]. Shear bond strength was determined using an Instron Universal testing machine with a "Flat-edge shear fixture" jig. The jig consisted of a metal holder with an adjustable screw to secure the specimen and an adjustable blade, which was used to shear the tube from the dentine. A 1 kN load cell at cross head speed of 1 mm/min was used.

The load at break was recorded and the bond strength τ calculated using Equation (7), in which F is the load at break, and A is the bonded area of the cylinder.

$$\tau = \frac{F}{A} \tag{7}$$

Each specimen was repeated 8 times, and 95% confidence interval was calculated from standard deviation using Equation (8).

$$CI = \frac{2SD}{\sqrt{n}} \tag{8}$$

3.6. Scanning Electron Microscopy

To visually assess micro-gap formation due to composite shrinkage, dentine blocks of 5 mm depth were produced containing cavities of 3 mm diameter. After drilling, the cavities were washed and then treated in the four different dentine pre-treatment methods above. Following treatment the cavities were restored by placing composites in 2 mm layers and curing as above. After 24 h in distilled water at room temperature, restored cavities were cut in half vertically to expose the interface on two sides and from top to bottom of the restoration. The surface of both dentine were polished using P2400 paper and sputter coated with gold/palladium before taking microscopic images of interfaces using a Scanning electron microscope (SEM)(Phillip XL-30, Eindhoven, The Netherlands).

4. Conclusions

In this study, dentine from different sources were analyzed. The ratio of hydroxyapatite to collagen in ivory was half that in human dentine and comparable to that of human dentine etched for 20 s with acid. Surface ivory and human dentine hydroxyapatite level decreased by half with every 23 s of acid etch. The bending strength of ivory was approximately half that of human dentine presumably due to lower hydroxyapatite content.

Additionally, to characterization of dentine the experimental and commercial composites bond strengths and micro-gap formation were investigated. The major factor increasing composite bond strength to dentine was the use of ibond. Acid etching and the addition of acidic monomers in the composites could increase bond strength in some circumstances.

Acknowledgments

Authors are grateful to UK Border Agency at Heathrow airport for providing ivory tusks. Authors would also like to thank Wendy Xia, Graham Palmer, George Georgiou, and Nicola J. Morden for their help and technical assistance. Saad Liaqat and Muhammad Adnan Khan also acknowledges financial support of Higher Education Commission, Government of Pakistan. Anas Aljabo and Hesham Ben Nuba would also acknowledge Libyan Ministry of Education. EPSRC have also provided support for this study (project EP/I022341/1). Authors would also acknowledge Schottlander, DMG and ESSCHEM for supply of monomers and glass.

Author Contributions

Saad Liaqat contributed to the whole process of sample preparation, data analysis, and preparation of manuscript. A. Aljabo contributed towards the mechanical testing, M. A Khan helped with the data analysis and sample preparation for Raman testing. H. B Nuba contributed in dentine collection, storage, and sample preparation for shear bond testing. P. Ashley, L. Bozec, and A.M Young supervised the whole procedure and also contributed to the data interpretation, and preparation of manuscript. Additionally, P. Ashley helped in getting ethical approval for human teeth, and its collection. L. Bozec, and A.M Young helped in arranging tusks for the research purpose.

Conflicts of Interest

The authors declare no conflict of interest.

References

1. Selwitz, R.H.; Ismail, A.I.; Pitts, N.B. Dental caries. *The Lancet* **2007**, *369*, 51–59.
2. Featherstone, J. The continuum of dental caries—Evidence for a dynamic disease process. *J. Dent. Res.* **2004**, *83*, C39–C42.
3. Bernardo, M. Survival and reasons for failure of amalgam versus composite posterior restorations placed in a randomized clinical trial. *J. Am. Dent. Assoc.* **2007**, *138*, 775–783.
4. Mehdawi, I.M.; Pratten, J.; Spratt, D.A.; Knowles, J.C.; Young, A.M. High strength re-mineralizing, antibacterial dental composites with reactive calcium phosphates. *Dent. Mater.* **2013**, *29*, 473–484.
5. Krifka, S.; Hiller, K.-A.; Bolay, C.; Petzel, C.; Spagnuolo, G.; Reichl, F.-X.; Schmalz, G.; Schweikl, H. Function of mapk and downstream transcription factors in monomer-induced apoptosis. *Biomaterials* **2012**, *33*, 740–750.
6. Erhardt, M.C.G.; Cavalcante, L.M.A.; Pimenta, L.A.F. Influence of phosphoric acid pretreatment on self-etching bond strengths. *J. Esthet. Restor. Dent.* **2004**, *16*, 33–40.
7. Wang, Y.; Spencer, P. Quantifying adhesive penetration in adhesive/dentin interface using confocal raman microspectroscopy. *J. Biomed. Mater. Res.* **2002**, *59*, 46–55.
8. Frankenberger, R.; Lopes, M.; Perdigão, J.; Ambrose, W.W.; Rosa, B.T. The use of flowable composites as filled adhesives. *Dent. Mater.* **2002**, *18*, 227–238.
9. Cardoso, M.V.; Moretto, S.G.; Carvalho, R.C.; Russo, E.M. Influence of intrapulpal pressure simulation on the bond strength of adhesive systems to dentin. *Braz. Oral Res.* **2008**, *22*, 170–175.

10. Earl, J.; Leary, R.; Perrin, J.; Brydson, R.; Harrington, J.; Markowitz, K.; Milne, S. Characterization of dentine structure in three dimensions using fib-sem. *J. Microsc.* **2010**, *240*, 1–5.
11. Jeon, J.-G.; Rosalen, P.; Falsetta, M.; Koo, H. Natural products in caries research: Current (limited) knowledge, challenges and future perspective. *Caries Res.* **2011**, *45*, 243–263.
12. De Munck, J.; Mine, A.; Poitevin, A.; van Ende, A.; Cardoso, M.V.; van Landuyt, K.L.; Peumans, M.; van Meerbeek, B. Meta-analytical review of parameters involved in dentin bonding. *J. Dent. Res.* **2012**, *91*, 351–357.
13. Skene, L. Ownership of human tissue and the law. *Nat. Rev. Genet.* **2002**, *3*, 145–148.
14. Wang, Y.; Spencer, P.; Walker, M.P. Chemical profile of adhesive/caries-affected dentin interfaces using raman microspectroscopy. *J. Biomed. Mater. Res. A* **2007**, *81A*, 279–286.
15. Ignjatović, Z.; Stojšin, I.; Brkanić, T.; Lončar, J. The effect of excessive coca-cola consumption on the development of dental erosions. *Stomatol. Glas. Srb.* **2012**, *59*, 148–153.
16. Pan, H.; Darvell, B.W. Effect of carbonate on hydroxyapatite solubility. *Cryst. Growth Des.* **2010**, *10*, 845–850.
17. Gupta, S.K.; Thangaraj, K.; Singh, L. Identification of the source of ivory idol by DNA analysis. *J. Forensic Sci.* **2011**, *56*, 1343–1345.
18. Poggio, C.; Lombardini, M.; Vigorelli, P.; Colombo, M.; Chiesa, M. The role of different toothpastes on preventing dentin erosion: An sem and afm study®. *Scanning* **2013**, *36*, 301–310.
19. Chandra, S.S.; Shankar, P.; Indira, R. Depth of penetration of four resin sealers into radicular dentinal tubules: A confocal microscopic study. *J. Endod.* **2012**, *38*, 1412–1416.
20. Chng, H.K.; Palamara, J.E.; Messer, H.H. Effect of hydrogen peroxide and sodium perborate on biomechanical properties of human dentin. *J. Endod.* **2002**, *28*, 62–67.
21. McKittrick, J.; Chen, P.Y.; Tombolato, L.; Novitskaya, E.E.; Trim, M.W.; Hirata, G.A.; Olevsky, E.A.; Horstemeyer, M.F.; Meyers, M.A. Energy absorbent natural materials and bioinspired design strategies: A review. *Mater. Sci. Eng. C* **2010**, *30*, 331–342.
22. Hayashi, M.; Koychev, E.V.; Okamura, K.; Sugeta, A.; Hongo, C.; Okuyama, K.; Ebisu, S. Heat treatment strengthens human dentin. *J. Dent. Res.* **2008**, *87*, 762–766.
23. Plotino, G.; Grande, N.M.; Bedini, R.; Pameijer, C.H.; Somma, F. Flexural properties of endodontic posts and human root dentin. *Dent. Mater.* **2007**, *23*, 1129–1135.
24. Vollenweider, M.; Brunner, T.J.; Knecht, S.; Grass, R.N.; Zehnder, M.; Imfeld, T.; Stark, W.J. Remineralization of human dentin using ultrafine bioactive glass particles. *Acta Biomater.* **2007**, *3*, 936–943.
25. Imbeni, V.; Nalla, R.K.; Bosi, C.; Kinney, J.H.; Ritchie, R.O. In vitro fracture toughness of human dentin. *J. Biomed. Mater. Res. A* **2003**, *66A*, 1–9.
26. Hayashi, M.; Okamura, K.; Koychev, E.V.; Furuya, Y.; Sugeta, A.; Ota, T.; Ebisu, S. Effects of rehydration on dentin strengthened by heating or uv irradiation. *J. Dent. Res.* **2010**, *89*, 154–158.
27. van Raaij, M.J.; Mitraki, A. Natural fibrous proteins: Structural analysis, assembly, and applications. In *Proteins in Solution and at Interfaces: Methods and Applications in Biotechnology and Materials Science*; Wiley: Hoboken, NJ, USA, 2013; pp. 219–232.
28. Wang, Q.; Wang, X.M.; Tian, L.L.; Cheng, Z.J.; Cui, F.Z. In situ remineralizaiton of partially demineralized human dentine mediated by a biomimetic non-collagen peptide. *Soft Matter* **2011**, *7*, 9673–9680.

29. Dibdin, G.; Poole, D. Surface area and pore size analysis for human enamel and dentine by water vapour sorption. *Arch. Oral Biol.* **1982**, *27*, 235–241.
30. Wenger, M.P.E.; Bozec, L.; Horton, M.A.; Mesquida, P. Mechanical properties of collagen fibrils. *Biophys. J.* **2007**, *93*, 1255–1263.
31. Roop Kumar, R.; Wang, M. Modulus and hardness evaluations of sintered bioceramic powders and functionally graded bioactive composites by nano-indentation technique. *Mater. Sci. Eng. A* **2002**, *338*, 230–236.
32. Profeta, A.C. Dentine bonding agents comprising calcium-silicates to support proactive dental care: Origins, development and future. *Dent. Mater. J.* **2014**, *33*, 443–452.
33. Burke, F.; Hussain, A.; Nolan, L.; Fleming, G. Methods used in dentine bonding tests: An analysis of 102 investigations on bond strength. *Eur. J. Prosthodont. Restor. Dent.* **2008**, *16*, 158–165.
34. Krifka, S.; Börzsönyi, A.; Koch, A.; Hiller, K.-A.; Schmalz, G.; Friedl, K.-H. Bond strength of adhesive systems to dentin and enamel—human *vs*. Bovine primary teeth *in vitro*. *Dent. Mater.* **2008**, *24*, 888–894.
35. Nicolae, L.C.; Shelton, R.M.; Cooper, P.R.; Martin, R.A.; Palin, W.M. The effect of udma/tegdma mixtures and bioglass incorporation on the mechanical and physical properties of resin and resin-based composite materials. *Conf. Pap. Sci.* **2014**, *2014*, 6143–6147.
36. Watanabe, T.; Tsubota, K.; Takamizawa, T.; Kurokawa, H.; Rikuta, A.; Ando, S.; Miyazaki, M. Effect of prior acid etching on bonding durability of single-step adhesives. *Oper. Dent.* **2008**, *33*, 426–433.
37. Breschi, L.; Mazzoni, A.; Ruggeri, A.; Cadenaro, M.; Di Lenarda, R.; De Stefano Dorigo, E. Dental adhesion review: Aging and stability of the bonded interface. *Dent. Mater.* **2008**, *24*, 90–101.
38. Oliveira, S.S.A.; Pugach, M.K.; Hilton, J.F.; Watanabe, L.G.; Marshall, S.J.; Marshall, G.W., Jr. The influence of the dentin smear layer on adhesion: A self-etching primer *vs*. A total-etch system. *Dent. Mater.* **2003**, *19*, 758–767.
39. Ozel Bektas, O.; Eren, D.; Akin, E.G.; Akin, H. Evaluation of a self-adhering flowable composite in terms of micro-shear bond strength and microleakage. *Acta Odontol. Scand.* **2013**, *71*, 541–546.
40. Vichi, A.; Margvelashvili, M.; Goracci, C.; Papacchini, F.; Ferrari, M. Bonding and sealing ability of a new self-adhering flowable composite resin in class i restorations. *Clin. Oral Investig.* **2013**, *17*, 1497–1506.
41. Moszner, N.; Salz, U.; Zimmermann, J. Chemical aspects of self-etching enamel–dentin adhesives: A systematic review. *Dent. Mater.* **2005**, *21*, 895–910.
42. Van Landuyt, K.L.; Snauwaert, J.; Peumans, M.; De Munck, J.; Lambrechts, P.; van Meerbeek, B. The role of hema in one-step self-etch adhesives. *Dent. Mater.* **2008**, *24*, 1412–1419.
43. Weinmann, W.; Thalacker, C.; Guggenberger, R. Siloranes in dental composites. *Dent. Mater.* **2005**, *21*, 68–74.
44. Braga, R.R.; Ballester, R.Y.; Ferracane, J.L. Factors involved in the development of polymerization shrinkage stress in resin-composites: A systematic review. *Dent. Mater.* **2005**, *21*, 962–970.
45. Kleverlaan, C.J.; Feilzer, A.J. Polymerization shrinkage and contraction stress of dental resin composites. *Dent. Mater.* **2005**, *21*, 1150–1157.
46. Braga, R.R.; Ferracane, J.L. Alternatives in polymerization contraction stress management. *Crit. Rev. Oral Biol. Med.* **2004**, *15*, 176–184.

47. Wei, Y.-J.; Silikas, N.; Zhang, Z.-T.; Watts, D.C. Hygroscopic dimensional changes of self-adhering and new resin-matrix composites during water sorption/desorption cycles. *Dent. Mater.* **2011**, *27*, 259–266.
48. Van Landuyt, K.L.; Snauwaert, J.; De Munck, J.; Peumans, M.; Yoshida, Y.; Poitevin, A.; Coutinho, E.; Suzuki, K.; Lambrechts, P.; van Meerbeek, B. Systematic review of the chemical composition of contemporary dental adhesives. *Biomaterials* **2007**, *28*, 3757–3785.
49. Wei, Y.-J.; Silikas, N.; Zhang, Z.-T.; Watts, D.C. Diffusion and concurrent solubility of self-adhering and new resin–matrix composites during water sorption/desorption cycles. *Dent. Mater.* **2011**, *27*, 197–205.
50. Obici, A.C.; Sinhoreti, M.A.C.; Frollini, E.; Correr Sobrinho, L.; Consani, S. Degree of conversion of z250 composite determined by fourier transform infrared spectroscopy: Comparison of techniques, storage periods and photo-activation methods. *Mater. Res.* **2004**, *7*, 605–610.
51. Xia, W.; Razi, M.M.; Ashley, P.; Neel, E.A.; Hofmann, M.; Young, A. Quantifying effects of interactions between polyacrylic acid and chlorhexidine in dicalcium phosphate–forming cements. *J. Mater. Chem. B* **2014**, *2*, 1673–1680.
52. International Organization for Standards. *Dentistry—Adhesion-Notched-Edge Shear Bond Strength Test*; International Organization for Standards: Geneva, Switzerland, 2013.

5

Prosthetic Meshes for Repair of Hernia and Pelvic Organ Prolapse: Comparison of Biomechanical Properties

Manfred M. Maurer [1,*], Barbara Röhrnbauer [1], Andrew Feola [2], Jan Deprest [2] and Edoardo Mazza [1,3]

[1] Institute of Mechanical Systems, ETH Zurich, Leonhardstrasse 21, Zurich 8092, Switzerland; E-Mails: barbara.roehrnbauer@gmx.de (B.R.); mazza@imes.mavt.ethz.ch (E.M.)

[2] Center for Surgical Technologies, Faculty of Medicine, Universitair Ziekenhuis "Gasthuisberg" Leuven, Katholieke Universiteit Leuven, Leuven 3000, Belgium; E-Mails: feolaaj12@gmail.com (A.F.); Jan.Deprest@uzleuven.be (J.D.)

[3] Empa—Swiss Federal Laboratories for Materials Science and Technology, Überlandstrasse 129, Dübendorf 8600, Switzerland

* Author to whom correspondence should be addressed; E-Mail: maurer@imes.mavt.ethz.ch;

Academic Editor: Amir A. Zadpoor

Abstract: This study aims to compare the mechanical behavior of synthetic meshes used for pelvic organ prolapse (POP) and hernia repair. The analysis is based on a comprehensive experimental protocol, which included uniaxial and biaxial tension, cyclic loading and testing of meshes in dry conditions and embedded into an elastomer matrix. Implants are grouped as POP or hernia meshes, as indicated by the manufacturer, and their stiffness in different loading configurations, area density and porosity are compared. Hernia meshes might be expected to be stiffer, since they are implanted into a stiffer tissue (abdominal wall) than POP meshes (vaginal wall). Contrary to this, hernia meshes have a generally lower secant stiffness than POP meshes. For example, DynaMesh PRS, a POP mesh, is up to two orders of magnitude stiffer in all tested configurations than DynaMesh ENDOLAP, a hernia mesh. Additionally, lighter, large pore implants might be expected to be more compliant, which was shown to be generally not true. In particular, Restorelle, the lightest mesh with the largest pores, is less compliant in the tested configurations than Surgipro, the heaviest, small-pore implant. Our study raises the question of defining a meaningful design target for meshes in terms of mechanical biocompatibility.

Keywords: mechanical biocompatibility; mesh prostheses; POP meshes; hernia meshes

1. Introduction

Mechanical biocompatibility of prosthetic meshes for hernia and pelvic organ prolapse (POP) is related to the ability of implants to display a mechanical behavior compatible with its function and favoring its integration into the surrounding native tissue [1–6]. This approach for implant assessment has received increased attention in recent years. While initial investigations focused on the ability of a mesh to provide sufficient strength and resistance to maximum loads [7–12], it recently became clear that the deformation behavior in a physiological range, also called "comfort zone", is of major importance [13,14]. A mismatch of mechanical properties of the implants compared to native tissue has been associated with clinical complications [15–19], although none of these works explicitly link mechanical properties with clinical outcome. It has recently been suggested that meshes designed to mimic the biomechanical properties of the area of application are advantageous [2,14,20]. These investigations are further motivated by an FDA safety communication [21] pointing at risks associated with existing prosthetic meshes and corresponding surgery procedures for repair of POP.

A wealth of studies has been conducted analyzing either hernia or POP meshes (see [3,22] for an extensive literature overview). However, little work was performed to compare the mechanical response of these two groups, which may shed light onto the prevalent clinical complications. The mechanical environment and loading conditions these implants are exposed to differ significantly between the abdominal wall and pelvic floor. Physiological loads in terms of membrane tension were calculated based on Laplace's law to be around 0.035 N/mm in the pelvic region and 0.136 N/mm at the abdominal wall at rest, but can be orders of magnitude higher at increased intra-abdominal pressures [5,7,23]. This gives an indication of the range of load at which mesh implants should work best in supporting and mimicking native tissue, thus ensuring mechanical biocompatibility.

Based on the experimental study presented in [22], the data analysis in the present investigation is extended to compare the mechanical properties of hernia and POP mesh implants with respect to physiological loading conditions.

2. Experimental Section

Nine mesh implants were investigated. They were grouped into hernia ($n = 5$) and POP ($n = 4$) implants based on the manufacturer's information, available on their respective websites and analyzed accordingly. All products are described in Table 1.

The mechanical testing procedure has been previously described in detail [22]. In short, each mesh type was tested in eight different configurations: 2 (uniaxial tension or biaxial tension) × 2 (dry or embedded) × 2 (0° or 90° direction).

Table 1. List of mesh types used for the present investigation, with their weight classified as ultralight, light or standard according to [24]. Principal directions of testing are marked in red. Scale bar (lower right): 5 mm. Their clinical application is listed as pelvic organ prolapse (POP) or hernia repair, as specified by the manufacturer.

Image	Mesh	Application	Material, Weight
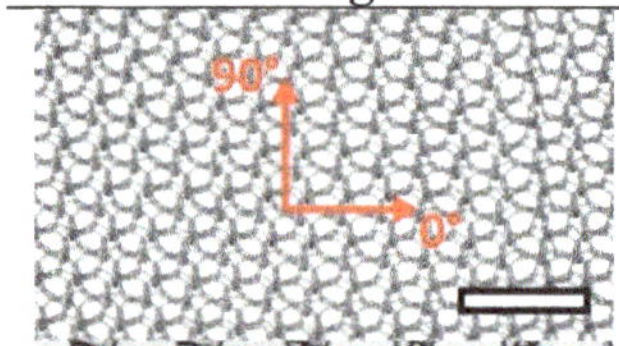	Bard™ Mesh Marlex (BM)	Hernia	Polypropylene, standard
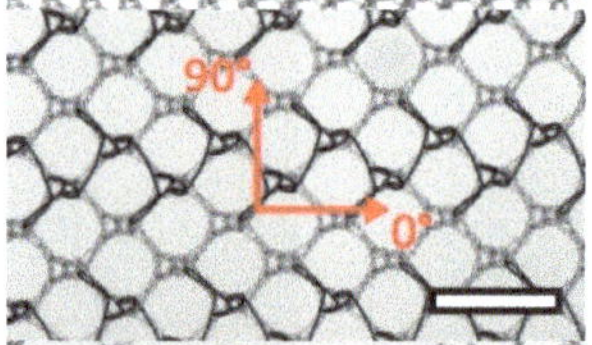	DynaMesh® ENDOLAP (DM)	Hernia	PVDF (polyvinylidene fluoride), standard
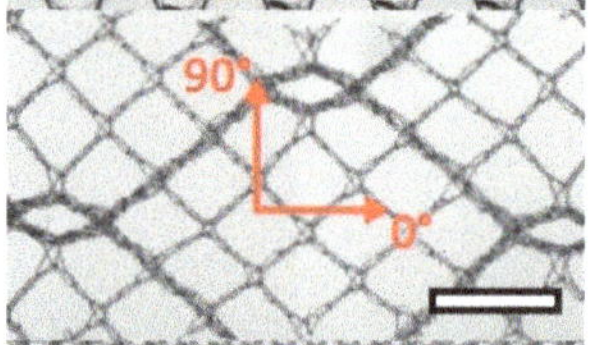	Ethicon Physiomesh® (PM)	Hernia	Polypropylene, ultralight
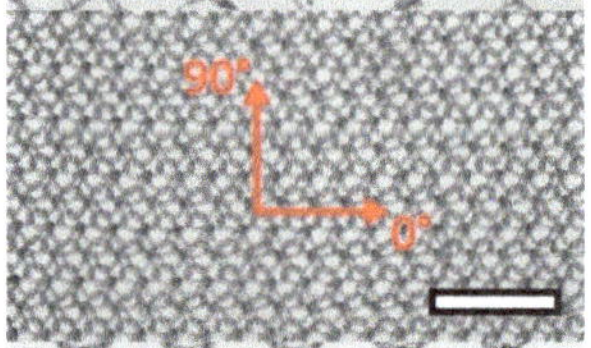	Surgipro™ Polypropylene Monofilament Mesh (SPMM)	Hernia	Polypropylene, standard
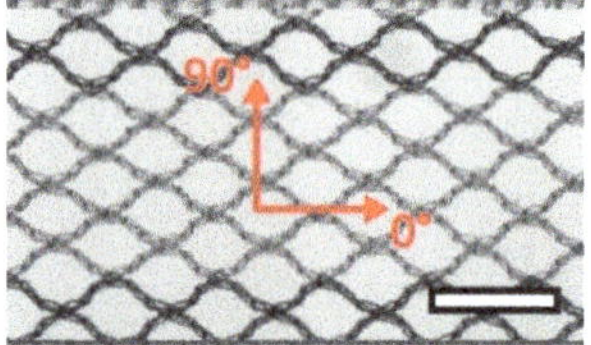	Ethicon Ultrapro™ (UP)	Hernia	Polypropylene, light
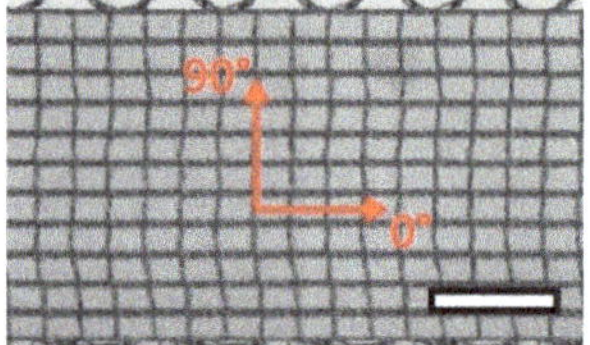	DynaMesh® PRS (DMPRS)	POP	PVDF, standard
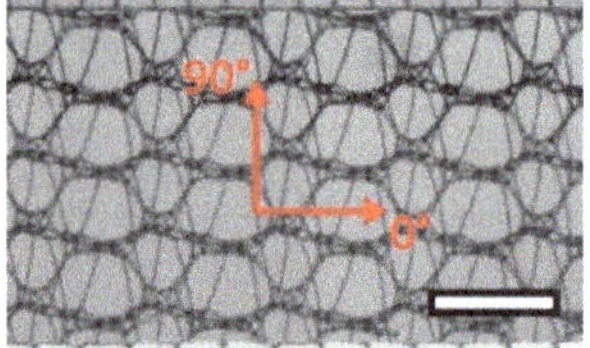	Gynecare PROLIFT™ (PE)	POP	Polypropylene, ultralight
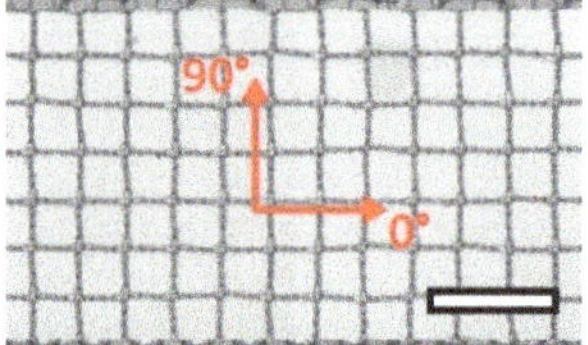	Coloplast Restorelle™ (Rest)	POP	Polypropylene, ultralight
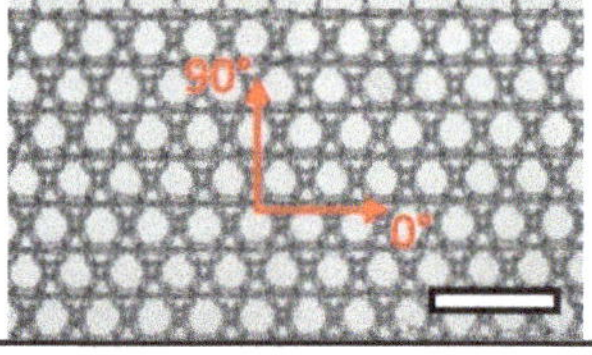	Parietex Ugytex® (UT)	POP	Polypropylene, light

These test configurations represent the *in vivo* loading and environmental conditions of the mesh implants. Long, narrow strips of meshes used in "line-type" suspensions are mainly loaded in uniaxial tension, whereas wider sheets, such as for hernia repair, are typically subjected to multiaxial tension states. Our earlier study examined the anisotropic behavior of these meshes along two perpendicular directions following the main knitting patterns. However, here, the focus is on the stiffer of the two directions on a per mesh basis. A dry mesh is tested as delivered, whereas embedded infers a specimen being embedded into a soft elastomer matrix (Young's modulus 0.0276 N/mm^2 [25]), mimicking *in vivo*, ingrown conditions.

Experiments with uniaxial tension and biaxial tension (realized as uniaxial strain test, also called "strip biaxial"; [22,26]) were performed on the same tensile test machine. In the uniaxial strain test, lateral contraction of the specimen is constrained, leading to stresses in the direction perpendicular to the loading axis, thus subjecting the sample to a biaxial state of tension. Test piece dimensions were selected to generate a free area of 30 mm × 15 mm (uniaxial) and 50 mm × 15 mm (biaxial). Each specimen was loaded to a maximum of 30% nominal strain (loading rate ~$10^{-3}s^{-1}$) and unloaded back to a pre-force threshold of 0.01 N for 10 cycles.

Deformation analysis was performed in an optical, non-contact procedure in the center of the specimen, allowing for extraction of local strains (ε_{loc}) as the result of an image analysis algorithm, thus avoiding edge and clamp effects at the specimen boundaries. Force measurements at the clamps were converted to nominal membrane tension (M_t (N/mm)) by dividing by the undeformed width of the sample. For a detailed description of the loading protocol and data extraction, refer to [22].

The resulting M_t–ε_{loc} curves of each of the 8 specimens of each type, as well as the area density and porosity measurements [27,28] form the basis for the analysis and comparison of mesh groups. Dry mesh samples of known dimensions were weighted before mechanical testing using a high-resolution balance, and their area density was calculated as weight per area (kg/m^2). Porosity is determined as the ratio of open area to the total area, including filaments of one undeformed unit cell of the knitting pattern [27,28].

From the M_t–ε_{loc}, curves the secant stiffness K (N/mm) in the stiffer direction at the reference membrane tension $M_t^{ref} = 0.035\ \mathrm{N/mm}$ (for hernia, as well as POP meshes) was extracted, for both the 1st and 10th cycle (see Figure 1). It is defined as:

$$K = \frac{M_t^{ref}}{\Delta\varepsilon} \tag{1}$$

where $\Delta\varepsilon$ is the difference of local strain at the reference membrane tension M_t^{ref} and at the beginning of the current cycle.

The specific value of M_t was chosen as a load representative of the membrane tension in the pelvic region under physiological intra-abdominal pressure (IAP) at rest [23]. Each mesh is thus characterized by 10 parameters, *i.e.*, a secant stiffness value for each of the tested configurations (uniaxial and biaxial tension, dry and embedded) in 1st and 10th cycle, as well as area density and porosity.

The implants are grouped into POP and hernia meshes as indicated on their official product insert. Each parameter is shown in a bar graph, as well as standard box plots in order to visualize the differences between the two groups. To determine the statistical significance, the Wilcoxon rank sum test (equivalent to the Mann–Whitney U-test) is applied for each parameter.

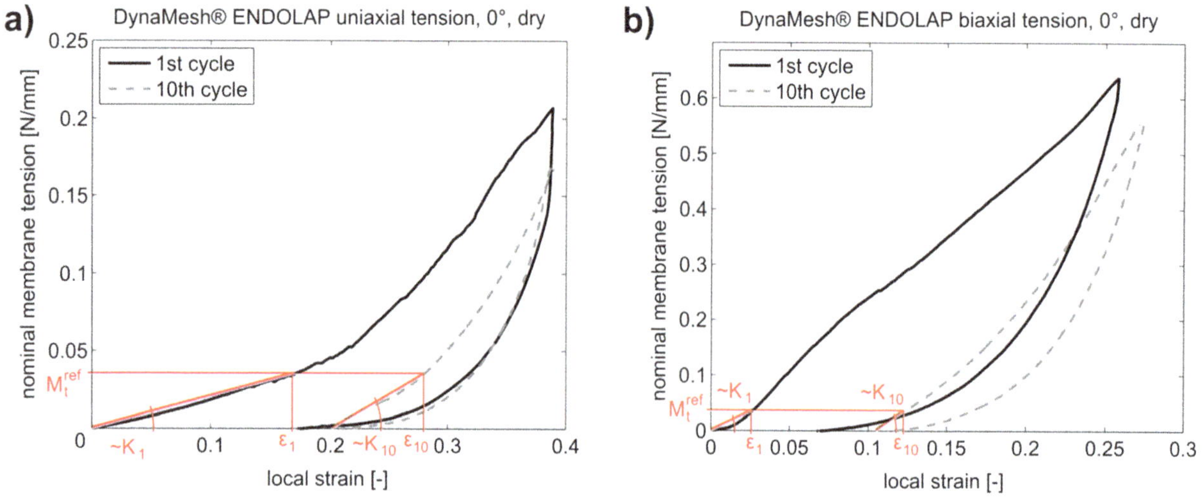

Figure 1. Exemplary tension-strain curves for DynaMesh ENDOLAP (DM) for the illustration of secant stiffness determination. (**a**) Uniaxial tension in the 0° direction, dry mesh, loading and unloading in 1st and 10th cycle; (**b**) Biaxial tension in the 0° direction, dry mesh, loading and unloading in 1st and 10th cycle. Secant stiffness in the 1st and 10th cycle is shown for both.

3. Results and Discussion

3.1. Results

Figure 2 shows the uniaxial secant stiffness K_{uni} (N/mm) of each specimen grouped according to the manufacturer indication into POP (red) and hernia (blue) meshes, with the mean of each group shown in darker red and blue, respectively. The specific testing conditions (dry/embedded and first/10th cycle) are indicated in each subgraph. Figure 3 represents the corresponding biaxial secant stiffness K_{bi} (N/mm). The respective stiffness values for each configuration are reported in Tables 2 (POP meshes) and 3 (hernia meshes).

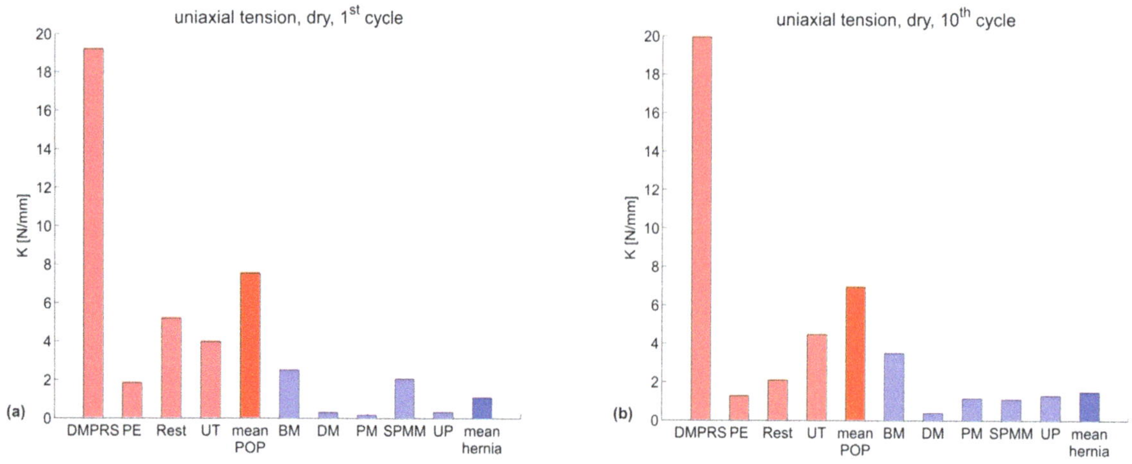

Figure 2. *Cont.*

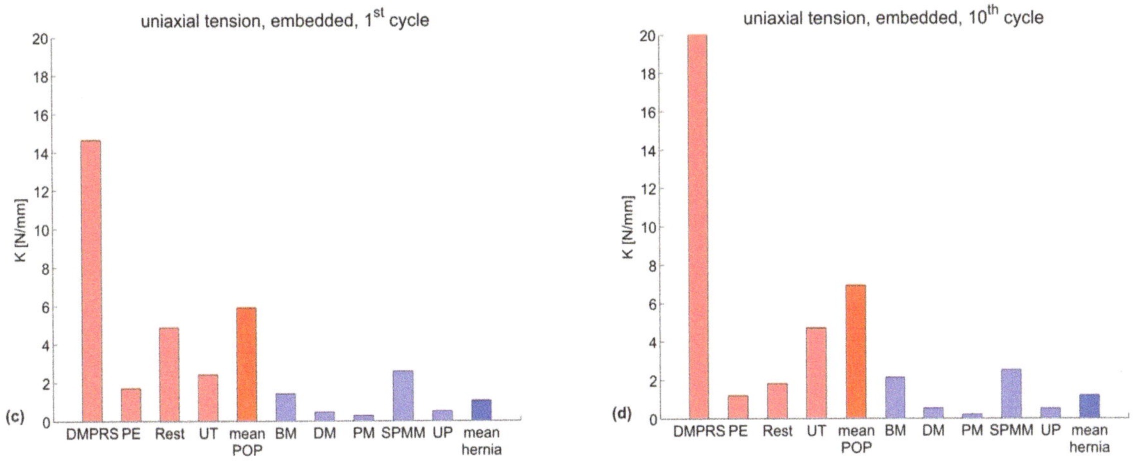

Figure 2. Uniaxial secant stiffness K_{uni} (N/mm) for all meshes in four configurations: (**a**) dry mesh, first cycle; (**b**) dry mesh, 10th cycle; (**c**) embedded mesh, first cycle; (**d**) embedded mesh, 10th cycle. POP meshes are shown in red; hernia meshes in blue. The mean of each group is plotted darker.

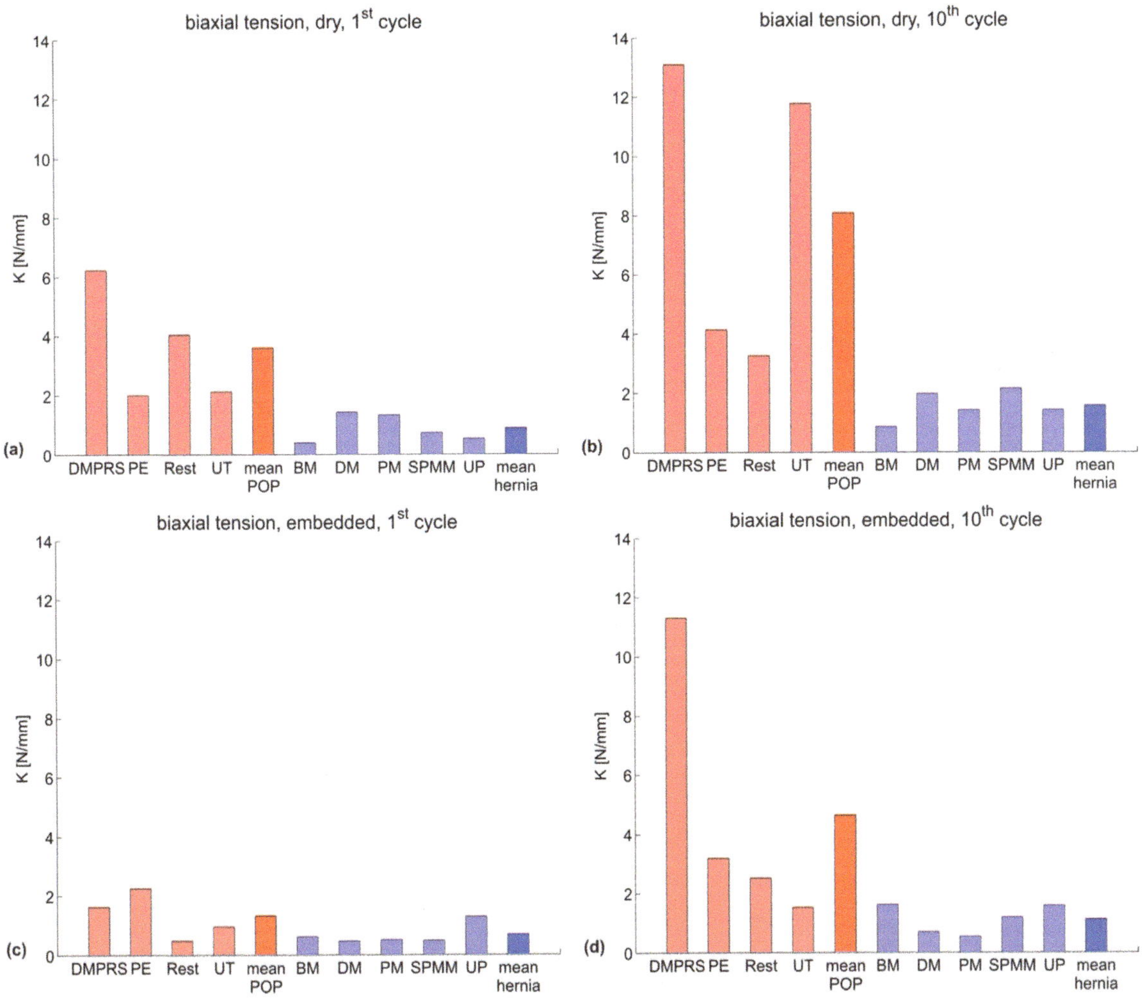

Figure 3. Biaxial secant stiffness K_{bi} (N/mm) for all meshes in four configurations: (**a**) dry mesh, first cycle; (**b**) dry mesh, 10th cycle; (**c**) embedded mesh, first cycle; (**d**) embedded mesh, 10th cycle. POP meshes are shown in red; hernia meshes in blue. The mean of each group is plotted darker.

Table 2. Numerical values of stiffness for all POP meshes and all tested configurations.

Testing configuration	*K* (N/mm) POP meshes				
	DMPRS	PE	Restorelle	UT	mean
uniaxial, dry, 1st cycle	19.2	1.8	5.2	4.0	7.5
uniaxial, dry, 10th cycle	19.9	1.3	2.1	4.5	6.9
uniaxial, embedded, 1st cycle	14.6	1.7	4.8	2.4	5.9
uniaxial, embedded, 10th cycle	20.0	1.2	1.8	4.7	6.9
biaxial, dry, 1st cycle	6.2	2.0	4.0	2.1	3.6
biaxial, dry, 10th cycle	13.1	4.1	3.2	11.8	8.1
biaxial, embedded, 1st cycle	1.6	2.2	0.5	0.9	1.3
biaxial, embedded, 10th cycle	11.3	3.2	2.5	1.5	4.6

Table 3. Numerical values of stiffness for all hernia meshes and all tested configurations.

Testing configuration	*K* (N/mm) hernia meshes					
	BM	DM	PM	SPMM	UP	mean
uniaxial, dry, 1st cycle	2.5	0.3	0.2	2.0	0.3	1.1
uniaxial, dry, 10th cycle	3.5	0.4	1.1	1.1	1.3	1.5
uniaxial, embedded, 1st cycle	1.4	0.5	0.3	2.6	0.5	1.0
uniaxial, embedded, 10th cycle	2.1	0.5	0.2	2.5	0.5	1.2
biaxial, dry, 1st cycle	0.4	1.4	1.3	0.7	0.5	0.9
biaxial, dry, 10th cycle	0.9	2.0	1.4	2.1	1.4	1.6
biaxial, embedded, 1st cycle	0.6	0.5	0.5	0.5	1.3	0.7
biaxial, embedded, 10th cycle	1.6	0.7	0.5	1.2	1.6	1.1

Figures 4 and 5 depict the summarizing box plots for the two groups, for uniaxial and biaxial stiffness in each configuration.

The variability for the POP group is very large for all parameters, thus affecting the statistical significance of the differences observed. The Wilcoxon rank sum test indicates a statistically-significant difference between the POP and hernia groups for the biaxial stiffness in the dry condition, both at the first and 10th cycle (p = 0.016 for both); see Figures 4a,b and 5a,b. The POP meshes were four- or five-fold stiffer than the hernia meshes in the first cycle and 10th cycle, respectively.

When comparing the mean and median stiffness for all configurations, POP implants are overall less compliant than hernia implants. Since the abdominal wall is known to be stiffer than vaginal tissue [9,29] and if mechanical biocompatibility were mainly dependent on similar properties to the implant area, one would expect a more compliant design for implants for POP compared to hernia. Embedding a mesh into a polymer matrix, thus reflecting the interaction with native tissue, affects the mechanical response of the implants. The differences between the groups are still evident also for this case (see Figures 2–4 and Figure 5c,d).

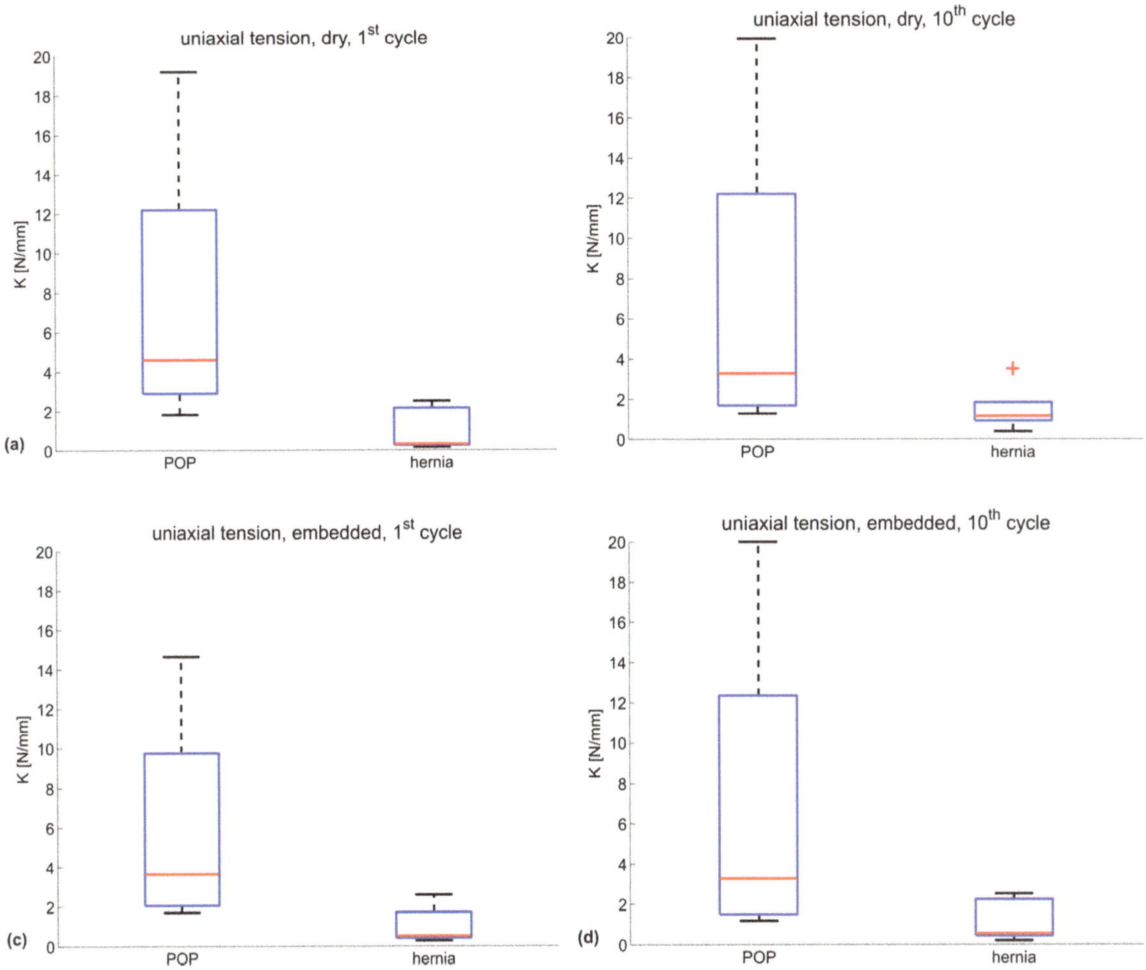

Figure 4. Box plots for uniaxial secant stiffness K_{uni} (N/mm) for all meshes in four configurations: (**a**) dry mesh, first cycle; (**b**) dry mesh, 10th cycle; (**c**) embedded mesh, first cycle; (**d**) embedded mesh, 10th cycle. The red line marks the median of the group; the box represents the 25th and 75th percentile; the extended whiskers the most extreme data points. The + sign indicates an outlier.

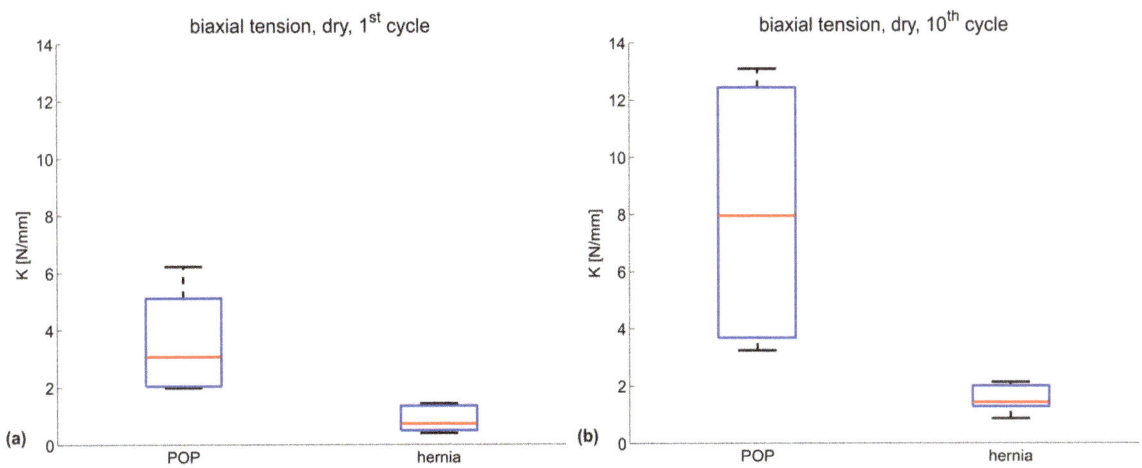

Figure 5. *Cont.*

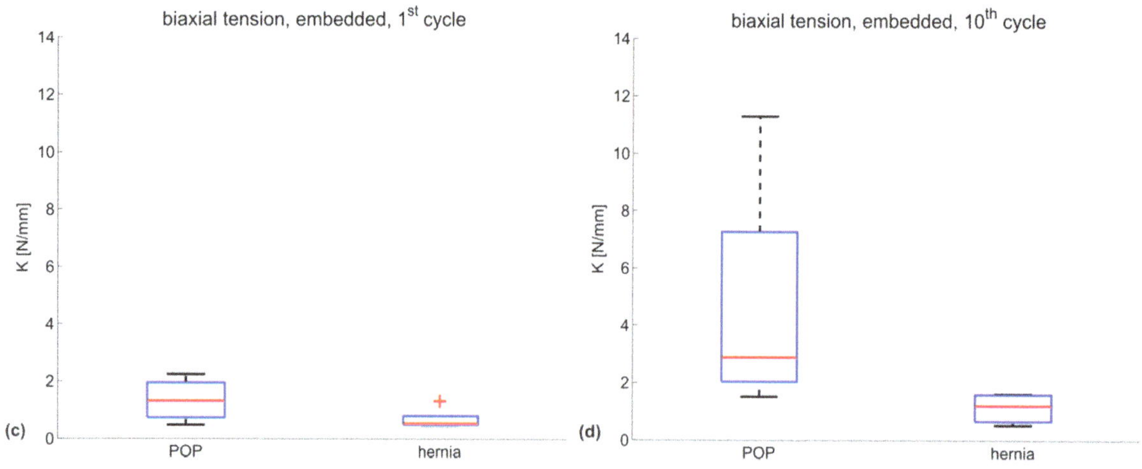

Figure 5. Box plots for biaxial secant stiffness K_{bi} (N/mm) or all meshes in four configurations: (**a**) dry mesh, first cycle; (**b**) dry mesh, 10th cycle; (**c**) embedded mesh, first cycle; (**d**) embedded mesh, 10th cycle. The red line marks the median of the group; the box represents the 25th and 75th percentile; the extended whiskers the most extreme data points. The + sign indicates an outlier.

High density and small pores are often linked to high stiffness in prosthetic meshes [30,31]. When comparing the POP and hernia groups, no statistically-significant difference can be found in these parameters (see Figures 6 and 7). However, tendencies can be seen with the hernia meshes being heavier (and similarly porous), while still being in general more compliant.

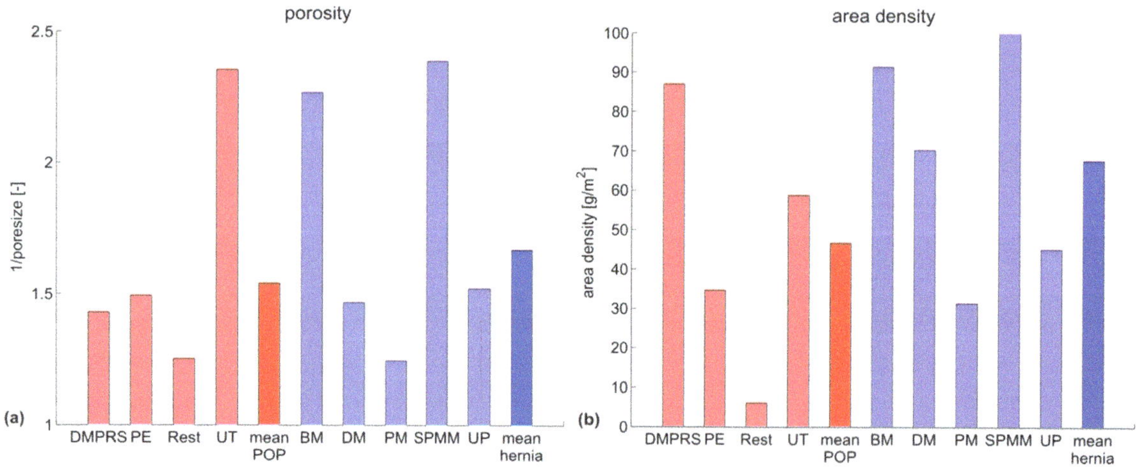

Figure 6. (**a**) Porosity and (**b**) area density of all meshes. POP meshes are shown in red, hernia meshes in blue, with their respective mean values plotted darker. Porosity is shown as the inverse of pore size, *i.e.*, higher values represent smaller pores.

3.2. Discussion

Better clinical outcome might be expected from meshes designed to mimic the physiologically-relevant deformation behavior of the underlying native tissue, thus ensuring mechanical biocompatibility. This entails a meaningful stiffness reference target for mesh design. However, the physiological loading configuration, as well as the range of load levels in terms of membrane tension in the abdominal and vaginal wall still remain largely uncertain. While membrane tensions in the abdominal wall are generally higher [7] than in the pelvic region (simply due to geometric reasons, as shown in [23]), increasing the level of membrane tension at which the secant stiffness is evaluated for the hernia meshes to a level of 0.136 N/mm (reported in [23] as a tension at rest in the abdominal wall) only marginally increases their stiffness and does not change the trends reported in Figures 2–7.

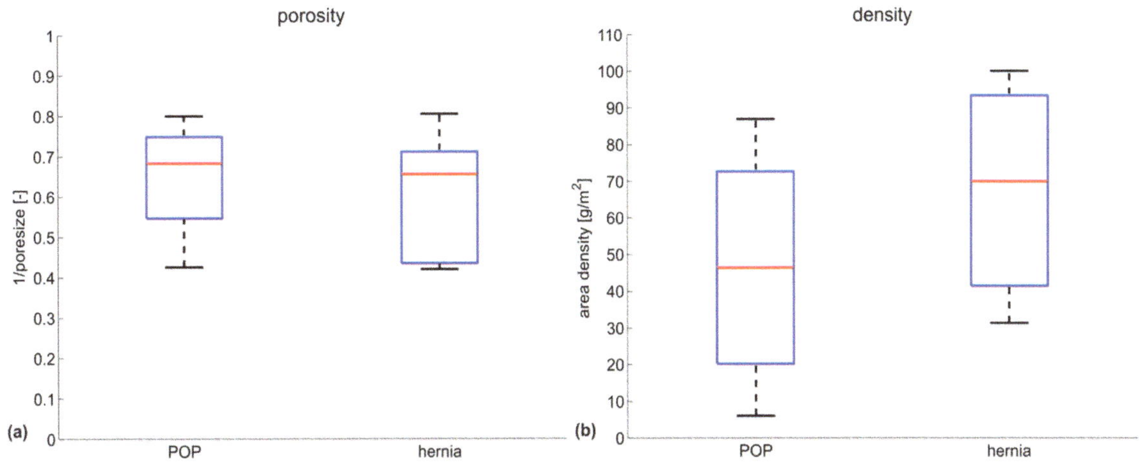

Figure 7. Box plots for (**a**) porosity and (**b**) area density of all meshes. The red line marks the median of the group; the box represents the 25th and 75th percentile; the extended whiskers the most extreme data points. Porosity is shown as the inverse of pore size, *i.e.*, higher values represent smaller pores.

The range of stiffness values for native tissue reported in the literature shows large variation and is mostly based on uniaxial tensile tests, while the predominant loading state *in vivo* is biaxial. Song *et al.* [32] report a Young's modulus of 0.042 N/mm^2 and 0.0225 N/mm^2 in the transverse and sagittal plane, respectively, for human abdominal wall during *in vivo* insufflation, which would translate to membrane stiffness values of 1.26 N/mm and 0.675 N/mm, respectively, multiplying by the reported thickness of around 30 mm [32]. Analyzing the uniaxial stress-strain graphs shown in [33], abdominal skin has a secant stiffness of 1.7 N/mm at a membrane tension reference of 0.136 N/mm, whereas vaginal wall stiffness is 1.45 N/mm at a reference of 0.035 N/mm membrane tension. Rabbits are one model system for mesh performance evaluation. Analysis of the stress-strain curves in [34] yields uniaxial stiffness values for the abdominal wall complex of rabbits of 0.87–0.98 N/mm, whereas [35] report 0.28 N/mm at the same reference membrane tension of 0.136 N/mm. Biaxial stiffness under inflation, however, increases by an order of magnitude to 2.41 N/mm. Similarly, vaginal wall tissue stiffness values at a reference membrane tension of 0.035 N/mm reach from 0.155 N/mm (prolapsed tissue; [29]), 0.675 N/mm (healthy tissue; [33]), 1.4 N/mm (prolapsed tissue; [11]), up to 6.47 N/mm (healthy tissue; [11]).

This variability is due to differences in experimental methodology, *in vivo vs. ex vivo* testing, cadaver testing, animal tissue *vs*. human tissue, pathological *vs*. healthy tissue, as well as the inherent variability of biological soft tissues. This scatter poses a significant problem in determining meaningful mechanical design targets for prosthetic mesh implants and warrants further investigation. Focus should be on the definition of consistent testing procedures based on physiological, *in vivo* loading and stress magnitude conditions.

It has to be noted that the approach of mimicking the ingrown state of the mesh using an elastomer matrix is only partially representative of the *in vivo* condition. This is mainly due to the differences between the embedding procedure and the process of tissue ingrowth. A mesh sample is simply laid into liquid elastomer and the elastomer left to cure. This results in an elastomer-mesh complex that is formed in an unloaded initial configuration, while *in vivo* ingrowth might be expected to happen in a loaded state. Due to this discrepancy, the definition of a secant stiffness between tension values chosen here (pre-force threshold as defined in the experimental protocol and reference membrane tension) might not be representative of the actual *in vivo* load range. This further highlights the need to investigate the expected *in vivo* loading conditions of mesh implants, so to define testing protocols that reproduce physiological states.

When comparing individual meshes in terms of their physical and mechanical parameters, an instructive example can be seen in two meshes manufactured by FEG Textiltechnik: DynaMesh ENDOLAP (DM), used in hernia repair, and DynaMesh PRS (DMPRS), used for pelvic prolapse repair. While their porosity is similar and DM is indeed the heavier of the two, as expected for a hernia mesh, DMPRS is much stiffer in all tested configurations; and up to two orders of magnitude in the case of uniaxial tension.

Similarly, Restorelle (a POP mesh) is the lightest implant with the largest pores; however, its stiffness is clearly above average for most tested configurations. Restorelle is less compliant in the configurations tested here than SPMM, the heaviest, small-pore implant. Note that in previous studies [13,36,37], Restorelle was shown to be more compliant than some of the meshes tested here. This can be attributed to the differences in the experimental protocols. In particular, the present experiments evaluate the response of meshes at a physiological tension level. Thus the low deformation regime determines the measured stiffness in the present work, whereas the ball burst test in [36] depends more on the higher deformation regime.

Lightweight and large porous meshes might be expected to be more compliant than heavy, small porous implants [36]. However, the present analysis shows that while POP meshes are on average indeed lighter and often have larger pores, they are generally stiffer in the physiological loading regime. In these loading configurations, porosity and density alone cannot be predictors for mesh stiffness. Their specific knitting pattern and microstructure can lead to mechanism-like behavior in a physiological loading range, effects that determine their compliance. This calls for a careful evaluation of the mechanical properties of each mesh on several length scales, in conditions representative of those expected *in vivo*.

The present experimental results do not take into account loads in directions other than the main knitting patterns. In fact, the meshes are usually implanted such that their knitting pattern aligns with principal loading directions, such as in line type suspensions (e.g., urethral slings) or sacrocolpopexy procedures. Some meshes, such as DM, Ultrapro (UP) and PE, even have colored filaments interweaved, guiding the physician during implantation. Deviation from this rule might lead to a mechanical response that strongly differs from the data reported here. Similarly, while each mesh was tested in two perpendicular directions, only the stiffer of the two is considered for the present analysis, being indicative of an upper bound of stiffness. Knitted meshes do indeed tend to behave in an anisotropic way, as investigated in [22], with anisotropy indices ranging from 1.0 for SPMM (similar stiffness in both evaluated directions) to 8.0 for DMPRS.

In addition, only one sample per configuration has been tested due to the limited availability of raw mesh material, which also limited the sample size. However, the level of variability for mesh implants reported in the literature [18,22,37,38] is low, justifying the analysis conducted.

4. Conclusions

The mechanical biocompatibility of prosthetic mesh implants for hernia and POP repair very likely is an important factor in ensuring their functionality and integration into the host tissue. We see matching mechanical properties in a physiological loading range as desirable and an important step towards reducing clinical complications.

This study has shown that some meshes designated as suited for POP repair tend to be stiffer than those used for hernia repair, even though the abdominal wall has been shown to be less compliant than the vaginal wall. Additionally, the expectation of lightweight, large pore meshes being more compliant than their counterparts was contradicted by the presented data and specific testing configurations, indicating that a biomechanical analysis of each product is necessary to determine its mechanical suitability. Knowledge of the physiological, *in vivo* mechanical environment in terms of loading configuration and magnitude is required in order to define a suitable design target for optimization of implants. Data reported in the literature show large variations in testing configurations and corresponding stiffness values for pelvic organs and abdominal wall tissue. The consensus for a standardized, physiological mechanical testing procedure is needed for native tissues and implants, opening the path for a conscious mechanical design of prosthetic meshes.

Acknowledgments

Manfred M. Maurer and Edoardo Mazza acknowledge internal funding by ETH Zurich for the support of this research.

Author Contributions

Manfred M. Maurer, Barbara Röhrnbauer, Jan Deprest and Edoardo Mazza conceived of and designed the study. Manfred M. Maurer, Barbara Röhrnbauer and Andrew Feola performed the experiments. Manfred M. Maurer and Edoardo Mazza analyzed the data. Barbara Röhrnbauer and Jan Deprest contributed materials. All authors contributed to writing the paper.

Conflicts of Interest

The authors declare no conflict of interest.

References

1. Röhrnbauer, B. *Mechanical Characterization and Modeling of Prosthetic Meshes*; ETH Zurich: Zurich, Switzerland, 2013.
2. Röhrnbauer, B.; Mazza, E. Uniaxial and biaxial mechanical characterization of a prosthetic mesh at different length scales. *J. Mech. Behav. Biomed. Mater.* **2014**, *29*, 7–19.

3. Mazza, E.; Ehret, A.E. Mechanical biocompatibility of highly deformable biomedical materials. *J. Mech. Behav. Biomed. Mater.* **2015**, *48*, 100–124.
4. Liang, R.; Abramowitch, S.; Knight, K.; Palcsey, S.; Nolfi, A.; Feola, A.; Stein, S.; Moalli, P.A. Vaginal degeneration following implantation of synthetic mesh with increased stiffness. *BJOG Int. J. Obstet. Gynaecol.* **2013**, *120*, 233–243.
5. Mangera, A.; Bullock, A.J.; Chapple, C.R.; Macneil, S. Are biomechanical properties predictive of the success of prostheses used in stress urinary incontinence and pelvic organ prolapse? A systematic review. *Neurourol. Urodyn.* **2012**, *31*, 13–21.
6. Konerding, M.A.; Chantereau, P.; Delventhal, V.; Holste, J.-L.; Ackermann, M. Biomechanical and histological evaluation of abdominal wall compliance with intraperitoneal onlay mesh implants in rabbits: A comparison of six different state-of-the-art meshes. *Med. Eng. Phys.* **2012**, *34*, 806–816.
7. Klinge, U.; Conze, J.; Limberg, W.; Brucker, C.; Ottinger, A.P.; Schumpelick, V. Pathophysiology of the abdominal wall. *Der Chirurg* **1996**, *67*, 229–233.
8. Choe, J.M.; Kothandapani, R.; James, L.; Bowling, D. Autologous, cadaveric, and synthetic materials used in sling surgery: Comparative biomechanical analysis. *Urology* **2001**, *58*, 482–486.
9. Junge, K.; Klinge, U.; Prescher, A.; Giboni, P.; Niewiera, M.; Schumpelick, V. Elasticity of the anterior abdominal wall and impact for reparation of incisional hernias using mesh implants. *Hernia* **2001**, *5*, 113–118.
10. Cosson, M.; Debodinance, P.; Boukerrou, M.; Chauvet, M.P.; Lobry, P.; Crépin, G.; Ego, A. Mechanical properties of synthetic implants used in the repair of prolapse and urinary incontinence in women: Which is the ideal material? *Int. Urogynecol. J.* **2003**, *14*, 169–178.
11. Rubod, C.; Boukerrou, M.; Brieu, M.; Jean-Charles, C.; Dubois, P.; Cosson, M. Biomechanical properties of vaginal tissue: Preliminary results. *Int. Urogynecol. J.* **2008**, *19*, 811–816.
12. Claerhout, F.; Verbist, G.; Verbeken, E.; Konstantinovic, M.; de Ridder, D.; Deprest, J. Fate of collagen-based implants used in pelvic floor surgery: A 2-year follow-up study in a rabbit model. *Am. J. Obst. Gyn.* **2008**, *198*, 91–96.
13. Jones, K.A.; Feola, A.; Meyn, L.; Abramowitch, S.D.; Moalli, P.A. Tensile properties of commonly used prolapse meshes. *Int. Urogynecol. J.* **2009**, *20*, 847–853.
14. Ozog, Y.; Konstantinovic, M.L.; Werbrouck, E.; de Ridder, D.; Edoardo, M.; Deprest, J. Shrinkage and biomechanical evaluation of lightweight synthetics in a rabbit model for primary fascial repair. *Int. Urogynecol. J.* **2011**, *22*, 1099–1108.
15. Fenner, D.E. New surgical mesh. *Clin. Obstet. Gynecol.* **2000**, *43*, 650–658.
16. Dietz, H.P.; Vancaillie, P.; Svehla, M.; Walsh, W.; Steensma, A.B.; Vancaillie, T.G. Mechanical properties of urogynecologic implant materials. *Int. Urogynecol. J.* **2003**, *14*, 239–243.
17. Cobb, W.S.; Kercher, K.W.; Heniford, B.T. The argument for lightweight polypropylene mesh in hernia repair. *Surg. Innov.* **2005**, *12*, 63–69.
18. Feola, A.; Abramowitch, S.; Jallah, Z.; Stein, S.; Barone, W.; Palcsey, S.; Moalli, P. Deterioration in biomechanical properties of the vagina following implantation of a high-stiffness prolapse mesh. *BJOG Int. J. Obstet. Gynaecol.* **2013**, *120*, 224–232.
19. Lowman, J.K.; Jones, L.A.; Woodman, P.J.; Hale, D.S. Does the prolift system cause dyspareunia? *Am. J. Obstet. Gynecol.* **2008**, *199*, 701–706.

20. Ozog, Y.; Konstantinovic, M.; Werbrouck, E.; de Ridder, D.; Mazza, E.; Deprest, J. Persistence of polypropylene mesh anisotropy after implantation: An experimental study. *BJOG Int. J. Obstet. Gynaecol.* **2011**, *118*, 1180–1185.
21. FDA Centre for Devices and Radiological Health. Fda safety communication: Update on serious complications associated with transvaginal placement of surgical mesh for pelvic organ prolapse. Avaiable online: http://www.fda.gov/medicaldevices/safety/alertsandnotices/ucm262435.htm (accessed on 23 March 2015).
22. Maurer, M.M.; Roehrnbauer, B.; Feola, A.; Deprest, J.; Mazza, E. Mechanical biocompatibility of prosthetic meshes: A comprehensive protocol for mechanical characterization. *J. Mech. Behav. Biomed. Mater.* **2014**, *40*, 42–58.
23. Ozog, Y.; Deprest, J.; Haest, K.; Claus, F.; de Ridder, D.; Mazza, E. Calculation of membrane tension in selected sections of the pelvic floor. *Int. Urogynecol. J.* **2014**, *25*, 499–506.
24. Coda, A.; Lamberti, R.; Martorana, S. Classification of prosthetics used in hernia repair based on weight and biomaterial. *Hernia J. Hernias Abdom. Wall Surg.* **2012**, *16*, 9–20.
25. Farine, M. *Instrumented Indentation of Soft Materials And Biological Tissues*; ETH Zurich: Zurich, Switzerland, 2013.
26. Hollenstein, M.; Ehret, A.E.; Itskov, M.; Mazza, E. A novel experimental procedure based on pure shear testing of dermatome-cut samples applied to porcine skin. *Biomech. Modeling Mechanobiol.* **2011**, *10*, 651–661.
27. Chu, C.C.; Welch, L. Characterization of morphologic and mechanical properties of surgical mesh fabrics. *J. Biomed. Mater. Res.* **1985**, *19*, 903–916.
28. Pourdeyhimi, B. Porosity of surgical mesh fabrics: New technology. *J. Biomed. Mater. Res.* **1989**, *23*, 145–152.
29. Peña, E.; Calvo, B.; Martínez, M.A.; Martins, P.; Mascarenhas, T.; Jorge, R.M.N.; Ferreira, A.; Doblaré, M. Experimental study and constitutive modeling of the viscoelastic mechanical properties of the human prolapsed vaginal tissue. *Biomech. Modeling Mechanobiol.* **2010**, *9*, 35–44.
30. Klinge, U.; Klosterhalfen, B.; Birkenhauer, V.; Junge, K.; Conze, J.; Schumpelick, V. Impact of polymer pore size on the interface scar formation in a rat model. *J. Surg. Res.* **2002**, *103*, 208–214.
31. Klosterhalfen, B.; Junge, K.; Klinge, U. The lightweight and large porous mesh concept for hernia repair. *Expert Rev. Med. Devices* **2005**, *2*, 103–117.
32. Song, C.; Alijani, A.; Frank, T.; Hanna, G.B.; Cuschieri, A. Mechanical properties of the human abdominal wall measured *in vivo* during insufflation for laparoscopic surgery. *Surg. Endosc.* **2006**, *20*, 987–990.
33. Gabriel, B.; Rubod, C.; Brieu, M.; Dedet, B.; de Landsheere, L.; Delmas, V.; Cosson, M. Vagina, abdominal skin, and aponeurosis: Do they have similar biomechanical properties? *Int. Urogynecol. J.* **2011**, *22*, 23–27.
34. Hernandez, B.; Pena, E.; Pascual, G.; Rodriguez, M.; Calvo, B.; Doblare, M.; Bellon, J.M. Mechanical and histological characterization of the abdominal muscle. A previous step to modelling hernia surgery. *J. Mech. Behav. Biomed. Mater.* **2011**, *4*, 392–404.
35. Rohrnbauer, B.; Ozog, Y.; Egger, J.; Werbrouck, E.; Deprest, J.; Mazza, E. Combined biaxial and uniaxial mechanical characterization of prosthetic meshes in a rabbit model. *J. Biomech.* **2013**, *46*, 1626–1632.

36. Feola, A.; Barone, W.; Moalli, P.; Abramowitch, S. Characterizing the *ex vivo* textile and structural properties of synthetic prolapse mesh products. *Int. Urogynecol. J.* **2013**, *24*, 559–564.
37. Shepherd, J.P.; Feola, A.J.; Abramowitch, S.D.; Moalli, P.A. Uniaxial biomechanical properties of seven different vaginally implanted meshes for pelvic organ prolapse. *Int. Urogynecol. J.* **2012**, *23*, 613–620.
38. Edwards, S.L.; Werkmeister, J.A.; Rosamilia, A.; Ramshaw, J.A.M.; White, J.F.; Gargett, C.E. Characterisation of clinical and newly fabricated meshes for pelvic organ prolapse repair. *J. Mech. Behav. Biomed. Mater.* **2013**, *23*, 53–61.

6

Quantitative Evaluation of Contamination on Dental Zirconia Ceramic by Silicone Disclosing Agents after Different Cleaning Procedures

Sebastian Wille [1,*], Vincent van Broeck [2], Thomas Strunskus [2], Franz Faupel [2] and Matthias Kern [1]

[1] Department of Prosthodontics, Propaedeutics and Dental Materials, School of Dentistry, Christian-Albrechts University, Arnold-Heller-Straße 16, Kiel D-24105, Germany; E-Mail: mkern@proth.uni-kiel.de

[2] Chair for Multicomponent Materials, Institute for Materials Science, Faculty of Engineering, Christian-Albrechts University at Kiel, Kaiserstr. 2, Kiel D-24143, Germany; E-Mails: vivb@tf.uni-kiel.de (V.B.); ts@tf.uni-kiel.de (T.S.); ff@tf.uni-kiel.de (F.F.)

* Author to whom correspondence should be addressed; E-Mail: swille@proth.uni-kiel.de;

Academic Editor: Andrew J. Ruys

Abstract: The aim of this study was to evaluate the effectiveness of cleaning procedures for air-abraded zirconia after contamination with two silicone disclosing agents. Air-abraded zirconia ceramic specimens (IPS e.max ZirCAD) were contaminated with either GC Fit Checker white or GC Fit Checker II. Untreated zirconia specimens were used as control. Afterwards the surfaces were cleaned either with waterspray or ultrasonically in 99% isopropanol or using a newly developed cleaning paste (Ivoclean). After cleaning X-ray photoelectron spectroscopy (XPS) was performed and the relative peak intensities of Zr, C and Si were used for a qualitative comparison of the residuals. There was no significant difference between the two different silicone disclosing agents. An additional cleaning step with isopropanol led to a significantly lower amount of residuals on the surface, but an additional cleaning process with Ivoclean did not reduce the amount of carbon residuals in comparison to the isopropanol cleaning. Just the silicone amount on the surface was reduced. None of the investigated cleaning processes removed all residuals from the contaminated surface. Standard cleaning processes do not remove all residuals of the silicone disclosing agent from the surface. This may lead to a failure of the resin-ceramic bonding.

Keywords: contamination; zirconia; silicone; cleaning; surface analysis

1. Introduction

In the last decade, zirconia restorations have become more and more important in dental applications. Due to its mechanical properties and also for esthetic reasons zirconia ceramics are widely used as fixed partial prostheses (FDPs), resin-bonded FDPs, full-coverage crowns, implant abutments, and endodontic posts [1]. In comparison to conventional cementation, which in most cases can also be used for luting zirconia restorations, durable resin-ceramic bonding is required for restorations with little or no mechanical retention such as resin-bonded retainers for fixed dental prostheses [2,3].

It has been shown in several *in vitro* studies that a strong, durable resin-zirconia bonding can be achieved by using a 10-methacryloxydecyl dihydrogenphosphate (MDP)-containing composite resin after air-abrasion with alumina particles [4–9]. These studies have been performed under well-controlled and clean conditions in the laboratory. However, in clinical situations zirconia ceramic bonding surfaces are contaminated by saliva, blood or silicone indicators during try-in of the restoration. As the set silicone shows a high silicone stability it might be assumed that no residuals are left on the surface after removal of the silicone disclosing agents. However, Yang *et al.* showed by XPS investigations that the surface is contaminated with silicone residuals after the use of silicone disclosing agent (Fit checker, GC) even after cleaning procedures including water rinsing or isopropanol rinsing [10]. In the same study the authors showed that contaminations on the zirconia surface lead to a dramatic drop in tensile bond strength. The reason for the contamination has been explained in several ways. Some older publications explain it by a chemical reaction between the silicone disclosing agent and the restoration surface leading to a stable contamination on the ceramic surface [11,12]. Some investigators also assumed that a not polymerized thin film is left on the surface that leads to a reduction in the bond strength [12,13].

However, meanwhile the company GC developed another product, the so-called Fit Checker II which is also on the market. The different compositions of both products are given in Table 1. Although both base materials are mainly using polymers from the group of the silicones, there are large differences in the compositions. The base material of Fit Checker white consists also of calcium carbonate whereas Fit Checker II contains silicon dioxide. The catalyst materials are completely different in their compositions. In the product description of Fit Checker II it is claimed that there will be a clean, residue free surface after peeling it away.

The aim of this study was to compare the older material Fit Checker white with Fit Checker II concerning the cleanability on zirconia surfaces. In this study cleaning procedures were chosen that are generally clinically available namely water spray and ultrasonic cleaning in isopropanol. In addition, also a newly developed cleaning medium (Ivoclean, Ivoclar Vivadent) was tested. Ivoclean consists of an alkaline suspension of zirconium oxide particles. As claimed by its manufacturer to the size and concentration of the particles in the medium, contaminants are much more likely to bond to them than to the surface of the ceramic restoration. Therefore, this cleaning medium is supposed to absorb contaminants and thus leaving behind a clean zirconia surface. Another aim of the study was to check, whether the easy cleaning procedures that are generally available are really efficient to remove contaminations in a way that the bond strength will not be significantly reduced.

Table 1. Used materials, their composition as given by the manufacturers in wt-% and their batch numbers.

Material	Company	Composition	Batch No.
IPS e.max ZirCad	Ivoclar Vivadent	ZrO_2: 87%–95%, Y_2O_3: 4%–6%, HfO_2: 1%–5%, Al_2O_3: 0%–0.5%	L48411
GC Fit Checker white—Base	GC Corporation	Dimethyl methylhydrogenpolysiloxane: 65% $CaCO_3$: 35%	1302221
GC Fit Checker white—Catalyst	GC Corporation	Petrolatum: 55%, Ethylsilicate: 40%, fatty acid salt: 5%	1302221
GC Fit Checker II Base	GC Corporation	SiO_2: 40%–60%, Polydimethyl siloxane: 38%–42%	1305151
GC Fit Checker II Catalyst	GC Corporation	SiO_2: 42%–45%, Vinyldimethylpolysiloxane: 50%–60%	1305151
Ivoclean	Ivoclar Vivadent	ZrO_2: 10%–15%, H_2O: 65%–80%, PEG: 8%–10%, NaOH: <1%, Pigments, Additives: 4%–5%	P66483

2. Results and Discussion

2.1. Results

The relative line intensities (areas) of the X-ray photoelectron spectroscopy (XPS) measurements of the C 1s, Si 2p and Zr 3d lines and the corresponding standard deviations for the different groups are shown in Table 2. There was no significant difference between the main groups FC1 and FC2 ($p = 0.350$). The two-way ANOVA with a post-hoc Tukey's HSD test also revealed that the cleaning process with an additional step of an ultrasonic bath in isopropanol showed a significant decrease in the relative peak intensities of C ($p = 0.006$) and Si ($p = 0.005$) for the previously contaminated specimens, whereas the additional cleaning with Ivoclean led only to a minor decrease of the Si amount on the surface as shown in Figure 1 and Figure 2. However, this decrease was statistically not significant ($p = 0.702$). Independently of the conducted cleaning processes, all previously contaminated specimens showed a significantly higher amount of C and Si on the surface than the uncontaminated reference groups ($p \leq 0.001$).

2.2. Discussion

The results of this study show that there are always some residuals of the silicone disclosing agent after the conducted cleaning processes left on the surfaces. As it has been shown by Yang *et al.*, these residuals can lead to a drastic decrease in the bond strength [14]. Also additional cleaning processes with Ivoclean and isopropanol do not remove all residuals from the silicone disclosing agents, although the amount of contamination is significantly decreased. In order to achieve a reliable bonding to zirconia ceramics one needs to use other cleaning processes that are more efficient, such as for example air abrasion with alumina particles [15]. However, this method might aggravate the degradation of zirconia [16].

Applying phosphoric acid etching did not lead to the removal of silicone residuals as Zhang *et al.* have shown on a nano-alumina coated zirconia surface [17]. Several studies concerning the use of plasma treatment on zirconia surfaces show also promising results that this method might be an efficient cleaning process for chair side application with less surface damaging effects [18,19].

Table 2. X-ray photoelectron spectroscopy (XPS) results: mean relative elemental intensities and corresponding standard deviations of Zr, C and Si of the test groups determined from the XPS line intensities (areas).

Test Group	Zr	C	Si
REF a	0.7723 ± 0.0207	0.2033 ± 0.0216	0.0234 ± 0.0057
REF b	0.7100 ± 0.0141	0.2683 ± 0.0147	0.0201 ± 0.0037
REF c	0.7350 ± 0.0152	0.2383 ± 0.0147	0.0243 ± 0.0024
FC1 a	0.5517 ± 0.0488	0.3900 ± 0.0522	0.0571 ± 0.0134
FC1 b	0.6633 ± 0.0163	0.2883 ± 0.0194	0.0514 ± 0.0071
FC1 c	0.6783 ± 0.0184	0.2800 ± 0.0141	0.0410 ± 0.0043
FC2 a	0.5117 ± 0.1277	0.4050 ± 0.1161	0.0826 ± 0.0334
FC2 b	0.6683 ± 0.0240	0.2867 ± 0.0216	0.0455 ± 0.0162
FC2 c	0.6367 ± 0.0513	0.3217 ± 0.0471	0.0411 ± 0.0099

Notes: REF: no contamination; FC1: contamination with Fit Checker white; FC2: contamination with Fit Checker II; a: water spray cleaning; b: additional ultrasonic bath with isopropanol; c: water spray, Ivoclean, water spray and ultrasonic bath with isopropanol.

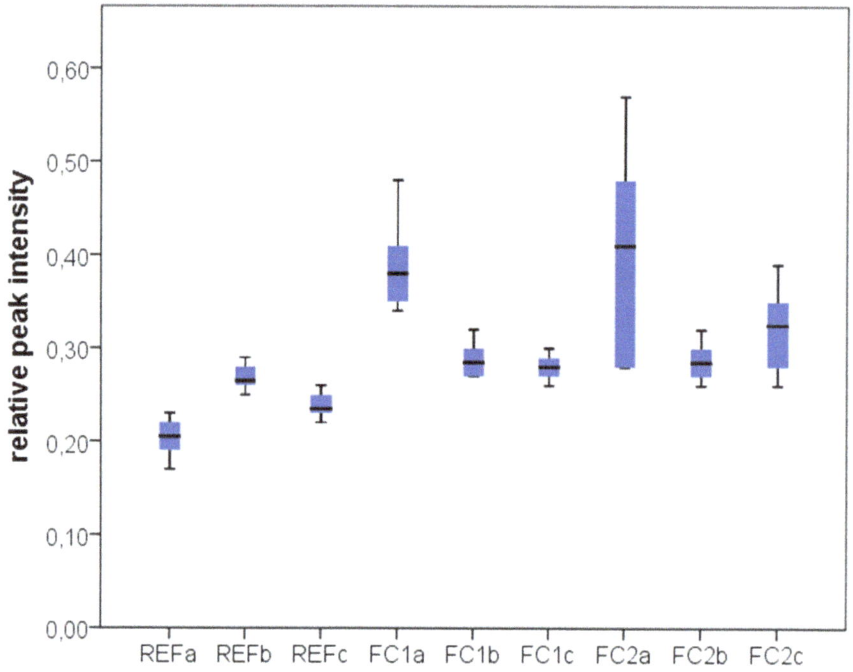

Figure 1. Relative peak intensity of carbon (group codes see Table 2).

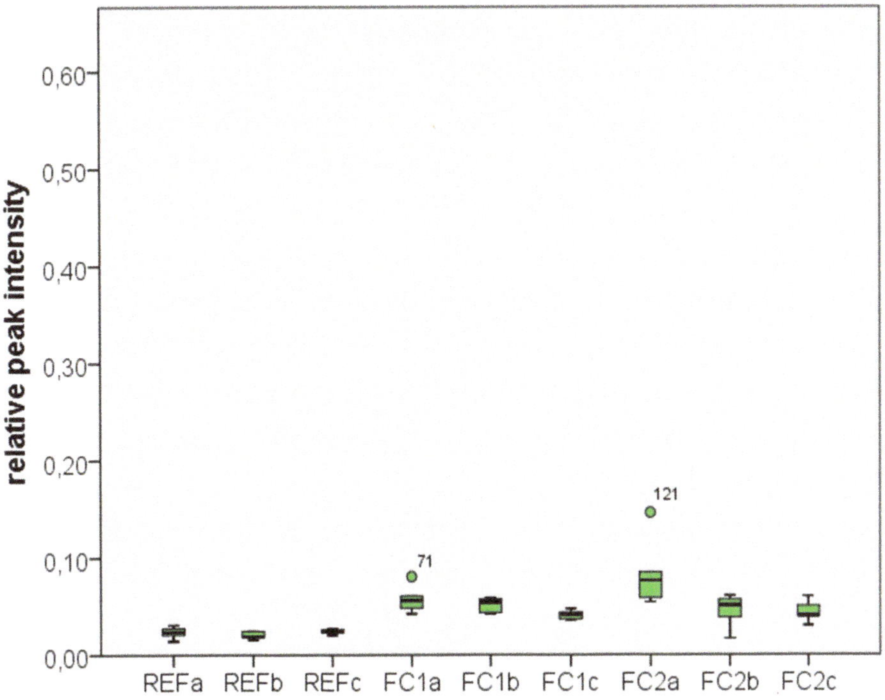

Figure 2. Relative peak intensity of silicone (group codes see Table 2).

No differences between the two silicone disclosing media GC Fit Checker white and GC Fit Checker II could be detected regarding the evaluated cleaning procedures. The XPS measurements show that the contamination of the surface with silicon and carbon based molecules can be reduced considerably by ultrasonic cleaning in 99% isopropanol in comparison to the simple water spray cleaning. However, the additional cleaning process with Ivoclean had only a small influence on the surface concentration of silicon. There was no significant change in the carbon concentration. Taking into account that Yang *et al.* have shown that carbon based residuals have a larger effect on the tensile bond strength than silicone residuals [14], the additional use of Ivoclean after the application of silicone disclosing agents might be questioned.

One limitation of this study is that the XPS measurements do not give any information about the structure of the contaminants on the surface. So it is unknown, whether the contaminations build a very thin film or single islands on the surface. The relatively large standard deviation in the first subgroup might be due to the fact that the water spraying was performed manually. This kind of cleaning was just a rough cleaning in order to remove visible contaminants. However small variations in the hardening and application process of the silicone disclosing agents might also have led to relatively large variations in this group. Another limitation of this study is that only silicone disclosing agents have been included. However, other clinical relevant contaminations such as e.g., saliva had not been considered.

3. Experimental Section

3.1. Specimen Preparation

Fifty-four pre-sintered IPS e.max ZirCAD specimens (Ivoclar Vivadent, Schaan, Liechtenstein) have been sectioned (10 mm × 10 mm × 1.5 mm) with a precision saw ISOMET 1000 (Buehler, Düsseldorf, Germany). Afterwards the zirconia specimens were ground with 600-grit silicon carbide paper before the final sintering process was performed according to manufacturer's instructions. Then the specimens were air-abraded with 50 μm alumina particles at 0.1 MPa using a P-G 400 (Harmisch+Rieth, Cologne, Germany) and cleaned in an ultrasonic bath with 99% isopropanol (Otto Fischer, Saarbrücken, Germany) for 3 min. After drying of the specimens with oil-free air at 0.33 MPa the specimens were divided into three main groups.

The specimens of two groups (FC1 and FC2 respectively) where treated with two different silicone disclosing agents. Therefore the components of the flowing silicone disclosing agents, GC Fit Checker white or GC Fit Checker II (GC Corporation, Tokyo, Japan) respectively, were mixed and applied to the specimens within 30 s. After the silicone disclosing agent had been specimen to one specimen, another specimen was slightly pressed on top of the covered surface and both zirconia specimens with the silicone disclosing agent in between were stored for 3 min. Finally the specimens were separated manually and the hardened silicone disclosing agent was removed. The third group was not contaminated before the cleaning processes were performed and are used as reference group (REF).

3.2. Cleaning Procedures

The cleaning procedures of the main groups FC1 and FC2 were performed subsequently after the contamination process. The specimens of all three main groups were divided into three subgroups with different cleaning procedures. Every wet cleaning step with isopropanol or water spray was followed by a drying step with air as it is described previously. The zirconia ceramic specimens of the first subgroup were only cleaned with water spray with a pressure of 0.15 MPa for 20 s (a), for the second subgroup additionally a cleaning step in an ultrasonic bath with 99% isopropanol was conducted for 3 min (b). The third cleaning procedure consisted of water spraying for 20 s, coating the bonding surface with Ivoclean (Ivoclar Vivadent, Schaan, Liechtenstein), dispensing Ivoclean with a microbrush for 20 s, water spraying for 20 s and ultrasonic bath with 99% isopropanol (c). The batch numbers of the used materials are given in Table 1.

3.3. XPS Analysis

All specimens were examined with X-ray photoelectron spectroscopy (XPS) to quantify the contamination on the surface of the zirconia specimens. For the measurements an Omnicron Full Lab (Omnicron NanoTechnology GmbH, Taunusstein, Germany) equipped with an Al/Mg Kα X-ray source and a VSW 100 hemispherical analyzer was used. All measurements were performed with an excitation energy of 1486.6 eV and in the constant analyzer energy (CAE) mode. A survey spectrum with a CAE of 100 eV was measured to monitor the characteristic lines of the elements present Afterwards three detailed measurements with a CAE of 20 eV were recorded for the peaks of carbon (C1s), silicon (Si2p) and zirconium (Zr3d). The areas under the peak curves were calculated with the EIS software (Omnicron NanoTechnology GmbH, Taunusstein, Germany). The oxygen O1s line dominating the spectrum was not used for the analysis because both the zirconia substrate as well as the residuals will contribute to the line intensity.

The collected data was checked for normal distribution and analyzed using two-way analysis of variance (ANOVA) with a post-hoc Tukey's HSD test (SPSS v20, Chicago, IL, USA) at a significance level of $p \leq 0.05$.

4. Conclusions

Taking the results and conditions of this study into account, the following conclusions can be drawn:

1. There is no difference between the two investigated silicone disclosing agents concerning the cleanability, with respect to the investigated cleaning processes.
2. Additional ultrasonic cleaning in 99% isopropanol will lead to a significant reduction of contamination.
3. Alternative cleaning processes need to be developed in order to achieve a better removal of contaminants in order to achieve optimal bonding conditions.

Acknowledgments

The authors wish to express appreciation for technical assistance and laboratory support of Frank Lehmann and Rüdiger Möller. They thank also Stefan Rehders and Matthias Wellmann for their help of the XPS measurements.

Author Contributions

Matthias Kern had the idea for the study. Franz Faupel, Matthias Kern, Thomas Strunskus, and Sebastian Wille conceived and designed the experiments. Vincent van Broeck performed the experiments. Vincent van Broeck, Thomas Strunskus and Sebastian Wille analyzed the data. Matthias Kern contributed the materials. Franz Faupel contributed the analysis tools. Sebastian Wille wrote the paper. All authors contributed to the discussion and proofread the paper.

Conflicts of Interest

The authors declare no conflict of interest.

References

1. Blatz, M.B.; Sadan, A.; Kern, M. Resin-Ceramic Bonding: A review of the literature. *J. Prosthet. Dent.* **2003**, *89*, 268–274.
2. Sasse, M.; Kern, M. CAD/CAM single retainer zirconia-ceramic resin-bonded fixed dental prostheses: Clinical outcome after 5 years. *Int. J. Comput. Dent.* **2013**, *16*, 109–118.
3. Sasse, M.; Kern, M. Survival of anterior cantilevered all-ceramic resin-bonded fixed dental protheses made from zirconia ceramic. *J. Dent.* **2014**, *42*, 660–663.
4. Kern, M.; Wegner, S.M. Bonding to zirconia ceramic: Adhesion methods and their durability. *Dent. Mater.* **1998**, *14*, 64–71.
5. Wegner, S.M.; Gerdes, W.; Kern, M. Effect of different artificial aging conditions on ceramic/composite bond strength. *Int. J. Prosthodont.* **2002**, *15*, 267–272.
6. Kern, M. Resin bonding to oxide ceramics for dental restorations. *J. Adhes. Sci. Technol.* **2009**, *23*, 1097–1111.

7. Blatz, M.B.; Phark, J.H.; Ozer, F.; Mante, F.K.; Saleh, N.; Bergler, M.; Sadan, A. *In vitro* comparative bond strength of contemporary self-adhesive resin cements to zirconium oxide ceramic with and without air-particle abrasion. *Clin. Oral. Investig.* **2010**, *14*, 187–192.
8. Kern, M. Letter to the editor: Bond strength of luting cements to zirconium oxide ceramics. *Int. J. Prosthodont.* **2000**, *13*, 350.
9. Wegner, S.M.; Kern, M. Long-term resin bond strength to zirconia ceramic. *J. Adhes. Dent.* **2000**, *2*, 139–145.
10. Yang, B.; Scharnberg, M.; Wolfart, S.; Quaas, A.C.; Ludwig, K.; Adelung, R.; Kern, M. Influence of contamination on bonding to zirconia ceramic. *J. Biomed. Mater. Res. Part B: Appl. Biomater.* **2007**, *81*, 283–290.
11. Sheth, J.; Jensen, M.; Tolliver, D. Effect of surface treatment on etched porcelain bond strength to enamel. *Dent. Mater.* **1988**, *4*, 328–337.
12. Sorensen, J.A. Improved seating of ceramic inlays with a silicone fit-checking medium. *J. Prosthet. Dent.* **1991**, *65*, 646–649.
13. Szep, S.; Schmid, C.; Weigl, P.; Hahn, L.; Heidemann, D. Effect of the silicone disclosing procedure on the shear bond strength of composite cements to ceramic restorations. *J. Prosthet. Dent.* **2003**, *89*, 60–65.
14. Yang, B.; Wolfart, S.; Scharnberg, M.; Ludwig, K.; Adelung, R.; Kern, M. Influence of contamination on zirconia ceramic bonding. *J. Dent. Res.* **2007**, *86*, 749–753.
15. Attia, A.; Kern, M. Effect of cleaning methods after reduced-pressure air abrasion on bonding to zirconia ceramic. *J. Adhes. Dent.* **2011**, *13*, 561–567.
16. Sato, H.; Yamada, K.; Pezzotti, G.; Nawa, M.; Ban, S. Mechanical properties of dental zirconia ceramics changed with sandblasting and heat treatment. *Dent. Mater. J.* **2008**, *27*, 408–414.
17. Zhang, C.; Degrange, M. Shear bond strengths of self-adhesive luting resins fixing dentine to different restorative materials. *J. Biomater. Sci.* **2010**, *21*, 593–608.
18. Han, G.J.; Chung, S.N.; Chun, B.H.; Kim, C.K.; Oh, K.H.; Cho, B.H. Effect of the applied power of atmospheric pressure plasma on the adhesion of composite resin to dental ceramic. *J. Adhes. Dent.* **2013**, *14*, 461–469.
19. Valverde, G.B.; Coelho, P.G.; Janal, M.N.; Lorenzoni, F.C.; Carvalho, R.M.; Thompson, V.P.; Weltemann, K.D.; Silva, N.R. Surface characterisation and bonding of Y-TZP following non-thermal plasma treatment. *J. Dent.* **2013**, *41*, 51–59.

7

Effect of Beverages on Viscoelastic Properties of Resin-Based Dental Composites

Muhammad Kaleem [1], Abdul S. Khan [2,3,*], Ihtesham U. Rehman [4] and Ferranti S. Wong [3]

[1] Army Medical College, National University of Sciences and Technology, Islamabad 44000, Pakistan; E-Mail: mkaleem-amc@nust.edu.pk

[2] Interdisciplinary Research Centre in Biomedical Materials, COMSATS Institute of Information Technology, Lahore 54000, Pakistan

[3] Centre for Oral Growth and Development, Institute of Dentistry, Barts and the London School of Medicine and Dentistry, Queen Mary University of London, London E1 2AD, UK; E-Mail: f.s.l.wong@qmul.ac.uk

[4] Kroto Research Institute, Department of Materials Science and Engineering, University of Sheffield, Sheffield S3 7HQ, UK; E-Mail: i.u.rehman@sheffield.ac.uk

* Author to whom correspondence should be addressed; E-Mail: draskhan@ciitlahore.edu.pk;

Academic Editor: Klara Hernadi

Abstract: The viscoelastic properties of three commercially available resin-based composites (Filtek™ P60, Filtek™ Supreme, and Filtek™ Z250; 3M ESPE, Bracknell, UK) were measured to determine the effect of beverages on their storage moduli and damping ratios. Rectangular samples of the three hybrid composites were immersed in three beverages at 37 °C for 1, 7, 30, and 60 days. At each time interval, these samples were subjected to three-point bend tests in temperature mode using a Perkin Elmer DMA7 Dynamic Mechanical Analyzer (Perkin Elmer Corp., Waltham, MA, USA) to measure the storage modulus and damping ratio. The immersion time had significant influence on the viscoelastic property of composites and it was found that generally for all samples the storage modulus was reduced, whereas damping values increased with immersion time. The viscoelastic behavior of tested materials seems to be related to the pH environment, hydrophilicity and the chemical composition of composites.

Keywords: dental composites; beverages; viscoelasticity; storage modulus

1. Introduction

Resin-based composite (RBC), as a dental restorative material, is heterogeneous and isotropic [1]. It is a polymer-based material, which transforms from a liquid to a rubbery-glassy state on polymerization. When subjected to an erosive condition, material may be lost due to monomers and oligomers leaving the matrix. Factors that may influence the speed of this reaction include the type of material's chemical bond, copolymer composition, hydrophilicity of the polymer matrix and the pH of the erosive agents [2]. The long-term stability as a restorative material is frequently related to its resistance to degradation under the sinusoidal masticatory load and effect of oral environment, especially with the increasing consumption of beverages [3]. Several researchers [4–8] have investigated the effects of food-simulating liquids on the properties of dental composites, however, to date, the effects of these agents on the viscoelastic properties of the restorative materials have not been tested.

Viscoelastic measurements can provide information on the polymeric system by measuring the propensity of the material to creep under load. Viscoelastic solids differ from elastic solids in that [9]: (i) on application of a constant stress, after the instantaneous strain, the strain increases with time (creep); (ii) on application of a constant strain, after the instantaneous stress, the stress decreases with time (stress relaxation); and (iii) if an alternating (sinusoidal) stress or strain is applied, stress and strain are out of phase. This phase angle is denoted by δ. Theoretically, the Young's modulus (E) of classical elasticity is replaced by the so-called complex modulus (E^*), where E^* is given by

$$E^* = E' + iE'' \tag{1}$$

where, E' is the storage modulus, and represents the elastic component of deformation; E'' is the loss modulus, and represents the viscous (inelastic) component; and $i = \sqrt{-1}$. These are related to δ:

$$\tan \delta = E''/E' \tag{2}$$

where, tan δ is a measure of energy loss, sometimes referred to loosely as damping capacity. It is related to the fraction of energy retained, resilience (R):

$$R = \exp(-\pi \tan \delta) \tag{3}$$

At the glass transition temperature (T_g), E' decreases dramatically over a short temperature range, E'' decreases initially and then increases, and tan δ go through maxima. Beyond T_g, the polymer is in the rubber-like state, where deformation is a function of entropy only.

Dynamic Mechanical Analysis (DMA) is now a standard method in measuring the viscoelastic properties and is suggested as a valuable tool for characterization of viscoelasticity of dental composites to give a better insight into their mechanical behavior at molecular level [10–15]. It measures the ratio of the amplitudes of stress to the applied strain, and the phase angle δ, and computes E', E'', and tan δ.

The aim of this study was to investigate the viscoelastic properties of three composite restorative materials after a period of storage in various beverages with different pH values. The working hypothesis for this study was that beverages could affect the viscoelastic stability of RBC restorative materials. This hypothesis was based on the assumption that acidic environment could cause breakage of the polymer-resin bonding with consequential change of the viscoelastic properties of the composite.

2. Results

The mean storage modulus (*E′*) and mechanical damping (tan δ) of all the groups along with their standard deviations are given in Figures 1 and 2, respectively. It was observed that the storage moduli for all samples were decreased with immersion time in media. The statistical analysis indicated that there was no significant difference after immersion for four week, whereas after eight week time period, all three composites showed a significant difference ($p = 0.037$) in all solutions. Samples immersed in apple juice (AJ) and Coca-Cola (CC) showed significant decrease in values as compare to deionized water (DW). Among samples, Filtek™ P60 and Filtek™ Z250 after eight week immersion in AJ, CC and DW showed significant difference ($p = 0.011$) with Filtek™ Supreme, whereas, insignificant difference was observed between Filtek™ P60 and Filtek™ Z250. The tan δ values for composites increased with immersion time, however, the insignificant difference was observed.

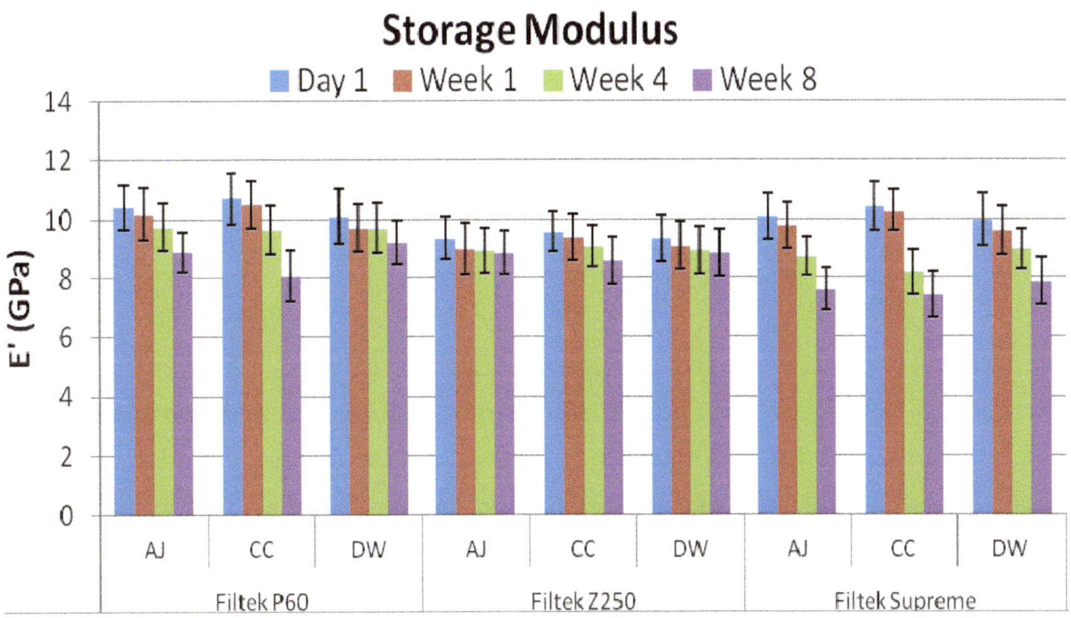

Figure 1. Comparative Storage Modulus values (GPa) of Z250, Filtek™ P60 and Filtek™ Supreme after immersion in apple juice (AJ), Coca-Cola (CC) and deionized water (DW) for one day, one week, four week, and eight week.

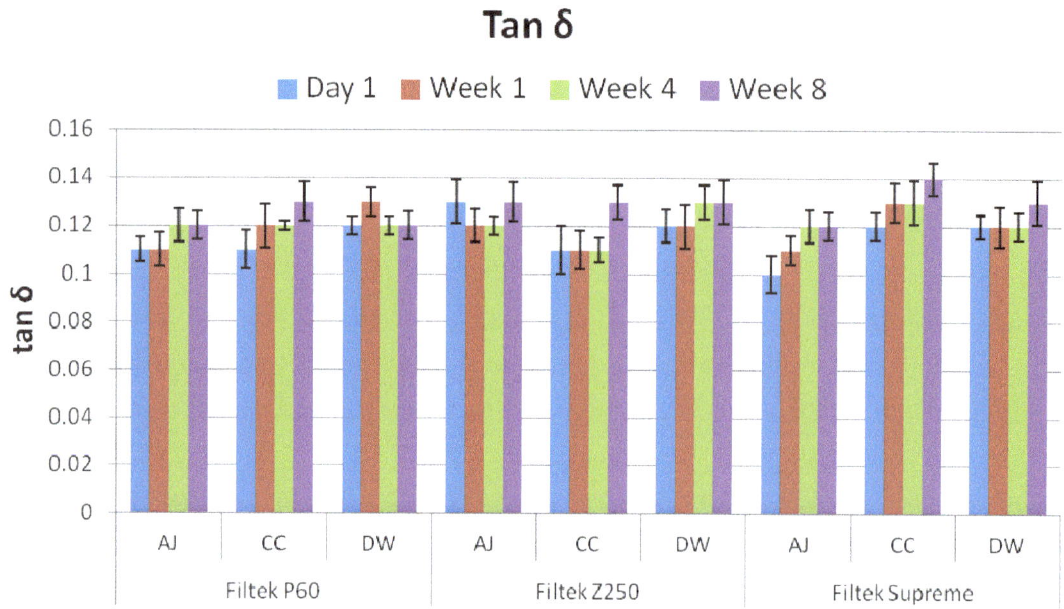

Figure 2. Comparative tan delta values of Z250, Filtek™ P60 and Filtek™ Supreme after immersion in apple juice (AJ), Coca-Cola (CC) and deionized water (DW) for one day, one week, four week and eight week.

3. Discussion

The properties of restorative material may degrade in the volatile oral environment. The mean pH of oral cavity is 6.28–7.34 [16]. Many foods and drinks (e.g., water, acidic soft drinks, alcoholic drinks, and food derivatives) affect the behavior of restorative materials. Previous studies have shown that fillers tend to fall out from resin materials [17], and the matrix component decomposes when exposed to low pH environments [18,19]. Many soft drinks are acidic and the pH is 3.0 or lower. This means that drinking acidic drinks over a long period and with continuous sipping can erode the tooth enamel—and the resin material as well. Therefore, the increase in use of RBC restorations made them more prone to be exposed to diverse conditions of oral environment, which can result in chemical degradation [20] and reduction in physical properties [21], subsequently effect on the clinical longevity of the material. There is lack of data about the viscoelastic stability of resin-composite restorative material after being subjected to various daily use beverages.

In the present study, different beverages promoted viscoelastic alterations, which are probably ascribed to a chemical reaction or dissolution of the composite when immersed in the tested beverages. These chemical reactions make the plasticization of polymer matrix and results in partial and complete debonding of filler particles. The variety of the results could be explained by the different testing conditions, especially with respect to storage medium, which has been shown to be important. The degrading effect of water on composites has been explained in the past with two different mechanisms. The first one is the plasticizing role of the water molecule, which creates more volume in the matrix and enhances the movement of the chain segments, resulting in the decrease of the stiffness of the material. The other mechanism is the leaching of composite's components in the water [22]. The comparative study showed that Filtek™ P60 and Filtek™ Z250 has higher storage modulus compared to Filtek™ Supreme. It could be due to presence of high molecular weight Bisphenol A-glycidyl methacrylate

(Bis-GMA), which restricts the molecular motion and decreases the dissipation of energy stored. In Filtek™ P60 and Filtek™ Z250 composites the low molecular weight monomer Triethyleneglycol dimethacrylate (TEGDMA) is replaced by Urethane dimethacrylate (UDMA) and Bisphenol A polyethylene diether dimethacrylate (Bis-EMA), so the reduction in the concentration of TEGDMA leads to higher viscosity. Both of these resins are higher molecular weight and therefore have fewer double bonds per unit of weight. The high molecular weight materials also impact the measurable viscosity and results in less shrinkage, reduced aging and a slightly softer resin matrix. Additionally, these resins impart a greater hydrophobicity and are less sensitive to changes in atmospheric moisture. Filtek™ Supreme composed of BisGMA, UDMA, Bis-EMA and TEGDMA. UDMA has low water absorption and solubility characteristics [23], whereas, TEGDMA is a hydrophilic monomer and able to absorb water [24]. It was found that TEGDMA containing composite showed low storage modulus with immersion time, which was due to increase in water uptake surface hydrophilicity. Hydrophilic groups such as the ethoxy group in TEGDMA are thought to show affinity with water molecule by hydrogen bonding to oxygen [25]. However, their comparative values with other composites were due to presence of UDMA-based monomers that causes low viscosity and low water absorption and successful polymerization with visible light [26,27]. On the basis of this, the specific composition and/or characteristics of RBCs that was used in this study may have favored the specific degradation. In respect to specimens immersed in Coca-Cola (pH 2.4), it was found that immersion media promoted degradation pattern on the composites. It is expected that Coca-Cola might damage the integrity of restorative materials because of presence of phosphoric acid. Previous studies [28–30] reported the dissolution of calcium phosphate from the tooth structure due to usage of Coca-Cola.

The DMA three-point bend test was used, therefore, the comparison of the modulus values for RBCs are limited to only in the axial direction. The direction of testing load means that the materials were subjected to tension in the lower surface and compression in the upper surface. The general trend for these composites showed that with increase in immersion time the storage modulus was reduced, whereas damping values increased which indicate that the RBCs absorbs less energy in repeated stress, and the external energy is likely to be dissipated within the material as residual stress. The acidic environment may cause the scission of polymeric chain and due to dynamic force the movement of molecules was expected to increase leading to movement of polymer chain, which reduces the stress and consequently the storage modulus. The force strike at one point and the energy dissipate to adjacent molecules and it enhances the molecular movement. The more the molecular movement, the lower is the stiffness. The pattern of stress distribution in a three-point bend test is similar to the common pattern of stress distribution in mastication and is clinically relevant [31].

Mostly in literature, the modulus values of composites were obtained at room temperature (21 ± 3 °C), whereas, the viscoelastic parameter is temperature dependent. Therefore, in this study the data were obtained at 37 °C, where testing results are very important in order to predict the clinical performance. The standard deviation of the measured values in the present study might be attributed to the variation in the roughness of the specimen surface because all the specimens were prepared manually, and the testing configuration is sensitive to surface flaws and defects. However, this feature appears to be consistent with previous mechanical and thermal studies [32–35]. It would be of interest to compare the viscoelastic moduli of these materials to those of natural tooth tissue. It has been shown that the relaxation modulus of human dentin has a linear dependence on the logarithm of time [36], indicating that dentin is also a viscoelastic material. The linear dependence of relaxation or creep on log time has been shown by Gent [37] to be applicable for elastomers and Braden and Wilson [38] for glass ionomer cements. The authors were unable to find any report that used DMA to measure the viscoelastic moduli for dentin in order to make direct comparison. The Young's modulus for dentin, measured by simple bending technique, was reported to be in the range of 11.4–19.3 GPa [39]. This value is lower than the storage modulus for intertubular dentin, 21 GPa (range 17–23 GPa), measured by combining a nano-indentation technique with Atomic Force Microscopy [40].

4. Clinical Significance

As these restorative materials have to be placed in the teeth, it is very important that these materials behave in the similar manner as the enamel and dentin, when they were subjected to occlusal forces. Therefore, it could be observed that beverages promoted a significant degradation of the resin matrix, even within the period of evaluation used in the present study, demonstrating that the clinicians must advise patients about the possible effects of these solutions over composites. According to this study, Filtek™ P60 showed better results compared to other restorative materials, however, further studies are necessary to verify the influence of beverages on other mechanical and physical properties.

5. Materials and Methods

Three different types of commercial photo-activated resin-based composite (RBC) restorative materials (Filtek™ P60, Filtek™ Supreme and Filtek™ Z250, 3M ESPE, Bracknell, UK) were used. The resin system and filler particle size with their coloring shades are given in Table 1. The resin matrices in these commercial products are BIS-GMA, BIS-EMA, UDMA, where, in Filtek™ Supreme there is small amount of TEGDMA along with UDMA. In Filtek™ Supreme, the filler contains a combination of a non-agglomerated/non-aggregated, 20 nm nano-silica filler, and loosely bound agglomerated zirconia/silica nano-cluster, consisting of agglomerates of primary zirconia/silica particles with size of 5–20 nm fillers. The cluster particle size range is 0.6 to 1.4 microns. The filler loading is 78.5% by weight. In Filtek™ P60 and Filtek™ Z250, the particle size distribution of filler is 0.01 μm to 3.5 μm with an average particle size of 0.6 μm. The fillers are zirconia/silica with 61% and 60% by volume without silane treatment, respectively.

Table 1. Composition of commercial dental composites with filler particle size and loading by volume.

<table>
<tr><th>Material</th><th>Shade</th><th>Fillers Particle size/
Loading by volume</th><th>Resin System †</th></tr>
<tr><td>Filtek™ P60
Batch No. 4720</td><td>A3</td><td>0.01–3.5 μm/
61% vol.</td><td rowspan="2">Bis-GMA, UDMA and
Bis-EMA</td></tr>
<tr><td>Filtek™ Z250
Batch No. 6020</td><td>A2</td><td>0.01–3.5 μm/
60% vol.</td></tr>
<tr><td>Filtek™ Supreme
Batch No. 3910</td><td>A2B</td><td>5–75 nm/
57%–60% vol.</td><td>Bis-GMA,
Bis-EMA and TEGDMA</td></tr>
</table>

Notes: † Resin System: Bis-GMA = Bisphenol A-glycidyl methacrylate; UDMA = Urethane dimethacrylate; BisEMA = Bisphenol A polyethylene diether dimethacrylate; TEGDMA = Triethyleneglycol dimethacrylate.

5.1. Sample Preparation

The RBCs were used without medication and were handled according to the manufacturer's instruction. The composite materials were injected in low light condition directly into PTFE molds (24 × 3 × 1.5 mm^3) with no bonding agents. The specimens were polymerized by using a blue visible light (ESPE Elipar® Trilight, Seefeld Bavaria, Germany, wavelength ≈ 470 nm). As the sample length was 24 mm, in order to obtain maximum polymerization, curing was carried out in sections and each section was cured for 60 s. The distance of blue visible light was constant (1 mm) for all samples.

To evaluate the effect of liquid diets on these materials, samples were immersed in two experimental media, Coca-Cola (Coke) and Apple Juice (Nestle), and a control medium, deionized water. The composition and pH values of these media are given in Table 2. Ninety specimens were prepared for each composite material. Ten specimens of each material were immersed in each media at 37 °C for 1, 7, 30, and 60 days (*i.e.*, day 1, week 1, week 4 and week 8). The media were changed daily to maintain constant pH. At each time period, the samples were dried and their viscoelastic properties were measures using DMA.

Table 2. Solvent used in the study along with contents and pH values.

Solution	Content	pH
Coca-Cola	• Carbonated Water • Sugar • Carbohydrate • Phosphoric Acid • Caffeine	2.5
Apple Juice	Pure Apple juice (nothing added or removed, no flavor and sugar)	4.0
Deionized Water		7.4

5.2. Dynamic Mechanical Analysis

A Perkin Elmer DMA7 (Perkin Elmer Corp., Waltham, MA, USA) in three-point bending mode was used to measure the viscoelastic properties of the materials. For a specimen of known geometry, if L = distance between the two supports, b = width, and t = depth, the oscillating strain (ε_o) is given by,

$$\varepsilon_o = 3ty^0/L_2 \quad (4)$$

where, y^0 is the displacement amplitude. The maximum oscillating stress (σ_o) occurs on the upper and lower surfaces and was given by,

$$\sigma_o = 3F_0L/2bt^2 \quad (5)$$

where, F_0 is the axial force amplitude. Therefore, by substituting for stress and strain, the complex modulus (E^*) was given by,

$$E^* = F_0L^3/2y^0bt^3 \quad (6)$$

The support separation in three-point bend test was 20 mm and the specimen length was about 24 mm. Prior to testing, the exact dimensions (width and depth) for each specimen were measured at three different points and averaged. Testing was performed in the temperature scan mode using the parameters and conditions shown in Table 3. The temperature was measured with a thermocouple positioned approximately 1 mm away from the sample. Helium gas at a rate of 30 mL min^{-1} was used in the furnace and cooling water maintained the isothermal environment outside the furnace. The measured data were automatically saved at the end of each test using the Pyris Manager software. One way ANOVA statistical test was used to check the significance level between samples for each time period. A p value ≤ 0.05 was regarded as being significant.

Table 3. The DMA parameters used in temperature mode.

Parameter	Conditions
Initial Temperature	26 °C
Final Temperature	50 °C
Static Force	400 mN
Dynamic Force	380 mN
Frequency	1 Hz
Heating Rate	10 °C/min

Acknowledgments

The authors would like to thank 3M ESPE, UK for providing the commercial composites for this study.

Author Contributions

The present paper is based on the research work of Muhammad Kaleem. Abdul Samad Khan provided help in Dynamic Mechanical Analysis and writing of manuscript. Planning of research strategy was carried out by Ihtesham ur Rehman and Ferranti Wong.

Conflicts of Interest

The authors declare no conflict of interest.

References

1. Dietschi, D.; Magne, P.; Holz, J. Recent trends in esthetic restorations for posterior teeth. *Quintessence Int.* **1994**, *25*, 659–677.
2. Gopferich, A. Mechanisms of polymer degradation and erosion. *Biomaterials* **1996**, *17*, 103–114.
3. Lingstrom, P.; Imfeld, T.; Birkhed, D. Comparison of three different methods for measurement of Plaque-pH in humans after consumption of soft bread and potato chips. *J. Dent. Res.* **1993**, *72*, 865–870.
4. Luiz, B.K.M.; Quintella, C.M.; Friedrich, L.A.; Slva, E.; Veiga, W.; Prates, L.H.M.; Bertolino, J.B.; Pires, A.T.N. Effect of drinks on the surface properties of dental resin composites. *Polym. Test.* **2007**, *26*, 855–861.
5. Wongkhantee, S.; Patanapiradej, V.; Maneenut, C.; Tantbirojn, D. Effect of acidic food and drinks on surface hardness of enamel, dentine, and tooth-coloured filling materials. *J. Dent.* **2006**, *34*, 214–220.

6. Luiz, B.K.M.; Amboni, R.D.M.C.; Prates, L.H.M.; Roberto, B.J.; Pires, A.T.N. Influence of drinks on resin composite: Evaluation of degree of cure and color change parameters. *Polym. Test.* **2007**, *26*, 438–444.
7. Sarrett, D.C.; Coletti, D.P.; Peluso, A.R. The effects of alcoholic beverages on composite wear. *Dent. Mater.* **2000**, *16*, 62–67.
8. Badra, V.V.; Faraoni, J.J.; Ramos, R.P.; Palma-Dibb, R.G. Influence of different beverages on the microhardness and surface roughness of resin composites. *Oper. Dent.* **2005**, *30*, 213–219.
9. Aklonis, J.J.; McKnight, W.J. *Introduction to Polymer Viscoelasticity*; Wiley: New York, NY, USA, 1983; pp. 10–24.
10. Emami, N.; Soderholm, K.J.M. Dynamic mechanical thermal analysis of two light-cured dental composites. *Dent. Mater.* **2005**, *21*, 977–983.
11. Whiting, R.; Jacobsen, P.H. Dynamic mechanical properties of resin-based filling materials. *J. Dent. Res.* **1980**, *59*, 55–60.
12. Wilson, T.W.; Turner, D.T. Characterization of polydimethacrylates and their composites by dynamic mechanical analysis. *J. Dent. Res.* **1987**, *66*, 1032–1035.
13. Papadogiannis, Y.; Lakes, R.S.; Petrou-Americanos, A.; Theothoridou-Pahini, S. Temperature dependence of the dynamic viscoelastic behavior of chemical- and light-cured composites. *Dent. Mater.* **1993**, *9*, 118–122.
14. Mesquita, R.V.; Geis-Gerstorfer, J. Influence of temperature on the visco-elastic properties of direct and indirect dental composite resins. *Dent. Mater.* **2008**, *24*, 623–632.
15. Sideridou, I.D.; Karabela, M.M.; Vouvoudi, E.C. Dynamic thermomechanical properties and sorption characteristics of two commercial light cured dental resin composites. *Dent. Mater.* **2008**, *24*, 737–743.
16. Aframian, D.J.; Davidowitz, T.; Benoliel, R. The distribution of oral mucosal pH values in healthy saliva secretors. *Oral Dis.* **2006**, *12*, 420–423.
17. Neamat, A.B.; Han, L.; Okamoto, A.; Iwaku, M. Effect of alcoholic and low pH soft drinks on fluoride release from compomer. *J. Esthet. Dent.* **2000**, *12*, 97–104.
18. Bayne, S.C.; Thompson, J.Y.; Swift, E.J., Jr.; Stamatiades, P.; Wilkerson, M. A characterization of first generation flowable composites. *J. Am. Dent. Assoc.* **1998**, *129*, 567–577.
19. Kitchens, M.; Owens, B.M. Effect of carbonated beverages, coffee, sports and high energy drinks, and bottled water on the *in vitro* erosion characteristics of dental enamel. *J. Clin. Pediatr. Dent.* **2007**, *31*, 153–159.
20. Bagheri, R.; Burrow, M.F.; Tyas, M. Surface characteristics of aesthetic restorative materials—An SEM study. *J. Oral Rehabil.* **2007**, *34*, 68–76.
21. Yap, A.U.J.; Tan, S.H.L.; Wee, S.S.C.; Lee, C.W.; Lim, E.L.C.; Zeng, K.Y. Chemical degradation of composite restoratives. *J. Oral Rehabil.* **2001**, *28*, 1015–1021.
22. Papadogiannis, D.Y.; Lakes, R.S.; Papadogiannis, Y.; Palaghias, G.; Helvatjoglu-Antoniades, M. The effect of temperature on the viscoelastic properties of nano-hybrid composites. *Dent. Mater.* **2008**, *24*, 257–266.
23. Ertaş, E.; Güler, A.U.; Yücel, A.C.; Köprülü, H.; Güler, E. Color stability of resin composites after immersion in different drinks. *Dent. Mater. J.* **2006**, *25*, 371–376.

24. Bagheri, R.; Burrow, M.F.; Tyas, M. Influence of food-simulating solutions and surface finish on susceptibility to staining of aesthetic restorative materials. *J. Dent.* **2005**, *33*, 389–398.
25. Arima, T.; Hamada, T.; McCabe, J.F. The effects of cross-linking agents on some properties of HEMA-based resins. *J. Dent. Res.* **1995**, *74*, 1597–1601.
26. Douglas, W.H.; Craig, R.G. Resistance to extrinsic stains by hydrophobic composite resin systems. *J. Dent. Res.* **1982**, *61*, 41–43.
27. Khan, Z.; von Fraunhofer, J.A.; Razavi, R. The staining characteristics, transverse strength and microhardness of a visible light cured tenture base material. *J. Prosthet. Dent.* **1987**, *57*, 384–386.
28. Aliping-Mckenzie, M.; Linden, R.W.A.; Nicholson, J.W. The effect of Coca-Cola and fruit juices on the surface hardness of glass ionomers and compomers. *J. Oral Rehabil.* **2004**, *31*, 1046–1052.
29. Villalta, P.; Lu, H.; Okte, Z.; Garcia-Godoy, F.; Powers, J.M. Effects of staining and bleaching on color change of dental composite resins. *J. Prosthet. Dent.* **2006**, *95*, 137–142.
30. Dos Santos, P.A.; Garcia, P.P.; de Oliveira, A.L.; Chinelatti, M.A.; Palma-Dibb, R.G. Chemical and morphological features of dental composite resin: Influence of light curing units and immersion media. *Microsc. Res. Tech.* **2010**, *73*, 176–181.
31. Gohring, T.N.; Gallo, L.; Luthy, H. Effect of water storage, thermocycling, the incorporation and site of placement of glass-fibers on the flexural strength of veneering composite. *Dent. Mater.* **2005**, *21*, 761–772.
32. Musanje, L.; Shu, M.; Darvell, B.W. Water sorption and mechanical behavior of cosmetic direct restorative materials in artificial saliva. *Dent. Mater.* **2001**, *17*, 394–401.
33. Tezvergil, A.; Lassila, L.V.J.; Vallitu, P.K. The effect of fiber orientation on the thermal expansion co-efficient of fiber-reinforced composite. *Dent. Mater.* **2003**, *19*, 471–477.
34. Khan, A.S.; Phillips, M.J.; Tanner, K.E.; Wong, F.S.L. Comparison of the visco-elastic behavior of a pre-impregnated reinforced glass fiber composite with resin-based composite. *Dent. Mater.* **2008**, *24*, 1534–1538.
35. Khan, A.S.; Tahir, M.T.; Khan, M.; Mian, S.A.; Rehman, I.U. An update on glass fiber dental restorative composites: A systematic review. *Mater. Sci. Eng. C* **2015**, *47*, 26–39.
36. Trengrove, H.G.; Carter, G.M.; Hood, J.A.A. Stress relaxation properties of human dentin. *Dent. Mater.* **1995**, *11*, 305–310.
37. Gent, A.N. Relaxation processes in vulcanized rubber. I. Relation among stress relaxation, creep, and hysterisis. *J. Appl. Polym. Sci.* **1962**, *6*, 433–441.
38. Braden, M.; Wilson, A.D. Relationship between stress relaxation and viscoelastic properties of dental materials. *J. Dent.* **1982**, *10*, 181–186.
39. Renson, C.E.; Braden, M. Experimental determination of the rigidity modulus Poisson's ratio and elastic limit in shear of human dentin. *Arch. Oral Biol.* **1975**, *20*, 43–47.
40. Balooch, M.; Marshall, G.W.; Marshall, S.J.; Warren, O.L.; Asif, S.A.S. Evaluation of a new modulus mapping technique to investigate microstructural features of human teeth. *J. Biomech.* **2004**, *37*, 1223–1232.

Polysaccharides for the Delivery of Antitumor Drugs

Bianca Posocco [1,2,†], Eva Dreussi [1,2,†], Jacopo de Santa [1,2,†], Giuseppe Toffoli [1], Michela Abrami [3], Francesco Musiani [4], Mario Grassi [2], Rossella Farra [2], Federica Tonon [2], Gabriele Grassi [3,*] and Barbara Dapas [3]

[1] Centro di Riferimento Oncologico, Via Franco Gallini 2, I-33081 Aviano (PN), Italy; E-Mails: bposocco@cro.it (B.P.); edreussi@cro.it (E.D.); jdesanta.@cro.it (J.S.); gtoffoli@cro.it (G.T.)

[2] Department of Engineering and Architecture, University of Trieste, Via Alfonso Valerio, 6/A, I-34127 Trieste, Italy; E-Mails: mario.grassi@di3.units.it (M.G.); rfarra@units.it (R.F.); effe.tonon@gmail.com (F.T.)

[3] Department of Life Sciences, Cattinara University Hospital, Trieste University, Strada di Fiume 447, I-34149 Trieste, Italy; E-Mails: mikystars@hotmail.com (M.A.); bdapas@units.it (B.D.)

[4] Laboratory of Bioinorganic Chemistry, Department of Pharmacy and Biotechnology, University of Bologna, I-40127 Bologna, Italy; E-Mail: Francesco.musiani@unibo.it

† These authors contributed equally to this paper.

* Author to whom correspondence should be addressed; E-Mail: ggrassi@units.it;

Academic Editor: Loo Say Chye Joachim

Abstract: Among the several delivery materials available so far, polysaccharides represent very attractive molecules as they can undergo a wide range of chemical modifications, are biocompatible, biodegradable, and have low immunogenic properties. Thus, polysaccharides can contribute to significantly overcome the limitation in the use of many types of drugs, including anti-cancer drugs. The use of conventional anti-cancer drugs is hampered by their high toxicity, mostly depending on the indiscriminate targeting of both cancer and normal cells. Additionally, for nucleic acid based drugs (NABDs), an emerging class of drugs with potential anti-cancer value, the practical use is problematic. This mostly depends on their fast degradation in biological fluids and the difficulties to cross cell membranes. Thus, for both classes of drugs, the development of optimal delivery materials is crucial. Here we discuss the possibility of using different kinds of polysaccharides, such as chitosan, hyaluronic acid, dextran, and pullulan, as smart drug delivery materials. We first describe the main features of polysaccharides, then a general overview about the aspects ruling drug release mechanisms and the pharmacokinetic are reported. Finally, notable examples of polysaccharide-based delivery of conventional anti-cancer drugs and NABDs are reported. Whereas additional research is required, the promising results obtained so far, fully justify further efforts, both in terms of economic support and investigations in the field of polysaccharides as drug delivery materials.

Keywords: polysaccharides; anticancer drugs; NABDs; delivery

---

1. Introduction

Polysaccharides can be defined as polymeric carbohydrate structures composed of repeating monosaccharide units joined by glycosidic bonds [1]. These natural polymers are, by far, the most abundant renewable resource on the Earth, with an annual formation rate exceeding the world production rate of synthetic polymers by some orders of magnitude. Interestingly, naturally designed polysaccharides can carry out various different functions. For example, they can accomplish structural tasks, such as cellulose and chitin, storage functions, such as starch and glycogen, and gel structures (made up of mono- and copolymers), such as mucopolysaccharides (glycosaminoglycans), agar, and pectins [2].

Polysaccharides found and continue to find wide application in the drug delivery field [3–8] for various and important reasons: (1) they can be obtained in a well characterized and reproducible way from natural sources [9]; (2) they can undergo a wide range of chemical and enzymatic reactions to give origin to different materials [10]; (3) they are biocompatible, biodegradable and have low immunogenic properties [11]; (4) ionic polysaccharides can, partially, or even totally, substitute synthetic polymers in the design of stimuli responsive drug delivery systems since they show pH- and ion-sensitiveness [12]; (5) ionic polysaccharides are muco-adhesive [13]; (6) they can be prepared as conjugates or complexes with proteins, peptides, and other bio-macromolecules [14]; (7) they can easily form gels [14]; and (8) they can give rise to interpenetrated polymeric networks (IPN) and semi-IPN that can show physico-chemical properties which are remarkably different from those of the macromolecular constituents [15]. All these characteristics make polysaccharides excellent materials for the realization of "smart" delivery systems capable to release, at the appropriate time and site of action, entrapped drugs in response to specific physiological stimuli [15].

The aim of this review is to present relevant polysaccharides and to show some of their most significant applications in the drug delivery fields. More in detail, in Section 2 we will provide a general overview of polysaccharides structure, types, and delivery/absorption mechanisms; in Section 3 and 4 we report examples of the use of polysaccharides as delivery materials for clinically used anticancer drugs and experimental nucleic acid based drugs (NABDs), chosen as relevant examples among existing drugs.

2. Polysaccharides

2.1. Polysaccharide Structure

Polysaccharide classification basically relies on their "primary or covalent structure" that represents the sequence of monomeric units along the chain [2,10]. These units are, usually, characterized by a heterocyclic structure, where one atom is oxygen and the remaining are carbon atoms. The most common repeating units, which naturally occur in the polysaccharide chains, present a six-membered ring structure. Despite the carbon atom numbers, the monomeric units are linked by chemical bonds of covalent nature that connect the carbon atom in position 1 (*C*(1)) through an oxygen bridge to a carbon atom of the following unit via the so-called "glycosidic linkages" (Figure 1).

As these covalent linkages are not completely flexible (they limit the monomers to a narrow range of relative orientations), an isolated polysaccharide chain can only adopt certain shapes (called "secondary structures"), which are dependent on their primary sequence. At a higher degree of organization, energetically favorable interactions between shaped chains may result in ordered specific structures known as "tertiary structures". Finally, such compact entities may interact among themselves to give higher levels of organization, which are described as "quaternary structures". Common examples of polysaccharide "secondary structures" (called conformations) are ribbons and helices. "Tertiary structures" are represented by the arrangement of multiple helices or aggregates of ribbons, as shown in Figure 2, in the case of cellulose and starch.

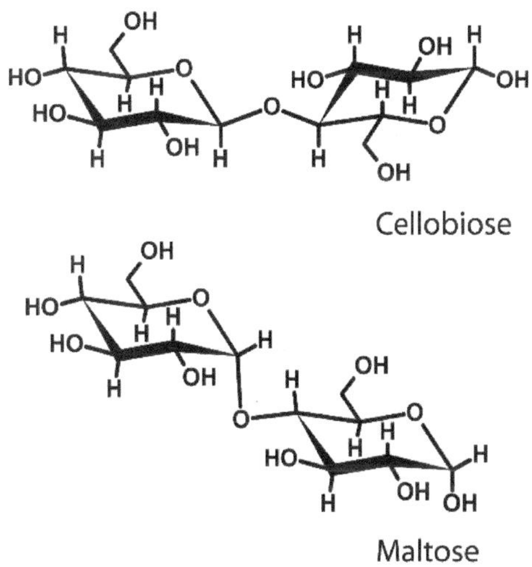

Figure 1. Glycosidic linkage between two different saccharide units. Schematic representation of cellobiose (**top**) and maltose (**bottom**). In the first case, the anomeric carbon (C2) is in the beta configuration (equatorial), while in the second case, it is in the alpha configuration (axial).

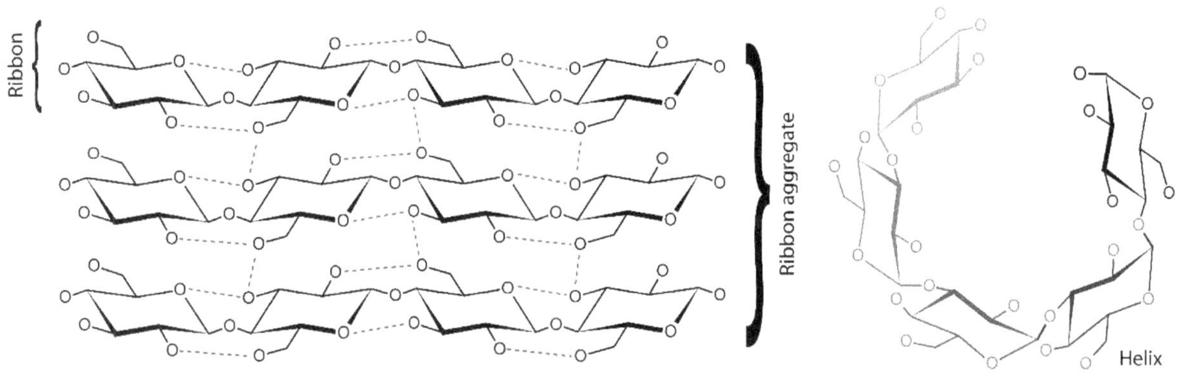

Figure 2. Schematic representation of ribbons and a ribbon aggregate (**left**), as found in cellulose, and of a helix (**right**), as found in starch. Hydrogen bonds are represented by dashed red lines.

2.2. Polysaccharides Types

As shown in Table 1, polysaccharides can have animal, vegetal (algae, plants, seeds, and exudates), and microbial/fungi origins.

Therefore, they hold a wide range of different functions. Polysaccharides can act as energy storage materials, such as starch, glycogen, and polysaccharides of some plant seed (locust bean gum and guar gum). Polysaccharides can also contribute to the structural integrity and mechanical strength of plant tissues by forming a hydrated cross-linked three-dimensional network, as in the case of pectins in land plants and carrageenans, agar, and alginate in marine species. In the animal world, hyaluronate, chondroitin sulfate and other related glycosaminoglycans play a fundamental part in governing the solution properties of some physiological fluids and in the structure of the intercellular matrix. Polysaccharides may also work as protective substances, as in the case of exudate gums from plants, where they provide a preventive function by sealing off the injured parts of the plant against bacterial infections.

Table 1. Origin of Polysaccharides.

Animal Polysaccharides	
Chitin	
Chitosan [16–28]	
Glycosamminoglycans [29–38]	
Vegetal Polysaccharides	
Land Vegetal	**Marine Vegetal**
Cellulose [39–41]	Alginate (Phaeophyceae—brown seaweed)
Pectins	Agar (Rhodophyceae—red seaweed)
Galactomannans	Carragenans (Rhodophyceae—red seaweed)
Acacia gum	–
Starch	–
Microorganisms/Fungi	
Alginate (*Pseudomonas aeruginosa* and *Azotobacter vinelandii*) [42–44]	
Dextran (*Leuconostoc spp.* and *Lactobacillus spp.*) [45–50]	
Gellan (*Pseudomonas elodea*)	
Pullulan (*Aureobasidium pullulans*) [51–53]	
Scleroglucan (*Sclerotium glucanicum*) *Xanthomonas campestris*	
Xanthan (*Xanthomonas campestris*)	

2.2.1. Animal Polysaccharides

Chitin, a high molecular weight polysaccharide, is a fundamental skeletal material in crustaceans, insects, and spiders that contributes to the structure of the exoskeleton, the lining of the gut, the tendons, the wing coverings, and the internal skeleton. The primary structure of chitin is closely related to that of cellulose (linear polymer of 1,4-linked β-D-glucopyranose), except for the replacement of the –OH group at each *C*(2) position by an acetamido group $–NHCOCH_3$. Chitosan is the name given to fully or partially deacetylated chitin.

Glycosaminoglycans (GAGs) are polysaccharides that contain amino sugars and are constituents of the extracellular space of many mammalian tissues. With the exception of hyaluronic acid, they are covalently linked to proteins in the natural state, and in this form, they are known as proteoglycans. Hyaluronic acid is a basic component of the extracellular matrix of connective tissue and it is present in large amounts in the vitreous humour of the eye, the synovial fluid of the joints, and in soft tissues, such as umbilical cord and rooster comb. In addition, hyaluronic acid has many peculiar chemical–physical and biological properties, which enable it to regulate and control the physiological functions of several tissues *in vivo*. In virtue of its extraordinary water-retaining capability, in aqueous solutions hyaluronic acid shows a high excluded volume since it behaves as an expanded random coil occupying large hydrodynamic volume. This is the reason why it is generally affirmed that it controls and regulates the trafficking of fluids and macromolecules within the interstitial space of tissues [54].

2.2.2. Vegetal Polysaccharides

Cellulose, a high molecular weight ($\approx 7 \times 10^5$–3.2×10^6 [10]) linear polymer of 1,4-linked β-D-glucopyranose, can be defined as the backbone of the vegetal kingdom as it is one of the most abundant and important component of land-growing plants.

Pectins represent a group of polysaccharides that occur as structural materials in all land-growing plants. They belong to the interrupted chain sequences family, being characterized by linear 1,4-linked α-D-galactopyranosyluronic acid sequences separated by α-L-rhamnopyranosyl residues. Galactomannans represent a class of polysaccharides that can be found in plant seeds such as carob or locust bean, guar, tara, and tamarind. These polysaccharides, or gums as they are often called, are derived from the endosperm of various leguminosae plant seeds, where they function as reserve materials utilized during germination. Most of these carbohydrate polymers share basic structural similarities consisting of linear 1,4-linked chains of β-D-mannopyranosyl residues to which single α-D-galactopyranosyl side chains are linked at position 6. Depending on the seed source, galactomannans can differ in their molecular weight, molecular weight distribution and galactose to mannose composition ratio (*G*/*M*). Moreover, at a given *G*/*M* ratio, they can also differ in terms of the side chain distribution along the mannan backbone. Gum Arabic, or acacia gum, is the most famous natural exudate, produced by more than 900 species of the Acacia trees. Natural Arabic gum is a mixture of potassium, magnesium and calcium salt of an acidic polysaccharide composed of six different carbohydrate units: D-galactopyranose, D-arabinopyranose, L-arabinofuranose, L-rhamnopyranose, D-glucopyranosyluronic acid and 4-*O*-methyl-D-glucopyranosyluronic acid.

A common function of polysaccharides that preferentially adopt hollow helix conformations is to act as energy reserves in biological systems. For these biopolymers, the conformations are relevant to their biological functions as they allow the macromolecules to be piled together and accumulated as concentrated deposits. These deposits can be organized in such a way that osmotic transport phenomena within the tissues are not affected and, yet, the molecules within them are easily accessible to be mobilized and employed when necessary. A typical example of energy reserve polysaccharide is represented by starch that, in nature, occurs as minute granules (2–100 μm particle size) in the roots, seeds, and stems of numerous types of plants including corn, wheat, rice, millet, barley, and potatoes. Starch is made up of a linear polymer, called amylose (1,4-linked α-D-glucopyranose units), and a branched polymer called amylopectin (linear backbone of 1,4-linked α-D-glucopyranose monomers bearing clusters of 1,6-α-linked glucopyranosyl branches with average length of 20–30 residues). Amylose constitutes only the 25% by weight of the total starch.

Polysaccharides can be also found in marine seaweeds or algae. Their task is similar to that of cellulose in land plants, with the difference that seaweeds need a more flexible structure to support the varying stresses due to currents and waves. The many varieties of seaweeds are usually subdivided on the basis of their predominant color pigments. Accordingly, Cyanophyceae (blue-green), Xanthophyceae (yellow-green), Chlorophyceae (green), Rhodophyceae (red), and Phaeophyceae (brown) can be mentioned. Among the many polysaccharides present in seaweeds and algae (see Table 1), from the drug delivery point of view, the most important are alginates and agar. The term alginate refers to a family of unbranched polysaccharides isolated from brown seaweeds [55]. They are composed of 1-4-linked β-D-mannuronic acid (*M*) and alpha-L-guluronic acid (*G*) arranged in a block-wise pattern with homopolymeric regions of *M*-blocks and *G*-blocks residues interspersed by regions of alternating structures (*MG*-blocks). The polysaccharide, known since 1881, has largely been used in biotechnological and industrial applications for its gel forming properties, which were described in detail around the 1970s by Grant [56] and Smidsrød [57].

The general term agar refers to a complex mixture of polysaccharide components which can be extracted from certain species of red seaweeds. Agar is composed of two major fractions: agarose, a neutral polysaccharide, and agaropectin, a charged, sulfated carbohydrate polymer. The percentage of agarose normally ranges from 50% to 90% of the total agar. The linear agarose backbone is based on a disaccharide repeating unit of the *AB* type, consisting of alternating 1,3-linked β-D-galactose (*A*) and 1,4-linked 3,6-anhydro-α-L-galactose (*B*). Depending on the algal sources, 6-*O*-methyl-D-galactose may also be present in variable amounts (1%–20%). Interestingly, agarose can give origin to thermo-reversible hydrogels.

Carrageenans are sulfated linear polysaccharides which can be extracted as matrix components of red seaweeds. Their chemical structure is based on a disaccharide repeating unit of the *AB* type so that they can be written generally as $(AB)_n$, where *A* is derived from a 1,3-linked β-D-galactose unit and *B* is usually, but not always, present as 1,4-linked 3,6-anhydro-β-D-galactose. Carrageenans sulfation takes place at position 2 and 4 in 1,3-linked galactose residues, whereas the 1,4-linked units occur as 2-sulfate, 6-sulfate, 2,6-disulfate and 3,6-anhydride-2-sulfate. No sulfation at *C*(3) apparently occurs.

2.2.3. Microorganisms/Fungi Polysaccharides

Microorganisms, such as bacteria and fungi, produce three distinct types of polysaccharides: (1) extracellular or exo-cellular polysaccharides, which can be found in the form of a discrete capsule that envelops the microbial cell and that is part of the cell wall itself (capsular polysaccharides), or as an amorphous mass secreted into the surrounding medium; (2) structural polysaccharides; and (3) intracellular storage polysaccharides.

The bacterial alginate is exo-cellularly produced (partially acetylated) by the bacteria *Pseudomonas aeruginosa* and *Azotobacter vinelandii* and its physical and chemical properties are very similar to their algal counterparts. *O*-acetylation invariably seems to be a characteristic feature in bacterial alginates that distinguishes them from the seaweed alginates.

Dextrans constitute a family of microbial polysaccharides produced by *Leuconostoc spp.* and *Lactobacillus spp.* Dextran is almost exclusively composed of 1,6-linked α-D-glucopyranose units with varying proportions of other bond types, such as α-1,3, α-1,2 and α-1,6 branch linkages. Gellan gum is produced by the bacterium *Pseudomonas elodea* and is composed of D-glucose, D-glucuronic acid and L-rhamnose in a molar ratio of 2:1:1. The primary structure of gellan is based on the tetra-saccharide repeating unit. Both *O*-acetyl and *O*-L-glyceryl substituents are present along the main chain, the former at the *C*(6) position and the latter at the *C*(2) position of the 3-linked glucose unit. The exo-cellular Pullulan, produced by the yeast-like fungus *Aureobasidium pullulans*, is a linear polymer composed of glucose residues polymerized into repeating malto-triose units. The malto-triose units are linked via α-1,6 bonds; within each unit the glucose residues are linked α-1,4. Nevertheless, some 1,6-α-D-maltotetraose repeating units and 1,3-linked glucose residues have also been found as components of pullulan primary structure. The molecular weight of this polysaccharide can vary in the range 1.5×10^3–4×10^6 depending on the culture conditions and strain of microorganisms used.

Scleroglucan, a neutral capsular polysaccharide, is exo-cellularly secreted by certain imperfect fungi, particularly of the genus Sclerotium. The primary structure of scleroglucan from *Sclerotium glucanicum* has been characterized as a linear chain of β-1,3-linked D-glucose units with single D-glucose side chains linked β-1,6 to every third unit of the main chain. Scleroglucan molecular weight varies in the range 3.2×10^5–1×10^6. Xanthan gum is an anionic polymer secreted exo-cellularly by the bacterium *Xanthomonas campestris*. Its primary structure is based on a linear backbone of 1,4-linked β-D-glucose. At the *C*(3) position of every alternate glucose residue there is a charged tri-saccharide side chain containing a glucuronic acid residue between two mannose units. The terminal β-D-mannose is linked β-1,4 to the glucuronic acid, which, in turn, is linked α-1,2 to the α-D-mannose. In approximately one half of the terminal mannose residues, a pyruvic acid moiety is joined by a ketal linkage to the *O*(4) and *O*(6) positions. Acetyl groups are present as substituents at the *O*(6) position of the nonterminal D-mannose.

2.3. Polysaccharide Gels

Undoubtedly, among the different interesting properties of polysaccharides, the possibility of forming gel structures is, from the drug delivery point of view, very interesting. Indeed, gels, defined as coherent systems made up of a three-dimensional polymeric network trapping a continuous liquid phase (water or physiological media in the case of hydrogels) [10] (see Figure 3), can efficiently and reliably control the drug release kinetics of embedded drugs [58]. The presence of inter-chains linkages, called crosslinks, prevents the solubilization of the network of the polymeric chains in the fluid phase. Interestingly, despite the very large amount of liquid phase that can be hosted inside the polymeric network (the polymeric network volume fraction in the gel can be lower than 1%), gels show mechanical properties in between those of solids and liquids [10], thus, mimicking living tissues very well [59].

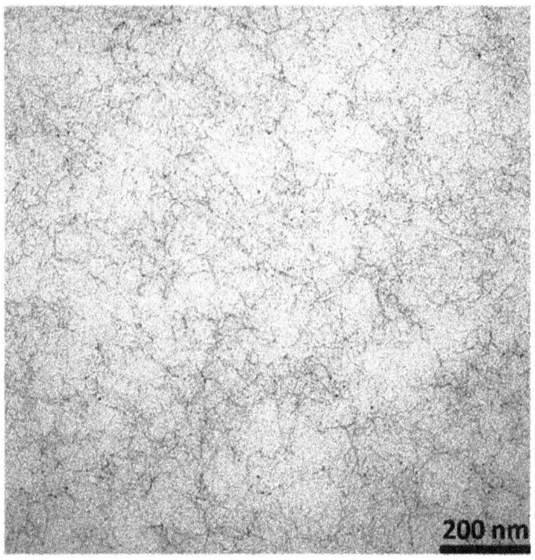

Figure 3. TEM image referring to a hydrogel obtained via the crosslinking of an alginate water solution (alginate mass fraction 2%) by means of a Ca^{++} solution (5 g/L). Alginate chains are represented by the dark wires while water occupies the gray zones. Details about the hydrogel treatment for the TEM analysis are reported in [60].

The chemical–physical properties of the polymeric network chains and the huge amount of absorbed water, make polysaccharide gels attractive candidates also for the release of proteins, peptides, vaccines and NABDs that can be embodied into the three-dimensional polymeric network meshes. The use of polysaccharide as delivery systems is also strengthened by the fact that hydrogel realization implies the use of safe procedures that do not imply proteins/NABDs denaturation. Thus, controlled release systems, based on polysaccharides hydrogels, can be a reliable tool to ensure the therapeutic proteins/peptides/NABDs concentration in tissues or in the blood for a long time [14].

Based on the crosslinked nature, (polysaccharides) hydrogels can be defined as chemical or physical. In the first case, crosslinks between different chains are strong, permanent, and punctual as they are due to covalent bonds. Conversely, in the second case, crosslinks are due to either polymer chains topological entanglements (spaghetti-like configuration, Figure 4) or physical interactions (this being typical of glucans and xanthan), such as H-bonds, ionic, Coulombic, van der Waals, dipole–dipole, and hydrophobic interactions.

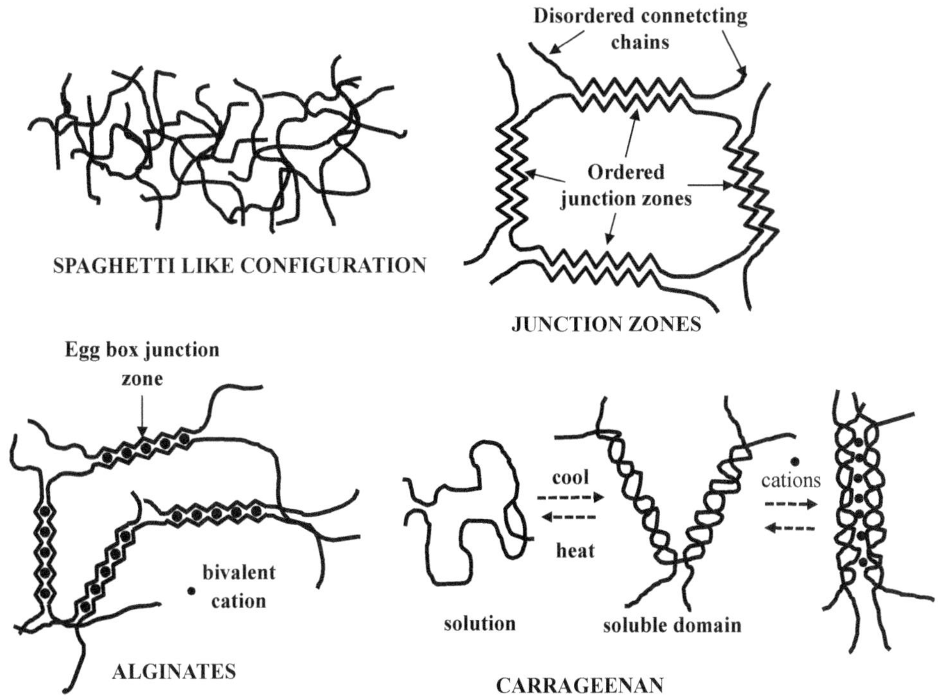

Figure 4. Structures occurring in physically cross-linked polysaccharides gels.

As a consequence, crosslinks are not punctual but they form junction zones where the regular coupling of chain portions belonging to different polymeric chains takes place. It is interesting to observe that the long chain segments departing from the ordered junction zones can form, with other chains, additional junction zones to form a polymeric three-dimensional network. As, in general, the physical interactions occurring in junction zones are not so strong, these junctions are often transient. Thus, they give origin to a statistical network characterized by a constant average crosslink density (moles of crosslinks per gel unit volume) and a time dependent distribution of crosslinks. These means that, due to Brownian motion of chains and segment of chains, network meshes continuously appear and disappear, being the average mesh number and dimension constant [10,58]. Figure 4 shows some typical situations occurring in the formation of polysaccharide gels.

Basically, drug delivery from hydrogels as those formed by polysaccharides, is ruled by physical processes, physicochemical processes, and system parameters [61]. In particular: (1) physical processes can include matrix swelling and erosion; (2) physicochemical processes can include matrix erosion, drug dissolution (recrystallization), drug transport (by diffusion and convection), drug interaction with hydrogel polymeric chains; and (3) system parameters can be represented by initial drug distribution and concentration inside the hydrogel, hydrogel geometry (cylindrical, spherical, *etc.*), and hydrogels size distribution (in the case of an ensemble of hydrogel particles). Of course, among all the above mentioned phenomena, those strongly affected by the hydrogel network topology are swelling, erosion, and drug diffusion. On the contrary, network physicochemical characteristics can play an important role for what concerns the drug interaction with the polymeric chains. In general, if the network topology highly affects the drug release kinetics from hydrogels ranging in the milli- or micro-range, polymer–drug interaction becomes the most important aspect for polymeric nano-delivery systems.

2.4. Polysaccharide Micro and Nanoparticles

Whereas polysaccharides can be used in the form of gel to deliver therapeutic drugs, they can be also arranged in the micro-nanoparticle shape. This last use has been limited in the past due to the harsh conditions necessary to prepare the particles, a fact that may compromise drug stability and loading efficiencies [62,63]. A recently developed approach to overcome this limitation is represented by the generation of super-hydrophobic surfaces [62]. This method is based on the preparation of a water solution containing the monomer (or polymer) and the drug at the desired concentration. Then, the aqueous solution is put on a super-hydrophobic surface constituted, for example, by surface modified copper, aluminum, or polystyrene substrates. Due to the high hydrophobicity of these surfaces, the aqueous solution assumes an almost perfect spherical shape whose diameter essentially depends on the solution volume deposited on the surface. The subsequent exposure to the polymerisation/crosslinking process leads to the formation of spherical particles [63]. Since the process takes place at the solid–air interface, no migration of the drug occurs, and, thus, 100% loading yield is possible [62]. Relying on this approach, Lima and co-workers [63] were able to realize spherical hydrogel particles, made up by interpenetrated dextran-methacrylated and poly(*N*-isopropylacrylamide) polymers, for the release of insulin and bovine serum albumin. Interestingly, they proved that the release kinetics could be tuned acting on particles composition and release environment temperature.

Whatever the technique used to prepare the drug loaded particles, the *in vivo* fate of the particles depends on some common steps reassumed by the ADME processes, *i.e.*, Absorption, Distribution, Metabolism and Elimination [64]. In the case of particulate delivery systems, such as those based on polysaccharide particles, cell membrane crossing (absorption) represent a critical step. In particular, the net superficial charge of the particles plays a key role. Whereas anionic complexes usually show good solubility and, thus, stability in the physiological environment, they cannot efficiently cross cell membranes. This is due to the electrostatic repulsion occurring with the negative electric charge of the cell membranes. Conversely, cationic complexes can efficiently bind to cell membrane due to the strong electrostatic interactions. Unfortunately, however, if the interaction is too strong, it can cause membrane disruption and consequent cell death. In addition, the presence of negatively charged blood proteins can induce the formation of complex-protein aggregates that are no longer soluble and precipitate. Whereas neutral complexes are not affected by the above mentioned problems, in the physiological environment they tend to associate each other resulting in a limited solubility. Thus the charge of the particles has to be carefully selected. Ideally the particle charge should guarantee a good stability in the physiological environment and should be able to allow an efficient interaction with the cell membrane without inducing significant damages.

Although very important, the surface properties are not the only variables affecting the *in vivo* particle/drug fate. In particular, the particle dimension and shape can affect the transport through biological barriers (e.g., microvascular walls, interstitial space, and cell membranes) [65]. It is well known that, in general, small nanoparticles (≤100 nm) are efficiently internalized by cells, while bigger particles are not [66]. However, nanoparticles suffer for the lack of the so called "margination effect" which consists in the movement of particles in flow toward the walls of a channel (blood vessel) [67]. Essentially, nanoparticles move inside the vessel (together with red blood cells) with a uniform radial distribution and limited near-wall accumulation. On the contrary micro-particles tend to preferentially accumulate next to the vessel walls, in a particle size and shape-dependent manner [65,67]. This localization facilitates micro-particle extravasation, thus favoring drug delivery to the surrounding tissue. In contrast, nanoparticles, which have a limited near-wall accumulation, have more difficulties to extravasate and thus to deliver the drug to the tissues. The margination phenomenon is particularly relevant in vessels nourishing tumor tissues due to their leaky nature (enhanced permeability and retention effect) which favors extravasation [67]. Thus, the extravasation of particles localized near the vessel wall (microparticles) is facilitated; in contrast, the extravasation of particles localized far from the vessel wall (nanoparticles) is disfavored.

3. Polysaccharide-Based Delivery Systems for Anti-Cancer Drugs

Most, if not all, the concepts expressed in Sections 2.3 and 2.4, with regard to the variables ruling drug delivery from polysaccharides, hold true for any drug delivery approach. This implies that the works reported in the next sections are the result of the complex interplay among all the aspects ruling drug release either from polysaccharide gels or particles. To make the text easily understandable, these aspects are generally omitted in the description of the experimental works; the focus is in contrast put on some other features peculiar of the specific polysaccharide-based delivery system.

The presented works have been subdivided into two parts: the first (Section 3.1) describing examples of the release of clinically relevant anticancer drugs, and the second (Section 3.2) describing examples of the release of nucleic acid based drug.

3.1. Polysaccharides for the Delivery of Clinically Relevant Anti-Cancer Drugs

In addition to the drug delivery aspects reported in Sections 2.3 and 2.4, other features specific to the delivery of clinically relevant anti-cancer drugs, have to be highlighted. We believe this is important to better understanding the rationale of the polysaccharide-based delivery examples described. For this purpose, in Sections 3.1.1 and 3.1.2 are reported some important aspects of drug delivery together with the main features of commonly used drugs, respectively. These two sections are then followed by the description of different polysaccharide-based delivery systems (Sections 3.1.3–3.1.8).

3.1.1. Critical Aspects of Systemic Drug Administration

The critical bottleneck of the administration of conventional cancer therapeutic molecules is represented by their high toxicity [68,69], mostly dependent on the indiscriminate distribution between diseased and healthy cells. Toxicity may also depend on the presence, in the pharmaceutical preparations, of organic solvents/detergents necessary to improve the poor solubility in water of many of these drugs [70]. Therefore, the development of targeted drug delivery systems with switchable (generally pH-dependent) drug release profiles has become a relevant issue in chemotherapy [71].

The conjugation of chemotherapy agents to macromolecules may offer several advantages (Figure 5).

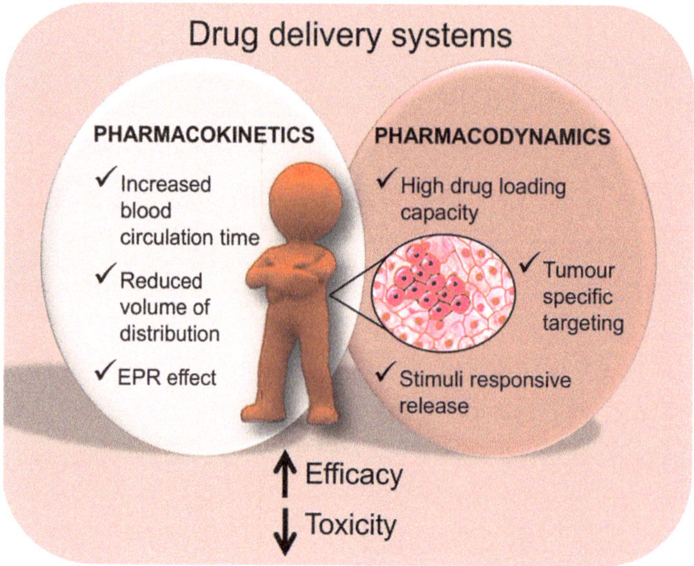

Figure 5. Advantages of chemotherapy agent conjugation with macromolecules.

Macromolecules can ameliorate drugs' pharmacokinetics (PK) and pharmacodynamics (PD) by different phenomena. Regarding PD, macromolecules have high drug loading capacity and, in addition, can be equipped with targeting molecules to increase drug specificity. In some cases, stimuli-responsive delivery has been studied to improve drug release in cancer tissues. In this regard, for example, pH-sensitive drug delivery systems take advantage of the different pH in normal (pH~7.4) *vs.* cancer (pH < 6) tissue [72]. With regard to PK, macromolecules can increase blood circulation time, reduce the volume of distribution and prolong the distribution/elimination phases [73]. It is also noteworthy that, in general, polymer–drug conjugates are not able to cross the endothelium surrounding healthy vessels. This mostly depends on the bigger size of polymer–drug conjugates compared to endothelium fenestration. In contrast, in pathological conditions like cancer, local vessels are more permeable due to the increased size of endothelium fenestration. Through these gaps, the delivery systems can extravasate and reach the tumor where they tend to reside due to the scarce lymphatic drainage. Collectively these phenomena go under the name of enhanced permeability and retention (EPR) effect [74]. The use of materials able to release the drug slowly, may further improve drug residence at the tumor site, possibly ameliorating drug efficacy and reducing side effects.

The use of macromolecule-based delivery systems also allows facing more complex clinical problems. For instance, it is possible to design multidrug delivery systems to improve, on one hand, the treatment efficacy and, on the other hand, the compliance of the patients. Another interesting option is related to the possibility to design so called "theragnostic" agents, that render possible the simultaneous cancer diagnosis and therapy. This approach typically combines the use of a therapeutic agent together with a targeting/marker agent, as below detailed.

Based on the above considerations, chemotherapeutic drugs have been encapsulated, conjugated or linked to different carriers [75]. Among these, biopolymers have been frequently used due to their biocompatibility, natural occurrence, often targeting ability, relatively inexpensiveness and the possibility of derivatization with different chemical groups [76].

3.1.2. Anticancer Drugs

Thus far, a great deal of effort has been spent to design polysaccharide-based systems to deliver anti-cancer drugs such as docetaxel (DTX), paclitaxel (PTX), cisplatin (CDDP), 5-fluorouracil (5-FU), and doxorubicin (DOX).

DTX and PTX are mitotic inhibitors belonging to the taxane family. These molecules promote the assembly of the free tubulin and stabilize the microtubules, thus, particularly impacting proliferating cells. According to FDA guidelines, DTX is indicated for the treatment of various solid tumors such as breast cancer, non-small cell lung cancer, hormone refractory prostate cancer, gastric adenocarcinoma, squamous cell carcinoma of head and neck. Among the numerous side effects reported, toxic deaths, hepatotoxicity, neutropenia, and hypersensitivity reactions can be mentioned. Another important limitation of DTX relates to its scarce water solubility that, together with the reported side effects, have lead numerous groups to try to ameliorate drug formulation and administration. As DTX, PTX is widely used in clinic for the treatment of breast cancer, ovarian cancer, and AIDS-related Kaposi sarcoma. The aforementioned side effects depend not only on the drugs, but also on the additives used to improve their poor solubility, like the surfactant polysorbate [77,78].

Another class of drugs commonly used in cancer treatment are DNA damaging agents. Among these, CDDP is employed for the therapy of many malignancies including metastatic testicular tumors, metastatic ovarian and advanced bladder cancer. Its administration is associated with many side effects due to its accumulation in the liver, kidney, and prostate.

5-FU is an antimetabolite that acts by interfering with the DNA and RNA synthesis. It is prescribed for the management of different malignancies such as carcinoma of the colon, rectum, breast, stomach, and pancreas. As for other chemotherapeutics, severe hematological and gastrointestinal toxicities and even toxic deaths are reported.

DOX, belonging to the class of anthracyclines, has mainly three different mechanisms of action. First, it binds to nucleic acids, intercalating the planar anthracycline nucleus with the DNA double helix; second, it interacts with topoisomerase II, an enzyme involved in DNA replication; third, it binds to the cellular membrane, affecting a variety of cellular functions. This drug is indicated for different cancers, such as bone sarcomas, Hodgkin's disease, ovary-, breast- and lung-cancer [79]. Among the side effects, it can be stressed in particular the cardiotoxicity and the possibility to develop acute myelogenous leukemia or myelodysplastic syndrome.

To improve drug effectiveness but especially to reduce side effects, different delivery options have been developed. In the next sections, we will concentrate on those based on the use of polysaccharides (Table 2). Most of the examples reported deal with the above-presented anticancer drugs; however, also other chemotherapeutics are reported and in these cases, their characteristics are briefly indicated.

Table 2. Polysaccharide-based delivery of chemotherapy drugs.

	System	Structure	Drug	Disease	*In vitro* Model	*In vivo* Model	Ref.
				Chitosan			
1	LMWC-PTX	LMWC	PTX	Melanoma, NSCLC	–	Mouse	[16]
2	LMWC-DTX	LMWC	DTX	NSCLC, GBM	NCI-H358, U87MG	Mouse	[17]
3	PTX-Cy5.5-CS	NPs	PTX	SCC	SCC7	Mouse	[18]
4	Cy5-CS	NPs	DOX	Sarcoma	HT1080	–	[19]
5	HGCS-Ce6 GCS-Ce6	NPs	Ce6	SCC, adenocarcinoma	SCC-7 HT-29	Mouse	[20]
6	FLT-CS	MPs	FLT	–	–	–	[21]
7	CS/GP CS/GP/GE	Hydrogel	CDDP	Oral cancer	–	–	[22]
				Hyaluronic Acid			
8	MSP-HA-DOX	Nano-conjugate	DOX	Breast cancer	MDA-MB-231	–	[29]
9	HA-DOX	Nano-conjugate	DOX	Breast cancer	MDA-MB-468NL	Rat	[30]
10	HA-CDDP	Nano-conjugate	CDDP	Lung cancer	A549	Rat	[31]
11	HA-CA-PTX	Nano-conjugate	PTX	Head and neck cancer	SCC7, NIH3T3	Mouse	[32]
12	HACD-AuNPs	Gold nanocluster	several	Breast cancer	MCF7, NIH3T3	–	[33]
				Dextran			
13	CMD-g-β-CD/(PAD-g-AD&AD-DOX))$_4$	Microsystem	DOX	–	HeLa	–	[45]
14	DEX-PDP	Nanosystem	DOX, CPT11	Breast cancer Colon cancer	MCF7, DLD1	–	[46]
				Pullulan			
15	P-PTX	NPs	PTX	Colon cancer	HCT116	Mouse	[51]
				Cellulose			
16	HPC-DNR	Conjugate	DNR	–	HeLa	Mouse	[39]
17	CMC-PABA, CMC-PAH	Nano-conjugate	Curcumin and derivates	Colon cancer	HT29, CRL1790	–	[40]
18	HPC-PAA-CdSe	Nanogel	TMZ	Melanoma	B16F10	–	[41]
				Other			
19	MR-5-FU	NPs	5-FU	Breast cancer	MCF7	–	[80]
20	FA-AG-MTX	NPs	MTX	–	AA8	–	[81]

3.1.3. Chitosan Based Delivery

Chitosan (CS) has been considered as delivery material because of its unique physicochemical and biological characteristics. The primary hydroxyl and amine groups located on the backbone of CS allow chemical modification to control its physical properties. The chemical modifications allow, for example, overcoming the low solubility at physiological pH and the poor buffering capacity of CS in its native form. Additionally, when a hydrophobic moiety is conjugated to a CS molecule, the resulting amphiphile may form self-assembled nanoparticles (NPs) that can encapsulate a variety of hydrophobic drugs and promote proper delivery [71].

Several examples of drug delivery systems based on CS have been reported in literature. Low molecular weight CS (LMWC) (MW < 10 KDa) conjugated with PTX (LMWC-PTX) through a succinate linker, was synthesized as a NP carrier for the oral delivery of PTX [16]. The antitumor efficacy of LMWC-PTX was evaluated in murine melanoma cells and in xenografted human non-small cell lung carcinomas (NSCLC) after oral administration. The strong antitumor activity was attributable to the improved water solubility, prolonged retention in the gastrointestinal tract and the ability to bypass the efflux pump of the gastrointestinal, cells as well as the metabolism in the intestine and liver. Moreover, LMWC could open quickly and reversibly the tight junctions between human epithelial colorectal adenocarcinoma cells (Caco-2), a useful characteristic that can improve intestine absorption [71,82].

Following an analogous approach, LMWC was also conjugated with DTX always through a succinate linker [17]. Cytotoxicity of LMWC-DTX conjugate was evaluated by MTT assay in human NSCLC-NCI-H358 and brain glioblastoma (U87MG) cell lines, showing similar effectiveness compared to the naked DTX. However, LMWC-DTX conjugate showed markedly enhanced (>200 times) water solubility compared to DTX alone. Moreover, following oral administration of LMWC-DTX in normal female BALB/c mice, blood half-life was increased by ~15-fold in comparison to the intravenously injected DTX. The *in vivo* antitumor efficacy was evaluated in nude mice bearing NCI-H358 and U87MG cells, respectively. The orally administered LMWC-DTX conjugate (10 mg DTX equivalent/kg) showed comparable antitumor efficacy to the same dose of the intravenously injected DTX for both NCI-H358 and U87MG models, but revealed much lower sub-acute toxicity as evaluated considering body weight loss and hematological toxicity in treated animals.

Another use of CS involves the generation of theragnostic NPs. In this case NPs carry a diagnostic and a therapeutic molecule enabling diagnosis, therapy, and monitoring of therapeutic response at the same time [83]. The possibility of exerting "Therapeutic Drug Monitoring" allows in principle the individual adjustment of drug dosing. This implies that it may be possible to optimize the therapy for any specific individual reaching a "personalized treatment" which is expected to be more effective and less toxic than available therapeutic protocols. CS-based theragnostic NPs, produced as spherical structures with approximately 260 nm in diameter, were loaded with PTX and linked with the fluorescent dye Cy5.5 for imaging [18]. NPs were then administered intravenously to mouse xenografted with head and neck squamous cell carcinoma cell line (SCC7). The high stability in serum, deformability and rapid uptake by tumor cells, allowed the NPs to localize to the tumor mass with minor uptake by normal tissues. Fluorescence imaging, used to directly monitoring tumor size, revealed that NPs were approximately two times more effective in tumor size reduction compared to naked PTX. The authors claimed that the superior tumor specificity of NPs was responsible for an increased drug accumulation in the tumor tissue, eventually resulting in a more potent antitumor effect. It would be interesting to see whether these NPs may exert the same targeting properties also with different tumor types and/or a different experimental set up. Despite this, the advantage of using CS NPs remains obvious compared to the administration of naked PTX.

More complex CS-based theragnostic NPs were generated with the possibility to monitor the drug release *via* Förster resonance energy transfer (FRET) [83]. A pH-responsive DOX-loaded NPs made of N-palmitoyl CS bearing a Cy5 fluorescent moiety (Cy5-CSNPs) were prepared [19]. The proper spatial positioning of DOX and Cy5 allowed FRET to occur. After endocytosis in the fibrosarcoma cell line HT1080, it was possible to observe that DOX fluorescence in the cytosol was barely seen when NPs were present in the slightly acidic early endosomes. However, after NPs moving to more acidic organelle (late endosomes/lysosomes), an improved release of DOX into the cytosol was detected. Consequently, a progressive accumulation into the cell nuclei occurred, paralleled by a significant DOX induced cytotoxicity. Notably, due to its mechanisms of action, DOX can exert its effect only reaching the nuclear compartment.

Photodynamic therapy (PDT) is a clinical treatment that employs photo-triggered chemical drugs as photosensitizers. To improve therapeutic efficacy and reduce the side effects in normal tissue, various nanoscale photosensitizer delivery systems have been developed [83]. As photosensitizers can generate both fluorescence and singlet oxygen upon laser irradiation, they can be used for both optical imaging and PDT at the same time [84]. In this regard, the hydrophobic photosensitizer chlorine e6 (Ce6) was loaded onto the hydrophobically-modified glycol CS NPs (average diameters of 300 to 350 nm) either by physical loading (HGCS-Ce6) or chemical conjugation (GCS-Ce6) [20]. Both HGCS-Ce6 and GCS-Ce6A displayed similar efficacy *in vitro* in the generation of singlet oxygen and a rapid cellular uptake in cultured SCC7. Compared to GCS-Ce6, HGC-Ce6 showed a burst of drug release *in vitro*, whereby 65% of physically loaded drugs were rapidly released from the particles within 6.5 h. However, following intravenous injection, HGCS-Ce6 did not accumulate efficiently in HT-29 human colon adenocarcinoma cells implanted in mice. In contrast, GCS-Ce6 had a prolonged circulation time and efficiently accumulated in the tumor, resulting in an excellent therapeutic effect. These findings suggest that the combined delivery of the two NP types may allow a fast delivery (HGC-Ce6) followed by slower release (GCS-Ce6), useful to maintain the pharmacological effects. The poor targeting ability of HGCS-Ce6 might be circumvented by adding targeting moieties on the NP themselves and/or including NPs in micro/macro delivery systems with targeting ability.

CS has been also utilized in form of microparticles (MPs) to deliver drugs. In the study of Elgindy *et al.* [21] CS-MPs (particle size range of 0.63–1.13 mm) were undertaken for the oral prolonged release of flutamide (FLT-CS-MPs). FLT is a nonsteroidal antiandrogenic agent recommended for prostate cancer monotherapy. FLT and its more potent active metabolite 2-hydroxyflutamide, competitively block dihydrotestosterone binding at androgen receptors, forming inactive complexes that cannot translocate into the cell nucleus. Formation of inactive receptors inhibits androgen-dependent DNA and protein synthesis, resulting in the growth arrest of tumor cell. Since FLT has low bioavailability after oral formulations due to low water solubility [85], Elgindy *et al.* [21] developed a water soluble formulation for a prolonged release. Simple ionic gelation and emulsification-ionic gelation techniques were used to prepare FLT-CS-MPs. Different formulations of FLT-CS-MPs were prepared and their ability to release the drug was tested *in vitro* at low pH to simulate the gastric environment. In general, a fast initial drug release at pH 1.2 was observed for the first two hours followed by a prolonged drug release for the remaining 10 h of the experiment. Notably, depending on the FLT-CS-MPs formulation, variants on this release behavior were possible. Although promising, this study is very preliminary and, thus, further studies in more complex experimental models are necessary to fully clarifying its efficacy.

In addition to the use in the form of NPs/MPs, CS has been also used as hydrogel for drug delivery. In this regard, there has been a great deal of interest in the use of hydrogels as chemotherapeutic drug delivery systems especially for oral administration. This is mainly due to the fact that the polymeric chains of the hydrogel can closely interact with the saliva glycoproteins, causing a muco-adhesion phenomenon [86]. For example, Moura *et al.* [22] investigated *in vitro* CDDP release from thermo-sensitive CS hydrogels cross-linked with glycerol phosphate disodium salt (CS/GP) (narrow pore size distribution centered around a mode of about 60 μm) and CS hydrogels that were ionic/covalently co-cross-linked with genipin (CS/GP/GE) (broad pore size distributions with larger pores which modes are around 200 μm). Both hydrogels could be produced *in situ* under physiologic conditions. Notably, the two types of hydrogels behaved differently as drug delivery systems. The amount of drug released was of about 20% for CS/GP and about 60%–70% for the CS/GP/GE hydrogel during the experimental time. Despite these differences, from the qualitative point of view the release profiles were similar for both hydrogel types being characterized by an initial burst, which reached a plateau following about 2–3 h. These systems are very attractive for the treatment of oral cancers because of the muco-adhesive property of the hydrogel, which can guarantee a prolonged permanence of the drug in the mouth. Moreover, the release kinetic can provide a first burst of the drug, useful for an immediate antitumor effect; then the slow release step is appropriate to maintain the therapeutic effects.

3.1.4. Hyaluronic Acid Based Delivery

In mammalian organisms, native Hyaluronic acid (HA) represents one of the main constituents of the extracellular matrix (ECM). Moreover, HA can modulate cellular fate by receptor-mediated uptake. For example, HA can bind cluster determinant 44 (CD44), receptor for hyaluronate-mediated motility (RHAMM), HA receptor for endocytosis (HARE) and lymphatic vessel endothelial hyaluronan receptor-1 (LYVE-1), in all cases inducing specific cellular consequences. The above properties together with the intrinsic hydrophilicity, biocompatibility, biodegradability and non-immunogenicity, make of HA an excellent candidate for biomedical needs.

DOX delivery by HA has been addressed in various works. Mesoporous silica (MPS) is a small nanomaterial known for its ability to host a large amount of drug and releasing it following changes in pH, temperature and by competitive binding [87,88]. The group of Chen [29] had chemically functionalized MPS surface with two chemicals (1-ethyl-3-(3-dimethylaminopropyl)carbodiimide) EDC and 3-(2-aminoethylamino) propyltrimethoxysilane NHS. The functionalization allows a homogeneous coating of MPS with HA, which results in high stability in physiological solution. Additionally, the HA coating of MPS allows the interaction with the receptor CD44, which is frequently over-expressed in tumors [89,90] and permits the complex internalization. Once in the cell, HA is digested by HA-specific enzymes leading to drug release into the cytosol from uncoated MPS. Based on this principle, Chen *et al.* [29] have found that MPS-HA loaded with DOX (MPS-HA-DOX) can decrease the viability of the human breast cancer cell line (MDA-MB-231) more effectively (15%) than DOX alone. Additionally, when administered to the non-tumor cell line NIH3T3, MPS-HA-DOX displayed 20% reduced unspecific toxicity compared to DOX alone. The authors explained this last feature with the reduced [91] expression of CD44 on NIH3T3, a fact which somehow protected the cells from particle binding and, thus, from DOX action.

Another smart approach to deliver DOX by HA is exploiting pH sensitive modifications. An example of this strategy has been developed by Cai's group [30]. Linking adipic acid hydrazide (ADH) to HA, the authors created a DOX (HA-DOX) conjugate. In this system the hydrazone bond (between HA and DOX) is hydrolyzed faster at pH 5 than pH 7.4, thus allowing the release of DOX in an acidic environment, such as that found in tumors and in internalization vesicles. The PK and toxicology features of the conjugate were studied in an MDA-MB-468NL (human breast adenocarcinoma cell line) orthotopic xenograft rat model. Compared to free DOX, HA-DOX, administered by subcutaneous injection, showed a reduced pick in plasma concentration (about 1/4). From the toxicology perspective, free DOX but not HA-DOX treatment induced renal and cardiac dysfunction. Moreover, HA-DOX conjugates reduced tumor growth more efficiently than naked DOX, with an increased animal survival rate of 50% over a time of 24 weeks. Even though the system has been tested only in one *in vivo* model, the results suggest its possible use in HA over-expressing tumors, a property which however needs to be further proven. Additionally, to test the CD44 specificity of the system, it would have been interesting to verify the effectiveness in CD44 non-over-expressing tumor cells.

The delivery of CDDP by HA based particles has been proposed by Xie *et al.* [31]. The authors tested a mono- and di-conjugate of CDDP and HA (HA-CDDP) in *in vitro* (A549 cell line, human lung adenocarcinoma epithelial cell line) and *in vivo* models (normal and xenografted rats) of lung cancer. The idea was to develop a local CDDP delivery to minimize the side effects that occur following CDDP intra veniously (*i.v.*) administration. Thus, the authors used the lung instillation route for HA-CDDP and compared the side effects with the classical *i.v.* administration of CDDP. Following lung instillation, HA-CDDP displayed better PK and reached higher concentrations in the lungs compared to *i.v.* administered naked CDDP. In particular, HA-CDDP reached 5.7- and 1.2-fold higher concentration than naked *i.v.* CDDP, 24 and 96 h after instillation, respectively. Notably, drug distribution to other tissues, such as brain, kidneys and liver was significantly more evident for naked *i.v.* CDDP compared to lung-instilled HA-CDDP. The more targeted localization of HA-CDDP resulted in reduced neuro- and nephron-toxic effects of HA-CDDP compared to CDDP. Whereas the investigation about the side effects are encouraging, future studies are required to examine the HA-CDDP therapeutic efficacy.

A curious and recent application of HA-coupled PTX has been undertaken by Thomas *et al.* [32]. The authors, by linking 5β cholanic acid (CA) to HA, produced micelle able to encapsulate and solubilize PTX in a ratio of 10:1 *w/w* (HA-CA-PTX). The study has been carried out *in vitro* in the cancer cell line SCC7 and in *in vivo* models (xenografted mice). The SCC7 cell line (CD44 over-expressing) and NIH3T3 (CD44 low-expressing) cells as control, were preliminary tested for the uptake using particles loaded with fluorescent dyes (Flamma™-552) but not PTX. In SCC7 cells, the uptake was superior for the dye containing HA-CA particles compared to naked dye; notably, the uptake was significantly superior in the CD44 over-expressing SCC7 cells compared to the CD44 low-expressing NIH3T3 cells, thus proving the targeting ability. With regard to the antitumor effect, HA-CA-PTX at various concentrations (0.01–100 μg/mL range), resulted to be from 10% to 40% more effective than naked PTX in SSC7. Bio distribution experiments *in vivo*, carried out using Flamma™-774 labeled particles administered intravenously, demonstrated that HA-CA particles accumulate in the liver from where they are cleared one day after administration; in contrast, the tumor localization was maintained up to three days. The harvested organs confirmed the particle accumulation in the tumor and in liver and evidenced some accumulation in the kidney. The *in vivo* efficacy tests conducted with 5 mg/kg of PTX in HA-CA micelles intravenously injected, showed a significant inhibition of tumor growth compared to naked PTX. Notably, whereas the maximum effect of naked PTX was observed at day 4 after administration, that of HA-CA-PTX occurred at day 8. This most likely reflects the different release kinetic, which may contribute to explain the improved therapeutic effect of HA-CA-PTX.

As a last example of HA mediated drug delivery, we mention the work of Li *et al.* [33]. This very promising, though not fully tested study, is based on the use of polysaccharide-gold nano-cluster. The system is constituted by a supramolecular carrier of gold NPs (AuNPs) bearing adamantine moieties and β-cyclodextrin-modified HA (HA-CD). The strong affinity between the β-CD cavity and the adamantane moieties allows the system to be stable [92]. Additionally, the system permits drug release in mild acidic environments. Besides these features, the system has other favorable properties: good biocompatibility [93,94], convenient synthesis and facile size control [95], robust stability under most *in vivo* conditions [96], tunable surface features and dense loading functionalities for specific cell targeting [97,98]. A further advantage of this carrier is represented by the possibility to be loaded with different drugs. In this regard, the authors demonstrated satisfactory loading abilities for both hydrophobic and hydrophilic drugs. A preliminary application of HACD-AuNP coupled with DOX in MCF7 breast cancer cells (over-expressing CD44), showed a significant reduction in cell viability, comparable to that observed with naked DOX. Notably, in NIH3T3 (CD44 low expressing), no significant cytotoxic effects were observed, suggesting that the CD44 targeting is the key determinant for HACD-AuNP/DOX effects. To confirm the targeting ability, it would have been interesting to test whether MCF7 pretreatment by HA receptor binding molecules could have prevented HACD-AuNP/DOX toxic effects. Despite this, the work is potentially interesting and constitutes the basis for further *in vivo* tests.

3.1.5. Dextran Based Delivery

Dextran is considered a convenient delivery material due to its hydrophilic character, biocompatibility, biodegradability and easy chemical derivatization.

A notable example of dextran-mediated delivery has been reported by the group of Luo [45]. The authors developed a process to produce microcapsules fabricated via host-guest interaction between polyaldehyde dextran-graft-adamantane and carboxymethyl dextran-graft-β-cyclodextrin. This kind of interaction has strong stability and it is pH responsive, due to the presence of acidic-sensitive hydrazone bonds in polyaldehyde dextran-graft-adamantane. Adamantane was used as DOX linking site giving rise to a complex structure named carboxymethyl dextran-graft-β-cyclodextran/polyaldenhyde dextran-graft-adamantane and adamantine-modified DOX $(CMD\text{-}g\text{-}\beta\text{-}CD/(PAD\text{-}g\text{-}AD \text{ and } AD\text{-}DOX))_4$. The presence of pH responsive bonds was introduced to confer the ability to release the drug in the acidic cancer tissues. The pH responsiveness was tested first *in vitro*, where it was proved the particle stability and the capacity to release the drug in relation to acidification. In particular, at pH 5.5 about 80% of the loaded drug was released, while at physiological pH only 18% of the drug was delivered. The cytotoxic potential of the DOX loaded microcapsules was then tested in cultured HeLa cells (human cervical cancer cell line), at different pHs (7.4 and 5.5). At pH 7.4 only a small amount of DOX was absorbed and thus no important effect on cell viability was observed. Under acidic conditions, about 50% of loaded drug was released with a dramatic reduction in cell viability. The pH responsive property together with the excellent biocompatibility of the unloaded microcapsules, make this particle type very promising. Future tests in *in vivo* model are necessary to confirm the obtained results and to determine the PK and PD profiles.

Another interesting and innovative strategy has been proposed by Pramod *et al.* [46], who reported a dextran-based method able to deliver two different drugs from the same particle. Dextran nano-vesicles were generated with a dextran backbone bound to a hydrophobic pentadecyl phenol (DEX-PDP) group via an aliphatic ester linkage. As this bond can be hydrolyzed only by the lysosomial enzymes esterases, outside the cell environment it has a high stability. The vesicles can be loaded with hydrophilic molecules in the core and hydrophobic ones in the external layer. The amphiphilic nature allowed the concomitant loading with the water-soluble DOX and the hydrophobic irinotecan (CPT11), an inhibitor of topoisomerase I, of which cytotoxicity is related to double-strand DNA damage. According to FDA, CPT11 is recommended for the treatment of metastatic colorectal cancer. In the study of Pramod *et al.* [46], three different vesicle types were prepared: the first loaded with DOX, the second with CPT11 and the third one with both of them. Cytotoxicity and drug release were tested in MCF7 and DLD1 (human colon cancer) cell lines. In both cancer cell lines, the combined administration of DOX and CPT11 in a 4:1 ratio had the best effect in terms of cytotoxicity, killing 80% and 60% of MCF7 and DLD1 cells, respectively. Notably, the activity of the particles loaded with the two drugs was always superior to that of the same particles loaded with the two drugs separately. Tests in more complex *in vivo* models are now necessary to prove the effectiveness of the vesicles. Additionally, it may be interesting to test the effectiveness of vesicles loaded with different combinations of drugs.

3.1.6. Pullulan Based Delivery

Pullulan (P) represents another potential interesting polysaccharide due to its biocompatibility, water solubility, lack of immunogenicity, and to the presence of multiple hydroxyl groups that can be functionalized. Notably, P can target the hepatic cells due to the ability to interact with the asialo-glycoprotein receptor present on hepatocytes. A notable example of P use is represented by the work of Lee *et al.* [51], where PTX-incorporated NPs have been synthesized using P acetate. NPs were spherical with a size from 80 nm for empty NPs and up to more than 100 nm for those loaded with PTX. *In vitro* it was demonstrated that PTX was released with an initial burst that lasted up to day two, followed by a reduced release over one week. In RAW264.7 macrophage cells, NPs alone were nontoxic and when loaded by PTX, they exerted a significant antitumor activity in the HCT116 human colon carcinoma cells. However, NPs-PTX were less effective than naked PTX, probably due to the slowed release kinetic. This features, which was detrimental in cultured cells was, on the contrary, the probable reason for the success *in vivo* in HCT116 xenografted in athymic nude BALB/c mice. Following *i.v.* administration and starting from ten days after drug administration, NPs-PTX demonstrated a superior activity than naked PTX in reducing tumor growth. Notably, compared to naked PTX, NPs-PTX induced a reduced body weight loss, suggesting a reduced systemic toxicity. The authors proposed that, in addition to the slowed release kinetic, the success of the increased effectiveness and reduced side effect of NPs-PTX was also dependent on the EPR effect. Whereas being fully acceptable, this hypothesis could have been verified by studying the distribution into the tumor tissue of NPs-PTX compared to naked PTX.

3.1.7. Cellulose Based Delivery

Cellulose, due to the biocompatibility and anti-microbial action, is one of the most promising materials for bio-applications. Although it is well known that cellulose is insoluble in many solvents and water, it can be chemically modified for improving its solubility. There are several commercial soluble derivatives of cellulose, such as hydroxypropyl methylcellulose (HPC), a non-ionic, thermoresponsive and biodegradable compound [99]. Metaxa and coworkers [39] used HPC conjugated to daunorubicin (DNR), a DOX analogue, via a glutathione-sensitive linker (*N,N'*-(disulfanediylbis(ethane-2,1-diyl))bis(2-methylacrylamide) (DSBMA). In the presence of glutathione, the disulfide bridges of the polymeric layer are transformed into thiol groups, resulting in the slow release of DNR. As glutathione is highly concentrated in tumor cells [100], the system is in principle able to have a cancer cell specific effect. This feature may be further improved by the covalent modification of HPC with a folic acid receptor (FR) targeting moiety. Being that FR scarcely expressed in normal cells and over-expressed in a range of cancer cells, the system has an additional cancer cell targeting ability. The HPC-DNR particles were able to significantly reduce the vitality of the cancer cell line HeLa, over-expressing FR. In contrast, negligible effects were observed in the HEK293 (human embryonic kidney cell line), which express FR at low levels. Over a period of three days, HPC-DNR displayed reduced activity compared to naked DNR, probably due to the slow release kinetic. This hypothesis is supported by the observation that HPC-DNR could localize into HeLa cells. In particular, two hours after administration, the amount of HPC-DNR was higher compared to naked DNR, and negligible amounts were seen in the FR negative NIH3T3 cells. *In vivo* (mouse model), HPC-DNR displayed rapid clearance from circulation following intravenous administration and an uptake by the mononuclear phagocyte system. In this work, no evidence of the effects of glutathione on DNR release from HPC-DNR were presented; additionally, no evidence on the *in vivo* activity and specificity have been provided. Despite these aspects, the possibility to bind FR and release drug in high glutathione concentration, make the system interesting for further investigation.

Another cellulose type, carboxymethylcellulose (CMC), can be used for drug delivery because of its hydrophilic nature. Plyduang *et al.* [40] explored this material to deliver tetrahydrocurcumin (THC). THC is the active metabolite of curcumin, a naturally occurring antioxidant, anti-inflammatory and anticancer agent [101]. However, due to its low water solubility, THC has only limited pharmacological activities. To overcome this drawback, Plyduang *et al.* [40] conjugated THC with CMC. In addition to being approved by FDA as a common excipient, CMC is inexpensive, has good compatibility to the skin and mucous membranes and presents suitable groups, which allow functionalization. The CMC derivative was conjugated with PABA (*p*-aminobenzoic acid) or PAH (aminohippuric acid)-diamine linkers, both of which can be cleaved by colonic bacteria. This property was chosen as the authors intended to prepare a colon-specific drug delivery material for the treatment of colonic cancer. To decrease side effects, the targeted drug delivery to the colon should not release the drug in the stomach and small intestine. After assessing the *in vitro* release of the conjugates (from 65% to 82% after 48 h), the analysis of cytotoxicity, revealed an increased toxicity in the human colon adenocarcinoma cell line HT-29 compared to the normal colon cell line CRL1790. This specificity, however, seemed to depend on the drug *per se* and not to any targeting ability of the delivery materials. The authors also showed that the incubation of the conjugates with gastric or small intestine homogenate resulted in a minor drug release; however, in the presence of cecal content, drug release had a burst. The authors explained this behavior with the action of colonic bacteria, which could have triggered the cleavage of PABA/PAH linkers. Despite being potentially interesting, these data need to be confirmed in physiological model considering the delivery behavior in living animals.

As mentioned before, the use of polysaccharides can be applied to integrate drug delivery and diagnostic/imaging in one system. These types of systems were, for example, investigated by Wu *et al.* [41]. The authors developed a class of multifunctional hybrid nanomaterials through the *in situ* immobilization of CdSe quantum dots (QDs) into the hydroxypropylcellulose-poly(acrylic acid) (HPC-PAA-CdSe) semi-interpenetrating (semi-IPN) nanogels. The fluorescent CdSe QDs is used for particles visualization; the HPC chains can provide rich-OH groups for sequestering CdSe ions into the gel network and stabilizing the in-situ formed CdSe QDs. Finally, the PAA chains are the pH sensitive portions that, depending on the pH, can induce reversible swelling/shrinking of the particles influencing drug delivery. Nanogels were successfully internalized into the mouse melanoma B16F10 cells as evaluated by fluorescence microscopy, giving no signs of morphological damage to the cells. The addition of the anticancer drug temozolomide (TMZ) into nanogel resulted in a significant reduction in B16F10 cell vitality compared to controls. Notably, nanogels did not alter the potency of TMZ as the drug either complexed in nanogels or in the naked form, had similar activity. TMZ is an oral anticancer agent approved for the treatment of newly diagnosed glioblastoma in combination with radiotherapy. The proposed nanogels may be useful to overcome some of the limitation of TMZ, such as the low solubility, the short half-life, and the considerable toxicity. In particular, TMZ toxicity may be reduced by adding to the nanogels cancer-cell targeting moieties. The developed nanogels are certainly promising for optimized drug delivery, but further *in vivo* studies are necessary to confirm their efficacy.

3.1.8. Other Polysaccharide Based Delivery

Other interesting, but thus far not yet deeply investigated, polysaccharides can be used as drug delivery materials. In this regard, a promising class is represented by sulfated polysaccharides. As most of them are negatively charged, they can bind positively charged molecules via electrostatic interaction. A recent study [80] considered the use of *Halomonas maura*, a halophile bacterial species, for its capability to produce huge amounts of sulfated exopolysaccharides particularly rich in sulfate residues. The numerous sulfate residues render this polysaccharide, named mauran (MR), an immune-modulating and anticancer agent itself. In the study, NPs composed of MR and CS were synthesized and filled with 5-FU (5-FU-MR). MR/CS NPs were synthesized by simple polyelectrolyte complexation of the anionic MR and the cationic CS. The obtained NPs, in the shape of spheres with a smooth surface, released 5-FU mainly via diffusion. No significant toxicity was observed in mouse connective tissue fibroblast cells (L929). Additionally, the authors could show that in MCF7 cells, the anti-proliferative capacity of 5-FU was enhanced following the release from NPs, compared to the naked drug. An explanation given for this observation was the increased 5-FU retention into the cells compared to the naked drug form. This feature together with the capacity to exert a sustained released of the drug (up to 10–12 days), indicate the interest for this kind of NPs, whose characterization should be further deepened to fully understand the effectiveness in drug release *in vivo*.

Another example of a not deeply studied polysaccharide is arabinogalactan (AG), a natural polymer extracted from Larix tree. The group of Pinhassi [81] created a smart system to ameliorate methotrexate (MTX) administration. MTX is a chemotherapeutic agent approved by FDA for the treatment of different cancers (breast cancer, acute lymphocytic leukemia). However, its administration causes different side effects. In the study of Pinhassi *et al.* [81], AG was conjugated with folic acid (FA), to allow the targeting to cancer cells which typically overexpress FA receptor. The drug loaded NPs displayed 6.3-fold increased cytotoxic activity in FR-over-expressing cells (AA8 Chinese hamster ovary cell line) compared to their FR-lacking counterparts. This interesting targeting ability will need to be confirmed in animal models; additionally it would be interesting to test whether the NPs could be conjugated with different targeting moieties to further improve the specificity of action.

3.2. Polysaccharides for the Delivery of Nucleic Acid Based Drugs

In addition to the clinically approved drugs, also other experimental molecules with potential therapeutic value can benefit from the polysaccharide-mediated delivery. Among the experimental molecules, we focussed on NABDs. These compounds, constituted by short sequences of either DNA or RNA, includes antisense oligonucleotides, decoys oligonucleotides, aptamers, triple helix forming oligonucleotides, DNAzymes, Ribozymes, small interfering RNAs (siRNAs) and micro interfering RNAs (miRNAs) [102–104]. As all NABDs are made of short nucleic acid molecules, they have comparable physical-chemical features. It follows that they display similar requirements in terms of the delivery material. Thus, whereas in this review we will focus on polysaccharide-based delivery systems for siRNAs/miRNAs, the same systems can be in principle effective also for the other class of NABDs.

In the next section, we will present the general biological features of siRNA/miRNA (Section 3.2.1) together with the specific problems related to the release of these molecules (Section 3.2.2). Then, we will present (Sections 3.2.3–3.2.7) the employment of representative and widely studied polysaccharides for NABD delivery.

3.2.1. Small and Micro Interfering RNA Molecules

The RNA interference (RNAi) is a cellular process characterized by the degradation, in a sequence dependent fashion, of a target RNA by short double stranded RNAs (dsRNAs) [105]. miRNAs and siRNAs are the short dsRNAs able to trigger RNAi. While miRNAs are synthesized within the cell, siRNAs are commonly of exogenous origin deriving, for example, from viruses and transposons. With regard to the action mechanism, of the two RNA filaments constituting miRNA/siRNA, one (the so-called antisense strand) is up-taken by the cellular protein complex termed RISC (RNA-induced silencing complex), while the other (called sense strand) is discarded and does not take part to the silencing process. The antisense strand drives RISC to the target RNA that is bound via either a perfect or a partial complementarity in the case of siRNA or miRNA, respectively. In the case of miRNAs, the binding triggers the down-regulation of target gene expression most often by interfering with the translational machinery [106] and, in some cases, by inducing target RNA degradation by RISC. In contrast, siRNAs repress target gene expression only via RISC-mediated target RNA degradation. Due to the short length, miRNAs and siRNA can be easily chemically synthesized and used to target the mRNAs of genes causing disease. In this regard, there are several examples of the potential therapeutic value of miRNAs and siRNAs [107–113].

3.2.2. The Delivery Problems

Whereas the therapeutic potential of miRNAs/siRNAs, which from now on will be collectively indicated as NABDs, is generally recognized, their practical use is limited. This is substantially due to the absence of optimal delivery systems able to efficiently protect and direct them to the target tissue. If administered systemically (vascular route) in the naked form (Figure 6), NABDs have to face with the obstacle of blood nucleases, which can rapidly trigger their degradation. Additionally, naked NABDs tend to be eliminated by the reticulo-endothelial system, by kidney filtration [114], and, in some cases, tend to activate the innate immunity [115,116]. The crossing of the vascular wall before reaching the diseased tissue, represents another obstacle to NABDs action. The efficiency of this process is strongly influenced by the size of the endothelium fenestration, a feature that can vary considerably among tissues. For example, the liver endothelium is highly fenestrated, thus favoring extravasation [117] and accumulation of NABDs in the liver.

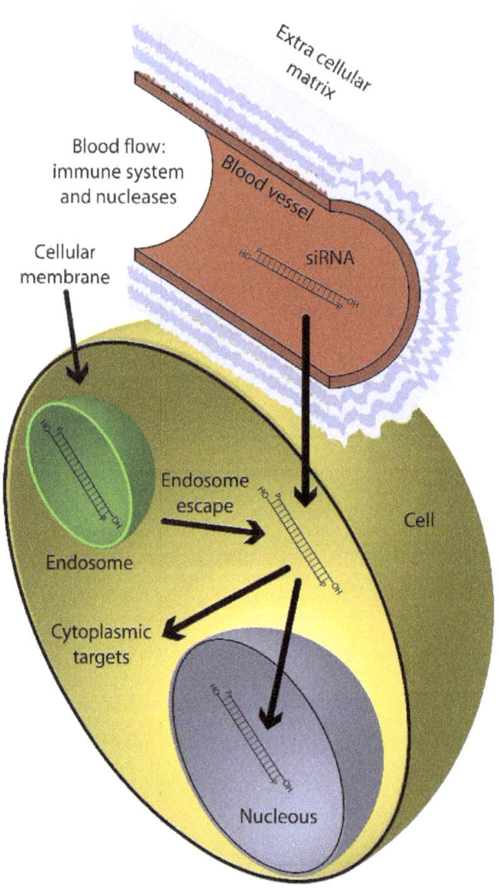

Figure 6. Biological obstacles in NABDs delivery.

Once in the desired tissue NABDs have to overcome the problem of the cell membrane crossing. This step is unfavorable due to the negatively charged surface of cellular membranes, which tend to repulse the negatively charged phosphate groups present in NABDs. Additionally, the hydrophilic nature of NABDs disfavors the crossing of the hydrophobic layer of the cell membrane. In the cells, NABDs are then susceptible to further degradation by cellular nucleases. In the cell, NABDs are also subjected to the cellular trafficking issue. This last aspect is relevant [118,119] as, following cellular uptake, NABDs can be sequestered into cytoplasmic vesicles named endosomes where they remain entrapped with no possibility to reach the target mRNA into the cell cytoplasm.

Together all the above considerations clearly indicate that naked NABDs have no chance to exploit their therapeutic action. A strategy undertaken to overcome NABDs degradation is based on the introduction of chemical modification into their structure to make them resistant to nucleases-mediated degradation. A second possibility is represented by the complexation/binding to different synthetic materials. While the discussion of the NABDs chemical modification [111] goes beyond the scope of the present work, the strategies adopted to complex NABDs with synthetic materials represents the topic of this review. In particular, among the several materials available [111], we concentrate on the use of polysaccharides as delivery substances due to their wide use and versatility. The reported examples are described below and summarized in Table 3.

Table 3. Polysaccharide-based delivery of NABDs.

	System	Structure	Drug	Disease	*In vitro* Model	*In vivo* Model	Ref.
			Chitosan				
1	Bio-CS-g-PEI	NPs	siRNA	Ovary cancer	HeLa OVCAR-3	–	[23]
2	CS-TAT CS-nonaArg	NPs	siRNA anti luciferase/ survivin	Brest cancer	MCF7	Mouse	[24,25]
3	CS-γPGA	NPs	siRNA anti luciferase/GFP	–	HT1080	–	[26,27]
4	psi-Pgp-GCt	NPs	siRNA anti Pgp	Brest cancer	MCF7	Mouse	[28]
			Hyaluronic Acid				
5	HA-Chol/2b	NPs	siRNA anti RFP	–	B16F10	–	[34]
6	PEG-HA-NP	NPs	siRNA anti Pgp siRNA anti GGCT	Breast cancer	MCF7	Mouse	[35,36]
7	CAP/dopa-HA	NPs	siRNA anti luciferase	–	HT29	Mouse	[37]
8	CAP-HA-DPA/Zn	NPs	siRNA anti luciferase	–	DU145 143B HCT116	Mouse	[38]
			Dextran				
9	Ac-DEX-spermine	NPs	siRNA anti luciferase	–	HeLa	–	[47,48]
10	DexS/pArg	NPs	siRNA anti GFP	–	HeLa	–	[49]
11	DexS/pArg/HY	NPs	siRNA anti SPARC	–	Fibroblast	–	[50]

Table 3. *Cont.*

	System	Structure	Drug	Disease	*In vitro* Model	*In vivo* Model	Ref.
			Pullulan				
12	P-PEI-FA	NPs	siRNA anti luciferase	–	HeLa	–	[52]
13	Ps	NPs	siRNA	–	MDA-MB-231	–	[53]
			Other				
14	CS/alginate	hydrogel	siRNA anti TNFα	Colitis	–	Mouse	[42]
15	CS/alginate	hydrogel	siRNA anti CD98	Colitis	Colon-26 RAW 264.7	Mouse	[43]
16	PEI/alginate	NPs	siRNA anti VEGFR-3	–	EPCs	–	[44]
17	PEG-Tf-CD	NPs	siRNA anti RRM2	–	–	Mouse Non-human primate Clinical trial	[120,121]

3.2.3. Chitosan Based Delivery

CS has been widely used as NABDs delivery agent mainly because it contains positively charged amino groups that permit the electrostatic interactions with the negatively charged NABDs. Darvishi *et al.* [23] have proposed the grafting of polyethyleneimine (PEI) on CS backbone. PEI is a synthetic polymer with a repeating unit of an amine group with two CH_2 spacers. Due to the presence of protonable amino groups in every third position, PEI has a high cationic density under physiological pH, which allows the spontaneous formation of polyelectrolyte complexes with nucleic acids. In addition to PEI, Darvishi *et al.* [23] added biotin (Bio-CS-g-PEI) whose receptor tends to be over-expressed in cancer cells. Bio-CS-g-PEI resulted to be water-soluble at physiological pH and had optimal transfection efficiency for siRNA at the Bio-CS-g-PEI/siRNA weight ratio of 30. Moreover, Bio-CS-g-PEI/siRNA transfection efficiency was 44% and 22% higher than that of CS-g-PEI/siRNA and PEI/siRNA complexes, respectively, in the HeLa tumor cells. Worth noting is the fact that, in OVCAR-3 cells, a model of human ovary cancer cells over-expressing biotin receptors on the cell membrane, the transfection efficacy was 74% and 23% higher, compared to CS-g-PEI and PEI complexes, respectively. Together, these data strongly support the targeting potential of Bio-CS-g-PEI/siRNA in cancer cells over-expressing biotin receptors. Confirmatory experiments are now necessary in *in vivo* model.

As an additional and interesting strategy to improve the efficiency of CS/siRNA transfection, cells penetrating peptides were used in combination with CS. In two different works it has been demonstrated that the grafting of TAT peptide [24] or nona-arginine peptide [25] ameliorated not only the quantity of internalized siRNA but also the silencing effect of siRNA compared to the siRNA delivered by unmodified CS. More in details, nona-arginine and TAT increased in MCF7 cells the uptake of 2.5 and 13 folds, respectively, compared to the unmodified CS. Notably, both for nona-argine or TAT grafted CS, the cyto-compatibility was definitively superior to unmodified CS. Additionally, the siRNA mediated silencing effect was excellent reaching 75%. TAT-grafted CS also showed its effectiveness *in vivo*, delivering a siRNA targeted against the mRNA of survivin, a protein known to inhibit cell apoptosis, thus favoring tumor cell survival. Following direct intra-tumor injection, TAT-CS-siRNA complexes were able to significantly reducing tumor growth in breast tumor-bearing mice in comparison to naked siRNA and saline treated mice. Interestingly, whereas mice treated by saline or naked siRNA developed lung metastasis, this was not the case for TAT-CS-siRNA treated animals [25], further showing the effectiveness of this delivery material.

Despite being the above reported success in the use of CS as siRNA delivery material very encouraging, it should be kept into consideration the fact that the strong interaction CS-siRNA, may negatively affect the siRNA release into the cell. This seems to be particularly true for high molecular weight (MW) CS [122], even if high MW CS can efficiently protect siRNAs from degradation. On the other hand, low MW CS is not optimal in siRNA protection against degradation in the extracellular environment, but it can efficiently release the siRNA in the cell [123]. Together, these observations indicate that the balancing between CS protection and release properties of siRNAs needs to be optimized [124]. Additionally, this balancing may need to be optimized also in relation to the target cell. For example, CS with intermediate chain lengths mediated an efficient siRNA delivery in H1299 (human lung carcinoma cell line), in the MCF7 and in HUVEC (human umbilical vein endothelial) [125]. However, very poor delivery was observed by us in the human hepatic cancer cell line JHH6 (Farra *et al.*, unpublished results), suggesting a cell-dependent efficiency of uptake.

With the purpose to improve CS-mediated delivery of siRNAs via the reduction of CS attraction towards siRNAs, Liao *et al.* conjugated to CS a negatively-charged poly(γ-glutamic acid), together with a γ-glutamyl unit at its N-terminal end (γPGA) [126] to form CS/γPGA complexes. As γPGA can be recognized by a membrane protein, γ-glutamyl transpeptidase (GGT), highly expressed in most mammalian cells, the entire CS/γPGA complexes were up taken by means of a receptor mediated endocytosis. Moreover, the authors showed that, using a double stranded DNA as a prototype of a nucleic acid molecule, the CS/DNA/γPGA complexes was gradually unpacked into the cell due to the electrostatic repulsion induced by the incorporated γPGA, which facilitated the intracellular release of DNA. This system was also used to deliver two disulfide bond-conjugated and PEGylated siRNAs (PEG, polyethylene glycol) directed against the mRNAs of luciferase and GFP (green fluorescence protein) [26]. To study the effects of the two siRNAs complexed with CS/γPGA (CS/siRNA/γPGA), the HT1080 cell line stably transfected with GFP/luciferase reporter gene plasmid was used. The authors could show that following endosomal escape, the negatively charged γPGA of the CS/siRNA/γPGA was able to disrupt the strength of electrostatic interactions between the conjugated siRNAs and the protonated CS. This facilitated the release of the two siRNAs into the cytosol where the intracellular glutathione could reduce the disulfide bond of the conjugated siRNAs making them free and, thus, able to silence the respective target mRNA (GFP and luciferase). The maximum suppression of gene expressions (80%) was observed at day five after administration and persisted, although progressively reduced in extent, up to day eleven. Notably, whereas at day one CS/siRNA/γPGA could reduce gene expression down to 28%, the CS/siRNA complex only reached 50%, showing the relevance of γPGA [27]. Despite that these data are very interesting, the comparison with a LF mediated delivery could have better defined the efficacy of the system.

An original CS-based delivery for siRNAs is the one proposed by Ji Young Yhee *et al.* [28]. The authors used glycol chitosan (GC) polymers bearing thiol groups (GCt) (Figure 7).

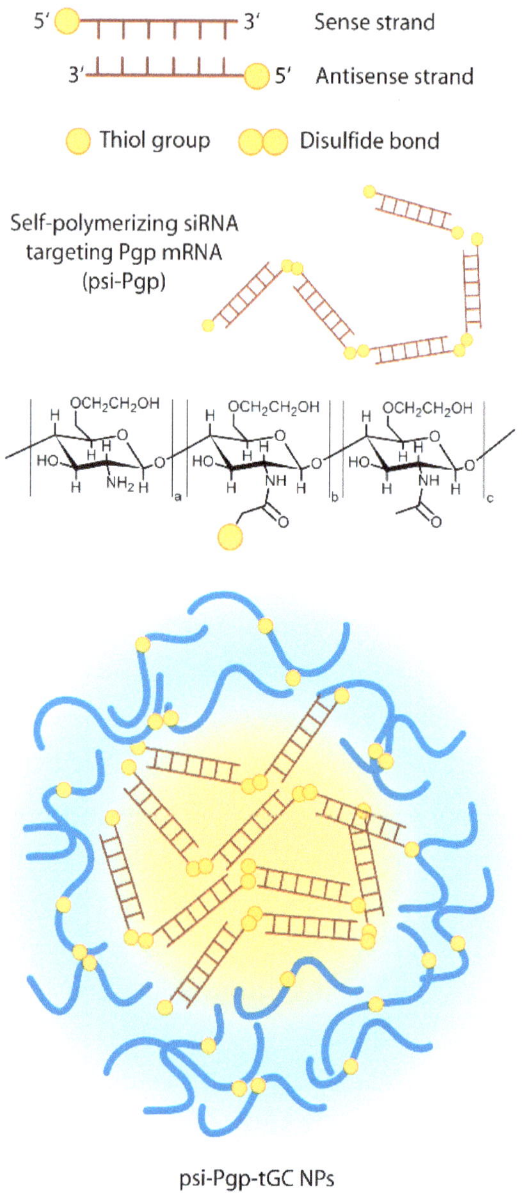

Figure 7. CS-based delivery for siRNAs proposed by Ji Young Yhee *et al.* [28].

The siRNA of interest was also thiolated at the extremities, thus inducing the formation of a sort of poly-siRNA (psi). GCt and psi were then mixed and allowed to bind by electrostatic interactions and by the chemical conjugation via thiol groups. The siRNA sequence used was directed against the mRNA of P-glycoprotein (Pgp), which mediates the therapeutic drug efflux from cancer cells, thus inducing cancer cell multi drug resistance (MDR). The resulting NPs named psi-Pgp-GCt, were spherical with an average hydrodynamic diameter of approximately 270 nm. The NPs could protect siRNA molecules from the degradation exerted by cellular and extracellular nucleases. At high particle concentrations (59.4 μg/mL, 400 nM of siRNA of irrelevant sequence), almost no reduction in the vitality of MCF7 was detected by MTT assay; in contrast, the siRNA of irrelevant sequence, released from lipofectamime (LF), displayed 50% reduction in cell vitality. Additionally, the psi-Pgp-GCt NPs were able to enter the cytoplasm at neutral to acidic pH (pKa of chitosan= 5–7) and to escape the lysosomes due to the change of endosomal pH. With regard to the functional effect, psi-Pgp-GCt NPs (100 nM of siRNA against Pgp) induced significant Pgp suppression in MCF7 cells which, however, was somewhat inferior to that obtained by the commercial available LF. Despite this, the reduction in Pgp levels reached by the produced NPs, was able to decrease the extrusion from the cells of the anti-cancer drug DOX. Thus, these data suggest that psi-Pgp-GCt may induce changes in drug sensitivity. This was proven by showing the reduced cell vitality following DOX administration in psi-Pgp-GCt transduced compared to non-transduced cells. *In vivo*, the bio-distribution of psi-Pgp-GCt delivered via the tail vein in MCF7 tumor-bearing mice, showed a predominant localization at the tumor site, which was about three times higher than in the control. Moreover, psi-Pgp-GCt co-treatment with DOX significantly enhanced the anti-cancer activity of DOX, as shown by the fact that the mean tumor volume of psi-Pgp-GCt + DOX group was significantly reduced compared to DOX treated animals. Thus, these data confirm, *in vivo*, the induction of drug sensitivity by psi-Pgp-GCt. Notably, the authors could also prove the significant reduction of DOX toxicity in animals treated by psi-Pgp-tGC NPs.

3.2.4. Hyaluronic Acid Based Delivery

The negative charge of HA does not allow an effective binding to the negatively charged NABDs. To overcome this limitation, HA has been used in combination with cationic polymers to form polyelectrolyte complexes by supramolecular self-assembly able to bind NABDs. In this regard, different molecules including PEI, Spermine [127], Poly L-Lysine [128], and other [1] have been tested. A notable example is represented by the work of Choi *et al.* [34]. The authors developed a siRNA delivery system based on HA-containing NPs with the HA backbone conjugated with cholesterol (HA-Chol NPs). HA-Chol NPs where generated with a variable number of cholesterol per 100 sugar residues (from 2.6 to 24). Then, the siRNA was combined with the 2b protein, a virus derived protein able to bind double stranded RNA (dsRNA). The author demonstrated that it was possible to encapsulate the 2b/siRNA complex into the HA-Chol NPs to form the HA-Chol/2b/siRNA complex. This particle showed a loading content of siRNA up to 11% *w/w* for NPs containing the higher amount of cholesterol. Using a siRNA directed against the red fluorescence protein (RFP), the authors showed that, in B16F10 cells expressing RFP, HA-Chol/2b/siRNA particles efficiently reduced RFP mRNA and protein levels. Worth noting, the commercially available LF 2000 was significantly less effective than the HA-Chol/2b/siRNA complexes. This elegant delivery system needs now to be tested in more complex scenario *in vivo*, using therapeutic siRNAs to fully demonstrate its delivery power.

HA can be also used as coating for positively charged NPs. The positive charge on the NP can interact with negatively charged serum protein, which can displace siRNA from the NP. This, in turn, results in the reduction of the amount of siRNA delivered to the target cells, and, to a certain extent, also in a possible off-targeting of the free siRNA with unpredictable effects in terms of toxicity. The coating with HA can minimize the siRNA displacement from the NPs thus improving cellular uptake, biocompatibility and colloidal stability. In this regard, an interesting example has recently been reported by Ran *et al.* [35]. The authors covalently linked HA to PEG molecules and used the resulting PEG-HA compound to coat liposome-siRNA NPs. Whereas liposome are commonly used as siRNA delivery molecules, they are not optimal for systemic delivery of siRNA and, thus, need to be conjugated with "smart molecules". The addition of PEG can efficiently shield the positive charge of liposome, thus reducing the siRNA displacement by negatively charged serum protein, from the NP (PEG-NP). However, PEG tend to reduce the cellular uptake and compromise endosome escape. HA can contribute to overcome these problems, as shown by Ran *et al.* [35] with the generation of liposome NPs coated by PEG-HA (PEG-HA-NP). PEG-HA-NP, with an average diameter of 190 nm and a negative surface charge, were stable in fetal bovine serum up to 24 h. Compared to NP, PEG-HA-NP carrying a siRNA directed against the mRNA of the P-glycoprotein, resulted in a more efficient down regulation of target mRNA in MCF7 cells. Because of the good stability in serum, the systemic administration of PEG-HA-NP could efficiently deliver the siRNA in tumor bearing mice reaching a 34% P-glycoprotein down regulation. In contrast, no activity was detected for siRNA delivered from either NP or HA-NP. As P-glycoprotein is involved in the efflux of drugs outside the cells, in so preventing drug accumulation into the cell its down regulation by the PEG-HA-NP delivered siRNA can contribute to reduce the problem of drug resistance in breast cancer cells. To further confirm the reliability of the PEG-HA-NP approach, in a different work by the same author [36], this system was successfully tested in MCF7 cells for the delivery of another siRNA targeted against γ-glutamylcyclotransferase (GGCT), a protein upregulated in various types of cancer cells. Since HA can selectively bind CD44, the use of PEG-HA-NP may be in principle extended to the targeting of the many different CD44 over-expressing tumor cells. However, the extent of the increased efficacy of PEG-HA-NP in CD44 over-expressing compared to under expressing cells remains unclear since the authors did not specifically address this aspect in their work.

HA has been combined with calcium phosphate (CAP), a natural inorganic material that can be found in bone and teeth. This material has been frequently used as gene delivery agent due to its biocompatibility, biodegradability, and ability to encapsulate negatively charged genetic material via the chelation of calcium ions during the process of calcium-phosphate crystals formation. The lack of delivery specificity and the uncontrollable growth of CAP crystals in a physiological solution has, however, limited its use as gene delivery system *in vivo*. Very recently, the conjugation of CAP with 3,4-dihydroxy-L-phenylalanine (DOPA), an unusual amino acid found in adherent threads of marine mussels, has been proposed as a novel device for the targeted siRNA delivery to tumors [37] in conjunction with HA. HA-DOPA was first synthesized and then used to stabilize the CAP/siRNA NPs, upon absorption on their surfaces. Due to the adhesive properties, the DOPA molecules have the role to bind to the surface of CAP particles while the hydrophilic HA backbone exerts protective functions and serves as a targeting moiety for CD44+ cancer cells. At the weight ratio DOPA-HA/siRNA of 60, the siRNA incorporation efficiency was about 84%; additionally, no significant cytotoxicity was observed as evaluated by MTT test. CAP/DOPA-HA containing an anti-luciferase siRNA (DOPA-HA to siRNA weight ratio of 60), reached in CD44+ human colon carcinoma cells (HT29-luc) over-expressing luciferase, 50% reduction of target gene expression. In addition to showing that the uptake mechanism of CAP/siRNA/ DOPA-HA was regulated by receptor-mediated endocytosis, the authors also demonstrated that the acidic endosome pH promotes CAP dissolution determining an increase in the concentration of calcium and phosphate ions. This, in turn, induces the swelling and disruption of the endosomal membrane due to increased internal osmotic pressure with the consequent siRNA release. In animals bearing HT-29-luc tumor xenografts, CAP/siRNA/DOPA-HA administered via the tail vein, accumulated predominantly in the tumor mass and, in part, also in the liver. This last phenomenon may be attributed to the presence of the HARE receptor in hepatocytes; however, no signs of hepatic toxicity were detected following the histological evaluation of the liver and the evaluation of the liver function tests. In our opinion, this observation needs further careful investigations. Notably, CAP/siRNA/DOPA-HA was able to significantly suppressing the expression of the target gene (luciferase) after a single dose (0.6 mg/kg) administered via the tail vein.

Another recent and interesting work based on the use of CAP and HA has been reported by Choi *et al.* [38] (Figure 8). The siRNA NP is composed by four major components: (i) a hydrophilic HA shell; (ii) a hydrophobic 5β-cholanic acid (CA) inner core; (iii) a phosphate receptor Zn(II)-dipicolylamine (DPA/Zn); (iv) a calcium phosphate layer (CAP). The HA shell confers the ability of NPs to bind CD44 positive tumor cells, CA offers the possibility to load into NPs hydrophobic drugs and DPA/Zn allows the binding to siRNA. In particular, DPA/Zn binds siRNAs via specific interactions between coordinated zinc ions of DPA and anionic phosphates of the RNA. Due to the significantly high phosphate concentration in the bloodstream, which can perturb the siRNAs-DPA/Zn binding, the author added CAP layers onto HA-DPA/Zn/RNA by *in situ* mineralization of sequentially added calcium and phosphate ions. In this way, the CAP layer can protect the DPA/Zn/RNA interaction in the biological environment. Additionally, as above discussed, at the low pH of endosome, the CAP is dissolved, favoring the endosome escape of DPA/Zn/RNA. Notably, the phosphate ions released from CAP competitively bind with DPA/Zn receptors triggering the HA-siRNA release. Subsequently, the HA backbone can be degraded by the enzyme hyaluronidase, abundant within tumor cells, eventually resulting in the release of the free siRNA. Worth note is the fact that this kind of NPs can allocate, in addition to siRNAs or other NABDs, also hydrophobic drugs due to the presence of CA. Thus, this NPs can be used to exert synergistic therapies mediated by siRNAs and hydrophobic drugs, such as those commonly used in cancer chemotherapy. Using red fluorescent labeled siRNA (Cy3-siRNA) Choi *et al.* [38] could detect in the prostate tumor cells DU145 and 143B, a strong fluorescence signal, which was about three times more intense than that obtained delivering the siRNA by liposomes. Moreover, releasing a siRNA directed against the luciferase gene in 143B tumor cell line over-expressing luciferase, the authors observed thirty times more activity of the NP-siRNA complexes compared to the liposome-mediated delivery of siRNA. In a HCT116 xenograft mouse model, NPs intravenously injected were able to dramatically reduce the expression of the target gene (Luciferase) over-expressed by the HCT116 tumor cell line. Interestingly, it was also demonstrated the feasibility of miRNA delivery by the prepared NPs. Finally, the authors also provided evidences of the possibility to co-deliver the anticancer drug PTX and siRNA to the HCT116 cancer cells. Notably, in an evolution of the above NPs, the authors [129] showed the feasibility to convert the functional group of Zn-DPA into other groups including a carboxylic or thiol group; moreover, the DPA analog can be covalently attached to different formulations suitable to generate multifunctional delivery materials via standard bio-conjugation techniques. Together, these data fully underline the great versatility of the proposed system.

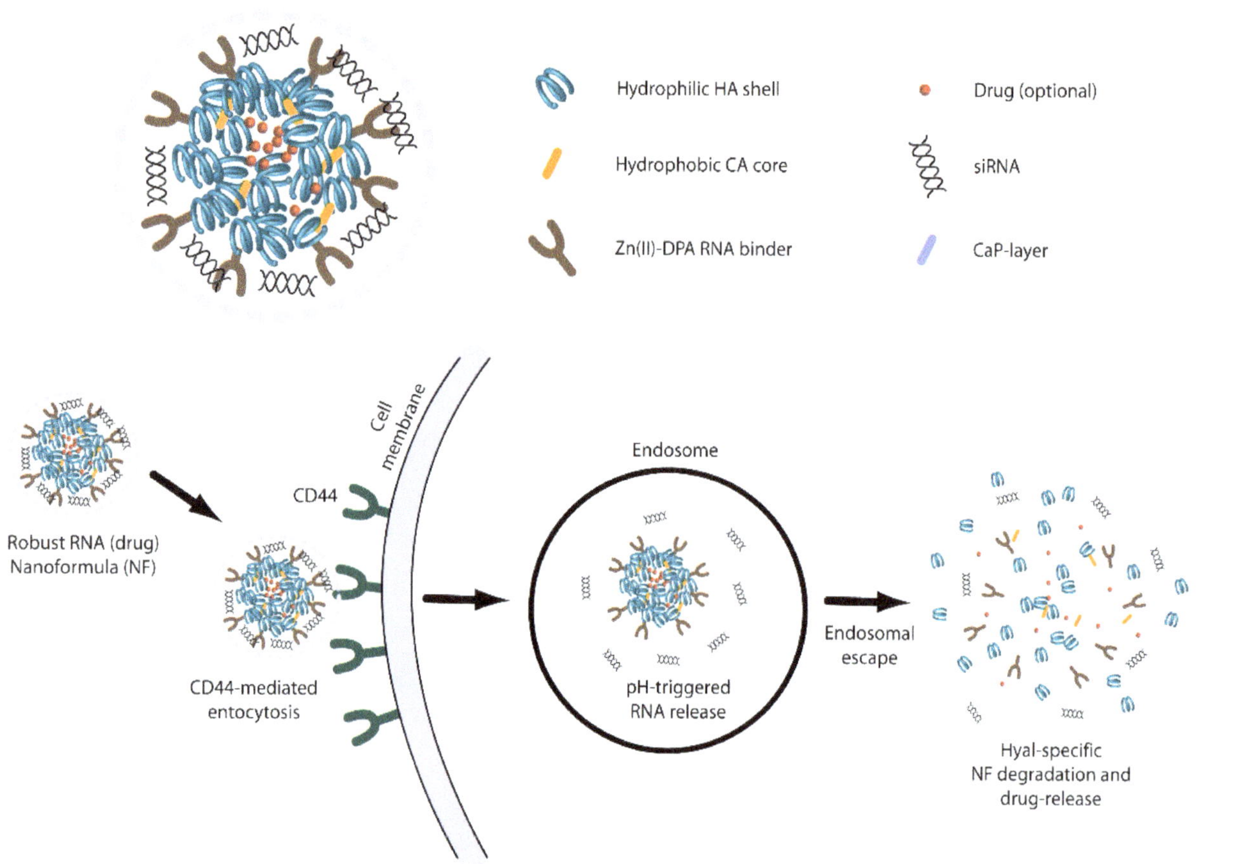

Figure 8. HA-based delivery for siRNAs proposed by Choi *et al.* [38].

3.2.5. Dextran Based Delivery

Thus far, several types of dextran-based NPs have been generated for NABD delivery. From the first attempt where dextran alone was bound via electrostatic interaction to NABDs, more complex solutions have been proposed. For example, Naeye *et al.* decorated the dextran/NABDs NPs with a hydrophilic PEG shell for the intravenous delivery route [130,131]. Whereas PEGylation did not prevent the partial dissociation of NABDs in human plasma, it was able to prevent NPs aggregation and the lysis of red blood cell. However, a clear charge dependent interaction with platelets and leukocytes was observed. On one hand, these results suggest the positive NPs features able to prevent a significant interaction with red blood cells; on the other hand, they indicate that further improvements aimed at the reduction of platelet interaction are necessary.

Another example of dextran NPs modification consisted in the conjugation of an acetal-modified dextran (Ac-DEX) with spermine as NABD delivery material [47]. The cationic nature of spermine improved the NPs complexation with the negatively charged siRNA and facilitated the cell membrane binding. Additionally, the Ac-DEX-spermine NPs resulted to be efficiently released from endosome, thus allowing the siRNA distribution with the cell cytoplasm [48]. This in turn resulted in an efficient silencing of the expression of the luciferase gene in HeLa cells expressing luciferase.

Very recently, dextran was used to encapsulate NPs containing NABD and PEI [49]. In particular, the authors encapsulated NABD/PEI with multiple layers of dextran sulfate (DexS) and poly-L-arginine (pArg). The idea below this approach is that the DexS/pArg layers can protect the NABD in the extracellular environment and prevent the known toxic effects of PEI. Once in the cellular environment, the DexS/pArg layers are progressively degraded, allowing PEI to mediate the known endosomal escape effect, thus favoring NABD translocation to the cytosol. As a cellular model it was used the HeLa cell line expressing the GFP against which the effect of an anti-GFP siRNA was tested. DexS/pArg capsule increased siRNA activity by about two to three times compared to the non-encapsulated PEI/siRNA. Notably, it was also shown that the minimal concentration of siRNA able to exert a visible effect was of 2.6 nM; in contrast, for the commonly available transfection reagents a concentration of about 100 nM was necessary. This data strongly support the potency of the system proposed that, in addition, displayed negligible cytotoxicity.

In a similar approach to that above reported, DexS and pArg were recently used to coat a core of hydroxyapatite (HY) particles [50]. The technique used is the layer-by-layer (LbL) self-assembly, a method based on the electrostatic interactions between positively (pArg) and negatively (DexS) charged polymers. A siRNA was inserted within the two most external pArg layers with an efficiency of about 97%. The resulting NPs yielded a size of 350 nm with a superficial charge of +43 mV. Initial studies performed using a fluorescently labeled siRNA indicated an efficient uptake into fibroblast with a clear cytoplasmic but not nuclear localization. The nuclear exclusion was explained with the size of the NPs, which was probably too big (350 nm) to enter the nucleus. NPs displayed a good endosome escape and a subsequent defoliation of the external pArg layers that allowed the siRNA homogeneous release into the cytoplasm. For the functional studies, a siRNA target against the secreted protein, acidic and rich in cysteine (SPARC) was chosen. SPARC is a calcium-binding matricellular protein that modulates cell/extracellular matrix and is involved in the generation of pathological tissue scarring and fibrosis. Thus, SPARC targeting may be a useful approach to modulate pathological fibrosis. The authors could show a significant reduction in the protein and mRNA levels of SPARC in fibroblasts; however, as it was not introduced a control containing a scrambled siRNA embedded into NPs, it is hard to evaluate the real effectiveness of the proposed system.

3.2.6. Pullulan Based Delivery

Due to its non-ionic nature, P functionalization with positive groups is necessary to allow the complexation with NABDs. P use as NABD delivery material has been recently undertaken by Wang *et al.* [52]. In this case, a low MW branched PEI (1000 Da) was grafted onto the backbone of P with succinic acid as a spacer; then PEI was conjugated with a peptide (FA) able to bind FR to generate the complex P-PEI-FA. FR is a tumor marker over-expressed on the cellular membrane of different malignant cells. P-PEI-FA showed significantly reduced toxicity compared to PEI in the HeLa cell line. P-PEI-FA complexation with nucleic acid molecules confirmed the modest toxicity in different cell tumor lines (Human hepatocellular carcinoma HepG2, HeLa and MCF7). The only exception occurred for the African green monkey kidney cell (COS-7) was, at the highest weight ratio tested, 50% reduction of cell viability was observed, thus suggesting a species-specific toxicity. In the HeLa cell line expressing the luciferase gene, P-PEI-FA complexed with an anti-luciferase siRNA, reduced luciferase expression by 82%, showing slightly higher gene silencing efficiency compared to P-PEI-siRNA (71%) and LF 2000-siRNA (76%). Notably, P-PEI-FA displayed also a targeting ability as it was highly effective in transducing the GFP expressing plasmid into the FR expressing HeLa cells but not in the FR non-expressing HepG2 cells. Together, these data suggest the feasibility of the use of P as NABD delivery material; tests in *in vivo* models will better define the efficacy of the proposed system.

In another recent work, it has been investigated the effects of serum on the transduction capability of NPs containing P chemically modified by spermine (pullulan-spermine, Ps) [53]. Serum is known to affect the transduction efficacy as it tends to be adsorbed by cationic molecules thus inducing the displacement of NABDs from NPs. Zhang *et al.* improved our knowledge about the effects of serum demonstrating that serum can significantly influence the NPs size and, thus, the transfection efficacy. In particular, the authors showed that low serum concentration (1.25% and 2.5%) in culture medium results in large particles of Ps-siRNA, while high serum concentration (10% and 40%) leads to small particles of Ps-siRNA. Compared to the small particles, large particles contain more amines eliciting a stronger proton sponge effect, which determines a more effective lysosomal escape of siRNA, and, thus, an improved siRNA biological effect. Notably, the large particles at the serum concentration of 1.25% display a transfection efficacy about seven times higher than that of the commercially available LF in the cancer cell line MDA-MB-231. Despite the higher transfection rate, large particles are more toxic than small ones. These interesting results prove the usefulness of P as NABDs delivery material and indicate the necessity to optimize NABDs also release in relation to serum-dependent effects.

3.2.7. Other Polysaccharide Based Delivery

Whereas the above-described polysaccharides have been often used as NABD delivery materials, other polysaccharides exist with a less frequent use. Among these, an interesting polysaccharide is represented by alginate. Due to its ability to form hydrogels, alginate has been mostly used for the controlled release of small molecule and protein. In this regard, a recent example is represented by the work of Laroui *et al.* [42]. Fab'-bearing TNFα-siRNA-loaded NPs were encapsulated into a matrix of CS/alginate and administered orally to mouse with induced colitis. Being colitis an inflammatory disease, the targeting of a pro-inflammatory cytokine, such as TNF-α, is considered beneficial to relieve inflammation. Moreover, the addition on the NPs of the Fab' portion of the F4/80 antibody, allows the targeting of macrophages, inflammatory cells sustaining colitis. The administration into the mouse colon of TNFα-siRNA-loaded NPs covered by Fab' was significantly more effective than the uncovered NPs in lessening colitis as shown by the attenuation of all parameters of colonic inflammation. Moreover, grafting Fab' onto the NPs improved the kinetics of endocytosis, as well as the macrophage-targeting ability. The use of alginates as releasing hydrogel was also considered in a similar recent work [43]. NPs were functionalized with molecules able to target CD98, an antigen whose over-expression on the surface of colonic epithelial cells and macrophages promotes the development and progression of colitis. NPs, loaded with an anti CD98 siRNA, were tested *in vitro* and *in vivo* embedded into CS and alginate hydrogel. *In vitro*, the NPs significantly reduced the levels of CD98 in the cell lines Colon-26 cells and RAW 264.7 macrophages, along with the reduction of inflammatory cytokines (TNFα, interleukin-6, and interleukin-12). Notably, administered orally to mouse with induced colitis, NPs also significantly reduced the severity of colitis. Collectively, these results show the relevant role of alginates hydrogel in delivering NABD loaded NPs.

Whereas alginates have been mainly considered as delivery hydrogel, only a limited number of works have used this molecule to prepare nanostructure for NABD release. In the past, alginate has been employed to reduce PEI cytotoxicity and Patnaik *et al.* showed successful siRNA transfection in various cell lines *in vitro* with PEI-alginate NPs containing PEI of different MW (25 kDa or 750 kDa) [132]. Very recently, NPs containing 25 kDa branched PEI and alginate were prepared and loaded with a siRNA targeting the mRNA of VEGFR-3 (vascular endothelial growth factor receptor-3) [44]. The VEGF-C/VEGFR-3 signaling pathway is involved in endothelial progenitor cells (EPCs) differentiation towards lymphoangiogenic cells; moreover, lymphangiogenesis is implicated in lymphatic metastasis of tumor cells. Thus, targeting VEGF-C/VEGFR-3 may be a useful approach to down regulate tumor lymphangiogenesis and lymphatic metastasis. NPs, with an average size and surface charge of 139.1 nm and 7.56 mV, respectively, were effective in reducing VEGFR-3 in purified EPCs. A limitation of these data is represented by the fact that only a semi-quantitative test was used to reveal the mRNA level of the siRNA target. Despite this fact, the authors could show the reduction in EPCs differentiation into lymphatic endothelial cells, together with a reduced attitude to proliferate and migrate. Confirmation in *in vivo* model is now necessary to deeply understanding the power of these alginate-based NPs.

Finally, among other polysaccharide able to release NABD, cyclodextrin has to be mentioned. CDs are naturally occurring cyclic oligosaccharides, consisting of α-1,4 linked D-glucopyranose units. They are characterized by an amphiphilic topology as they form structures hydrophilic on the exterior part and hydrophobic in the inner part. This configuration allows CD to carry hydrophobic molecules inside and outside hydrophilic compounds. CD are also indicated as drug delivery materials as they improve cellular delivery, resistance to endonucleases and can reduce immunogenicity. In NABD delivery, CD can be used as either backbone polymer or as derivative grafted on other polymers. Whereas a detailed description of CD as NABD delivery has been given recently elsewhere [1], here we will just concentrate on some notable examples. With regard to the use of CD as backbone polymer, it should be mentioned the works of Davis' group [133–135]. In this case, PEG and transferrin (Tf) were linked to the CD backbone. Tf was introduced to allow the targeting of cancer cells, which often overexpress the Tf receptor. The resulting PEG-Tf modified CD complexes, loaded with siRNA targeting the ribonucleotide reductase subunit M2 (RRM2), showed significant tumor growth inhibition in murine models [120]. Moreover, dose-escalating studies in non-human primates demonstrated that PEG-Tf modified CD complexes were well tolerated. This delivery approach named RONDEL™ ("RNAi/Oligonucleotide Nanoparticle Delivery", Calando Pharmaceuticals) was tested in the first in-human phase I clinical trial administered to patients with solid cancer refractory to standard therapies [121]. NPs were detected in post-treatment tumor biopsy sections together with a significant down-regulation of RRM2.

With regard to the use of CD as derivative molecule grafted on other polymers, we believe the works of Kulkarni *et al.* [136,137] represent relevant examples. CD was used as derivatization molecule for a neutral polymer, *i.e.*, poly(vinyl alcohol) (PVA) polymer bearing PEG and cholesteryl (Chol) or adamantane (Ad) modified grafts. The resulting CD-Chol/Ad-PVA-PEG NPs were able to form complexes with siRNA via electrostatic interactions. Worth note, the siRNA delivered by this NPs had comparable efficacy as the same siRNA released from PEI or LF 2000, while being far less toxic.

4. Conclusions

A significant limitation in the administration of conventional anticancer drugs is related to their high toxicity mostly dependent on the impossibility to discriminate between cancer and normal cells. Toxicity may also depend on the presence, in the pharmaceutical preparations, of molecules necessary to improve the poor solubility in water of many anticancer drugs. With regard to NABDs, an emerging class of drug with potential therapeutic value, despite evidence of their effectiveness, the practical use is limited. This mainly depends on the lack of materials able to overcome the delivery problems such as the fast degradation in biological fluids, the difficulties to cross cell membrane and the sequestration into cellular lysosomes. Thus, both for conventional anti-cancer drugs and NABDs, the development of optimal delivery materials may radically improve their therapeutic effectiveness. Additionally, appropriate delivery materials may allow the combined administration of conventional drugs and NABDs, possibly potentiating their therapeutic effects.

Among the several delivery materials studied thus far, polysaccharides represent very attractive molecules as they can be obtained in a reproducible way from natural sources, can undergo a wide range of chemical and enzymatic reactions, are biocompatible, biodegradable and have low immunogenic features. Together, these characteristics make polysaccharides excellent candidates for the realization of "smart" delivery systems capable to release, at the appropriate time and site of action the entrapped drugs. However, the development of such delivery systems is not an easy task due to the several aspects to be considered. First, it is important to optimize the polysaccharide-drug interaction, a variable that depends on the physical–chemical properties of the polysaccharide and of the drug. Second, it is necessary to optimize pharmacokinetics and pharmacodynamics through a careful evaluation of the ADME phenomenon. Third, it is necessary to confer a targeting ability to the polysaccharide carrier to limit the off targeting to healthy tissue. Finally, for the applications considering drug delivery from a gel, it is important to study the release mechanisms of the drug from the gels. Together, these considerations clearly indicate that only a multidisciplinary approach can successfully afford the task of the selection of the optimal delivery materials for the drug and application of interest. Whereas an ideal delivery system is not yet available, the examples reported in the present review indicate that many interesting options, based on the use of polysaccharide, are emerging. Thus, whereas additional research is required, the promising results obtained thus far fully justify further efforts in terms of both economic support and investigations in the field.

Acknowledgments

This work was in part supported by the "Fondazione Cassa di Risparmio of Trieste", by the "Fondazione Benefica Kathleen Foreman Casali of Trieste", by the "Beneficentia Stiftung" of Vaduz Liechtenstein and by the Italian Minister of Instruction, University and Research (MIUR), PRIN 2010–2011, [20109PLMH2]; FM was funded by CIRMMP (Consorzio Interuniversitario di Risonanze Magnetiche di Metallo-Proteine).

Author Contributions

Barbara Dapas, Mario Grassi and Gabriele Grassi proposed to the other authors the general topic of the manuscript. All the authors participated in the discussion for the selection of the main topics to be introduced in the manuscript giving advices in relation to their field of expertise.

Francesco Musiani took care about the drawing of the figures, the preparations of the tables and wrote the introduction.

Mario Grassi and Rossella Farra organized Section 2 with the Subsections 2.1–2.4.

Bianca Posocco, Eva Dreussi, Jacopo de Santa and Giuseppe Toffoli took care about Section 3.1. In particular, Bianca Posocco organized Sections 3.1.1–3.1.2; Eva Dreussi organized Sections 3.1.3–3.1.4; Jacopo de Santa organized Sections 3.1.5–3.1.6 and Giuseppe Toffoli organized Sections 3.1.7–3.1.8. Bianca Posocco, Eva Dreussi and Jacopo de Santa also worked on the assembly of the different sections of the manuscript.

Federica Tonon, Gabriele Grassi and Barbara Dapas took care about Section 3.2. In particular, Gabriele Grassi organized Sections 3.2.1–3.2.3; Federica Tonon organized Sections 3.2.4–3.2.5 and Barbara Dapas organized Sections 3.2.6–3.2.7. Barbara Dapas also worked on the whole manuscript suggesting stylistic and substantial improvements of the manuscript.

Michela Abrami organized the bibliography, the abbreviation list and wrote the conclusion.

Abbreviations

ATP binding cassette transporter B1	ABCB1
Absorption, Distribution, Metabolism and Elimination	ADME
Arabinogalactan	AG
Biotin	Bio
Calcium phosphate	CAP
Cluster determinant 44	CD44
Cisplatin	CDDP
Cyclodextrins	CDs
Carboxymethylcellulose	CMC
Irinotecan	CPT-11
Chitosan	CS
Daunorubicin	DNR
Doxorubicin	DOX
Double stranded RNAs	dsRNAs
Docetaxel	DTX
Extracellular matrix	ECM
Endothelial progenitor cells	EPCs
Enhanced permeability and retention effect	EPR
Folic acid	FA
5-fluorouracil	5-FU
Flutamide	FLT
Folic acid receptor	FR
Förster resonance energy transfer	FRET
Glycosaminoglycans	GAGs
Green fluorescence protein	GFP
γ-glutamylcyclotransferase	GGCT
Hyaluronic acid	HA
HA receptor for endocytosis	HARE
Hydroxypropyl methylcellulose	HPC
Hydroxyapatite	HY

Interpenetrated polymeric networks	IPN
Lipofectamine	LF
Low molecular weight chitosan	LMWC
Lymphatic vessel endothelial hyaluronan receptor-1	LYVE-1
Multi drug resistance	MDR
micro interfering RNAs	miRNAs
Mesoporous silica	MPS
Microparticles	MPs
Mauran	MR
Methotrexate	MTX
Molecular weight	MW
Nucleic acid based drugs	NABDs
Nanoparticles	NPs
Pullulan	P
Pullulan-spermine	Ps
p-aminobenzoic acid	PABA
Aminohippuric acid-diamine	PAH
Pharmacodynamics	PD
Photodynamic therapy	PDT
Polyethylene glycol	PEG
Polyethyleneimine	PEI
P-glycoprotein	Pgp
Pharmacokinetics	PK
Paclitaxel	PTX
Red fluorescence protein	RFP
Receptor for hyaluronate-mediated motility	RHAMM
RNA-induced silencing complexes	RISC
RNA interference	RNAi
Ribonucleotide reductase subunit M2	RRM2
small interfering RNAs	siRNAs
Acidic and rich in cysteine	SPARC
Transferrin	Tf
Tetrahydrocurcumin	THC
Temozolomide	TMZ
Vascular endothelial growth factor receptor-3	VEGFR-3

Cell lines

brain glioblastoma (human)	U87MG
breast cancer cell line (human)	MDA-MB-468NL
breast cancer cell line (human)	MDA-MB-231
breast cancer cells (human)	MCF7
cervical cancer cell line (human)	HeLa
colon adenocarcinoma (human)	HT29
colon cancer (human)	DLD1

colon carcinoma cells (human)	HCT116
colon cell line (human)	CRL1790
colon cells (human)	Colon-26
colorectal adenocarcinoma cells (human)	Caco-2
connective tissue fibroblast cells (mouse)	L929
fibroblast (murine)	NIH3T3
fibrosarcoma cell line (human)	HT1080
head and neck squamous cell carcinoma (human)	SCC7
hepatic cancer cell line (human)	JHH6
hepatocellular carcinoma cell (human)	HepG2
kidney cell (African green monkey)	COS-7
lung adenocarcinoma epithelial cell line (human)	A549
lung carcinoma cell line (human)	H1299
macrophage cells (murine)	RAW264.7
melanoma cell line (mouse)	B16F10
non-small-cell lung carcinoma (human)	NSCLC-NCIH358
ovary cancer cells (human)	OVCAR-3
ovary cell line (chinese hamster)	AA8
prostate tumor cells (human)	DU145
prostate tumor cells (human)	143B
umbilical vein endothelial (human)	HUVEC

Conflict of Interest

The authors declare no conflict of interest.

References

1. Raemdonck, K.; Martens, T.F.; Braeckmans, K.; Demeester, J.; De Smedt, S.C. Polysaccharide-based nucleic acid nanoformulations. *Adv. Drug Deliv. Rev.* **2013**, *65*, 1123–1147.
2. *Polysaccharides: Structural Diversity and Functiojnal Versatility*; Marcel Dekker: New York, NY, USA, 2005.
3. Goodarzi, N.; Varshochian, R.; Kamalinia, G.; Atyabi, F.; Dinarvand, R. A review of polysaccharide cytotoxic drug conjugates for cancer therapy. *Carbohydr. Polym.* **2013**, *92*, 1280–1293.
4. Mizrahy, S.; Peer, D. Polysaccharides as building blocks for nanotherapeutics. *Chem. Soc. Rev.* **2012**, *41*, 2623–2640.
5. Liu, Z.; Jiao, Y.; Wang, Y.; Zhou, C.; Zhang, Z. Polysaccharides-based nanoparticles as drug delivery systems. *Adv. Drug Deliv. Rev.* **2008**, *60*, 1650–1662.
6. Kosaraju, S.L. Colon targeted delivery systems: Review of polysaccharides for encapsulation and delivery. *Crit. Rev. Food Sci. Nutr.* **2005**, *45*, 251–258.
7. Lemarchand, C.; Gref, R.; Couvreur, P. Polysaccharide-decorated nanoparticles. *Eur. J. Pharm. Biopharm.* **2004**, *58*, 327–341.
8. Wen, Y.; Oh, J.K. Recent strategies to develop polysaccharide-based nanomaterials for biomedical applications. *Macromol. Rapid Commun.* **2014**, *35*, 1819–1832.

9. Alvarez-Lorenzo, C.; Blanco-Fernandez, B.; Puga, A.M.; Concheiro, A. Crosslinked ionic polysaccharides for stimuli-sensitive drug delivery. *Adv. Drug Deliv. Rev.* **2013**, *65*, 1148–1171.
10. Lapasin, R.; Pricl, S. *Rheology of Industrial Polysaccharides: Theory and Applications*; Blackie Academic & Professional, Chapman & Hall: London, UK, 1995.
11. Buschmann, M.D.; Merzouki, A.; Lavertu, M.; Thibault, M.; Jean, M.; Darras, V. Chitosans for delivery of nucleic acids. *Adv. Drug Deliv. Rev.* **2013**, *65*, 1234–1270.
12. Wang, Y.M.; Wang, P.G. Polysaccharide-based systems in drug and gene delivery. *Adv. Drug Deliv. Rev.* **2013**, *65*, 1121–1122.
13. Chen, M.C.; Mi, F.L.; Liao, Z.X.; Hsiao, C.W.; Sonaje, K.; Chung, M.F.; Hsu, L.W.; Sung, H.W. Recent advances in chitosan-based nanoparticles for oral delivery of macromolecules. *Adv. Drug Deliv. Rev.* **2013**, *65*, 865–879.
14. Ganguly, K.; Chaturvedi, K.; More, U.A.; Nadagouda, M.N.; Aminabhavi, T.M. Polysaccharide-based micro/nanohydrogels for delivering macromolecular therapeutics. *J. Control Release* **2014**, *193*, 162–173.
15. Matricardi, P.; Di, M.C.; Coviello, T.; Hennink, W.E.; Alhaique, F. Interpenetrating polymer networks polysaccharide hydrogels for drug delivery and tissue engineering. *Adv. Drug Deliv. Rev.* **2013**, *65*, 1172–1187.
16. Lee, E.; Lee, J.; Lee, I.H.; Yu, M.; Kim, H.; Chae, S.Y.; Jon, S. Conjugated chitosan as a novel platform for oral delivery of paclitaxel. *J. Med. Chem.* **2008**, *51*, 6442–6449.
17. Lee, E.; Kim, H.; Lee, I.H.; Jon, S. *In vivo* antitumor effects of chitosan-conjugated docetaxel after oral administration. *J. Control Release* **2009**, *140*, 79–85.
18. Kim, K.; Kim, J.H.; Park, H.; Kim, Y.S.; Park, K.; Nam, H.; Lee, S.; Park, J.H.; Park, R.W.; Kim, I.S.; *et al*. Tumor-homing multifunctional nanoparticles for cancer theragnosis: Simultaneous diagnosis, drug delivery, and therapeutic monitoring. *J. Control. Release* **2010**, *146*, 219–227.
19. Chen, K.J.; Chiu, Y.L.; Chen, Y.M.; Ho, Y.C.; Sung, H.W. Intracellularly monitoring/imaging the release of doxorubicin from pH-responsive nanoparticles using F+Ârster resonance energy transfer. *Biomaterials* **2011**, *32*, 2586–2592.
20. Lee, S.J.; Koo, H.; Jeong, H.; Huh, M.S.; Choi, Y.; Jeong, S.Y.; Byun, Y.; Choi, K.; Kim, K.; Kwon, I.C. Comparative study of photosensitizer loaded and conjugated glycol chitosan nanoparticles for cancer therapy. *J. Control Release* **2011**, *152*, 21–29.
21. Elgindy, N.; Elkhodairy, K.; Molokhia, A.; Elzoghby, A. Biopolymeric microparticles combined with lyophilized monophase dispersions for controlled flutamide release. *Int. J. Pharm.* **2011**, *411*, 113–120.
22. Moura, M.J.; Gil, M.H.; Figueiredo, M.M. Delivery of cisplatin from thermosensitive co-cross-linked chitosan hydrogels. *Eur. Polymer J.* **2013**, *49*, 2504–2510.
23. Darvishi, M.H.; Nomani, A.; Amini, M.; Shokrgozar, M.A.; Dinarvand, R. Novel biotinylated chitosan-graft-polyethyleneimine copolymer as a targeted non-viral vector for anti-EGF receptor siRNA delivery in cancer cells. *Int. J. Pharm.* **2013**, *456*, 408–416.
24. Yang, F.; Huang, W.; Li, Y.; Liu, S.; Jin, M.; Wang, Y.; Jia, L.; Gao, Z. Anti-tumor effects in mice induced by survivin-targeted siRNA delivered through polysaccharide nanoparticles. *Biomaterials* **2013**, *34*, 5689–5699.
25. Park, S.; Jeong, E.J.; Lee, J.; Rhim, T.; Lee, S.K.; Lee, K.Y. Preparation and characterization of nonaarginine-modified chitosan nanoparticles for siRNA delivery. *Carbohydr. Polym.* **2013**, *92*, 57–62.

26. Liao, Z.X.; Hsiao, C.W.; Ho, Y.C.; Chen, H.L.; Sung, H.W. Disulfide bond-conjugated dual PEGylated siRNAs for prolonged multiple gene silencing. *Biomaterials* **2013**, *34*, 6930–6937.
27. Liao, Z.X.; Peng, S.F.; Chiu, Y.L.; Hsiao, C.W.; Liu, H.Y.; Lim, W.H.; Lu, H.M.; Sung, H.W. Enhancement of efficiency of chitosan-based complexes for gene transfection with poly(gamma-glutamic acid) by augmenting their cellular uptake and intracellular unpackage. *J. Control Release* **2014**, *193*, 304–315.
28. Yhee, J.Y.; Song, S.; Lee, S.J.; Park, S.; Kim, K.; Kim, M.G.; Son, S.; Koo, H.; Kwon, I.C.; Jeong, J.H.; *et al*. Cancer-targeted MDR-1 siRNA delivery using self-cross-linked glycol chitosan nanoparticles to overcome drug resistance. *J. Control Release* **2014**, *198C*, 1–9.
29. Chen, Z.; Li, Z.; Lin, Y.; Yin, M.; Ren, J.; Qu, X. Bioresponsive hyaluronic acid-capped mesoporous silica nanoparticles for targeted drug delivery. *Chem. Eur. J.* **2013**, *19*, 1778–1783.
30. Cai, S.; Thati, S.; Bagby, T.R.; Diab, H.M.; Davies, N.M.; Cohen, M.S.; Forrest, M.L. Localized doxorubicin chemotherapy with a biopolymeric nanocarrier improves survival and reduces toxicity in xenografts of human breast cancer. *J. Control. Release* **2010**, *146*, 212–218.
31. Xie, Y.; Aillon, K.L.; Cai, S.; Christian, J.M.; Davies, N.M.; Berkland, C.J.; Forrest, M.L. Pulmonary delivery of cisplatin-hyaluronan conjugates via endotracheal instillation for the treatment of lung cancer. *Int. J. Pharm.* **2010**, *392*, 156–163.
32. Thomas, R.G.; Moon, M.; Lee, S.; Jeong, Y.Y. Paclitaxel loaded hyaluronic acid nanoparticles for targeted cancer therapy: *In vitro* and *in vivo* analysis. *Int. J. Biol. Macromol.* **2015**, *72*, 510–518.
33. Li, N.; Chen, Y.; Zhang, Y.M.; Yang, Y.; Su, Y.; Chen, J.T.; Liu, Y. Polysaccharide-gold nanocluster supramolecular conjugates as a versatile platform for the targeted delivery of anticancer drugs. *Sci. Rep.* **2014**, *4*, doi:10.1038/srep04164.
34. Choi, K.M.; Jang, M.; Kim, J.H.; Ahn, H.J. Tumor-specific delivery of siRNA using supramolecular assembly of hyaluronic acid nanoparticles and 2b RNA-binding protein/siRNA complexes. *Biomaterials* **2014**, *35*, 7121–7132.
35. Ran, R.; Liu, Y.; Gao, H.; Kuang, Q.; Zhang, Q.; Tang, J.; Huang, K.; Chen, X.; Zhang, Z.; He, Q. Enhanced gene delivery efficiency of cationic liposomes coated with PEGylated hyaluronic acid for anti P-glycoprotein siRNA: A potential candidate for overcoming multi-drug resistance. *Int. J. Pharm.* **2014**, *477*, 590–600.
36. Ran, R.; Liu, Y.; Gao, H.; Kuang, Q.; Zhang, Q.; Tang, J.; Fu, H.; Zhang, Z.; He, Q. PEGylated hyaluronic acid-modified liposomal delivery system with anti-gamma-glutamylcyclotransferase siRNA for drug-resistant MCF-7 breast cancer therapy. *J. Pharm. Sci.* **2014**, *104*, 476–484.
37. Lee, M.S.; Lee, J.E.; Byun, E.; Kim, N.W.; Lee, K.; Lee, H.; Sim, S.J.; Lee, D.S.; Jeong, J.H. Target-specific delivery of siRNA by stabilized calcium phosphate nanoparticles using dopa-hyaluronic acid conjugate. *J. Control Release* **2014**, *192*, 122–130.
38. Choi, K.Y.; Silvestre, O.F.; Huang, X.; Min, K.H.; Howard, G.P.; Hida, N.; Jin, A.J.; Carvajal, N.; Lee, S.W.; Hong, J.I.; *et al.* Versatile RNA interference nanoplatform for systemic delivery of RNAs. *ACS Nano* **2014**, *8*, 4559–4570.
39. Metaxa, A.F.; Efthimiadou, E.K.; Boukos, N.; Fragogeorgi, E.A.; Loudos, G.; Kordas, G. Hollow microspheres based on—Folic acid modified—Hydroxypropyl Cellulose and synthetic multi-responsive bio-copolymer for targeted cancer therapy: Controlled release of daunorubicin, *in vitro* and *in vivo* studies. *J. Colloid Interface Sci.* **2014**, *435*, 171–181.

40. Thipapun Plyduang, L.L. Carboxymethylcellulose-tetrahydrocurcumin conjugates for colon-specific delivery of a novel anti-cancer agent, 4-amino tetrahydrocurcumin. *Eur. J. Pharm. Biopharm.* **2014**, *88*, 351–360.
41. Wu, W.; Aiello, M.; Zhou, T.; Berliner, A.; Banerjee, P.; Zhou, S. In-situ immobilization of quantum dots in polysaccharide-based nanogels for integration of optical pH-sensing, tumor cell imaging, and drug delivery. *Biomaterials* **2010**, *31*, 3023–3031.
42. Laroui, H.; Viennois, E.; Xiao, B.; Canup, B.S.; Geem, D.; Denning, T.L.; Merlin, D. Fab'-bearing siRNA TNFalpha-loaded nanoparticles targeted to colonic macrophages offer an effective therapy for experimental colitis. *J. Control. Release* **2014**, *186*, 41–53.
43. Xiao, B.; Laroui, H.; Viennois, E.; Ayyadurai, S.; Charania, M.A.; Zhang, Y.; Zhang, Z.; Baker, M.T.; Zhang, B.; Gewirtz, A.T.; *et al*. Nanoparticles with surface antibody against CD98 and carrying CD98 small interfering RNA reduce colitis in mice. *Gastroenterology* **2014**, *146*, 1289–1300.
44. Li, T.; Wang, G.D.; Tan, Y.Z.; Wang, H.J. Inhibition of lymphangiogenesis of endothelial progenitor cells with VEGFR-3 siRNA delivered with PEI-alginate nanoparticles. *Int. J. Biol. Sci.* **2014**, *10*, 160–170.
45. Luo, G.F.; Xu, X.D.; Zhang, J.; Yang, J.; Gong, Y.H.; Lei, Q.; Jia, H.Z.; Li, C.; Zhuo, R.X.; Zhang, X.Z. Encapsulation of an adamantane-doxorubicin prodrug in pH-responsive polysaccharide capsules for controlled release. *ACS Appl. Mater. Interfaces* **2012**, *4*, 5317–5324.
46. Pramod, P.S.; Shah, R.; Chaphekar, S.; Balasubramanian, N.; Jayakannan, M. Polysaccharide nano-vesicular multidrug carriers for synergistic killing of cancer cells. *Nanoscale* **2014**, *6*, 11841–11855.
47. Thomas, T.; Thomas, T.J. Polyamines in cell growth and cell death: Molecular mechanisms and therapeutic applications. *Cell Mol. Life Sci.* **2001**, *58*, 244–258.
48. Cohen, J.L.; Schubert, S.; Wich, P.R.; Cui, L.; Cohen, J.A.; Mynar, J.L.; Frechet, J.M. Acid-degradable cationic dextran particles for the delivery of siRNA therapeutics. *Bioconjug. Chem.* **2011**, *22*, 1056–1065.
49. Ganas, C.; Weiss, A.; Nazarenus, M.; Rosler, S.; Kissel, T.; Rivera, G.P.; Parak, W.J. Biodegradable capsules as non-viral vectors for *in vitro* delivery of PEI/siRNA polyplexes for efficient gene silencing. *J. Control Release* **2014**, *196*, 132–138.
50. Tan, Y.F.; Mundargi, R.C.; Chen, M.H.; Lessig, J.; Neu, B.; Venkatraman, S.S.; Wong, T.T. Layer-by-layer nanoparticles as an efficient siRNA delivery vehicle for SPARC silencing. *Small* **2014**, *10*, 1790–1798.
51. Lee, S.J.; Hong, G.Y.; Jeong, Y.I.; Kang, M.S.; Oh, J.S.; Song, C.E.; Lee, H.C. Paclitaxel-incorporated nanoparticles of hydrophobized polysaccharide and their antitumor activity. *Int. J. Pharm.* **2012**, *433*, 121–128.
52. Wang, J.; Dou, B.; Bao, Y. Efficient targeted pDNA/siRNA delivery with folate-low-molecular-weight polyethyleneimine-modified pullulan as non-viral carrier. *Mater. Sci. Eng. C Mater. Biol. Appl.* **2014**, *34*, 98–109.
53. Zhang, W.; Liu, J.; Tabata, Y.; Meng, J.; Xu, H. The effect of serum in culture on RNAi efficacy through modulation of polyplexes size. *Biomaterials* **2014**, *35*, 567–577.
54. De Rosa, E.; Borselli, C.; Netti, P. Transport of large molecules in hyaluronic acid-based membranes and solution. *J. Membr. Sci.* **2006**, *273*, 84–88.

55. Turco, G.; Donati, I.; Grassi, M.; Marchioli, G.; Lapasin, R.; Paoletti, S. Mechanical spectroscopy and relaxometry on alginate hydrogels: A comparative analysis for structural characterization and network mesh size determination. *Biomacromolecules* **2011**, *12*, 1272–1282.
56. Grant, G.; Morris, E.; Rees, D.; Smith, P.; Thom, D. Biological interactions between polysaccharides and divalent cations: The egg-box model. *FEBS Lett.* **1973**, *32*, 195–198.
57. Smidsrod, O. Molecular basis for some physical properties of alginates in the gel state. *Faraday Discuss.* **1974**, *57*, 263–274.
58. Grassi, M.; Grassi, G.; Lapasin, R.; Colombo, I. *Understanding Drug Release and Absorption Mechanisms: A Physical and Mathematical Aproach*; CRC Press: Boca Raton, FL, USA, 2007.
59. Lin, C.C.; Metters, A.T. Hydrogels in controlled release formulations: Network design and mathematical modeling. *Adv. Drug Deliv. Rev.* **2006**, *58*, 1379–1408.
60. Abrami, M.; D'Agostino, I.; Milcovich, G.; Fiorentino, S.; Farra, R.; Asaro, F.; Lapasin, R.; Grassi, G.; Grassi, M. Physical characterization of alginate-Pluronic F127 gel for endoluminal NABDs delivery. *Soft Matter* **2014**, *10*, 729–737.
61. Grassi, M.; Grassi, G. Application of mathematical modeling in sustained release delivery systems. *Expert Opin. Drug Deliv.* **2014**, *11*, 1299–1321.
62. Alvarez-Lorenzo, C.; Concheiro, A. Bioinspired drug delivery systems. *Curr. Opin. Biotechnol.* **2013**, *24*, 1167–1173.
63. Lima, A.C.; Song, W.; Blanco-Fernandez, B.; Alvarez-Lorenzo, C.; Mano, J.F. Synthesis of temperature-responsive dextran-MA/PNIPAAm particles for controlled drug delivery using superhydrophobic surfaces. *Pharm. Res* .**2011**, *28*, 1294–1305.
64. Hildebrand, M. Pharmakokinetic und Arzneimittelentwicklung. *Deutske Apotheker* **1994**, *Z.4*, 23–32. (In German)
65. Muller, K.; Fedosov, D.A.; Gompper, G. Margination of micro- and nano-particles in blood flow and its effect on drug delivery. *Sci. Rep.* **2014**, *4*, doi:10.1038/srep04871.
66. Gratton, S.E.; Ropp, P.A.; Pohlhaus, P.D.; Luft, J.C.; Madden, V.J.; Napier, M.E.; DeSimone, J.M. The effect of particle design on cellular internalization pathways. *Proc. Natl. Acad. Sci. USA* **2008**, *105*, 11613–11618.
67. Carboni, E.; Tschudi, K.; Nam, J.; Lu, X.; Ma, A.W. Particle margination and its implications on intravenous anticancer drug delivery. *AAPS PharmSciTech* **2014**, *15*, 762–771.
68. Sikes, R.A. Chemistry and pharmacology of anticancer drugs. *Br. J. Cancer* **2007**, *97*, 1713.
69. Nussbaumer, S.; Bonnabry, P.; Veuthey, J.L.; Fleury-Souverain, S. Analysis of anticancer drugs: A review. *Talanta* **2011**, *85*, 2265–2289.
70. Torchilin, V.P. Targeted polymeric micelles for delivery of poorly soluble drugs. *Cell Mol. Life Sci.* **2004**, *61*, 2549–2559.
71. Park, J.H.; Saravanakumar, G.; Kim, K.; Kwon, I.C. Targeted delivery of low molecular drugs using chitosan and its derivatives. *Adv. Drug Deliv. Rev.* **2010**, *62*, 28–41.
72. Schmaljohann, D. Thermo- and pH-responsive polymers in drug delivery. *Adv. Drug Deliv. Rev.* **2006**, *58*, 1655–1670.
73. Cheng, Y.; Xu, T. The effect of dendrimers on the pharmacodynamic and pharmacokinetic behaviors of non-covalently or covalently attached drugs. *Eur. J. Med. Chem.* **2008**, *43*, 2291–2297.

74. Sriraman, S.K.; Aryasomayajula, B.; Torchilin, V.P. Barriers to drug delivery in solid tumors. *Tissue Barriers* **2014**, *2*, doi:10.4161/tisb.29528.
75. Liechty, W.B.; Kryscio, D.R.; Slaughter, B.V.; Peppas, N.A. Polymers for drug delivery systems. *Annu. Rev. Chem. Biomol. Eng.* **2010**, *1*, 149–173.
76. *Biopolymers In Drug Delivery: Recent Advances and Challenges*; Bentham Science Publishers: Sharjah, United Arab Emirates, 2012.
77. Hennenfent, K.L.; Govindan, R. Novel formulations of taxanes: A review. Old wine in a new bottle? *Ann. Oncol.* **2006**, *17*, 735–749.
78. Weiszhár, Z.; Czúcz, J.; Révész, C.; Rosivall, L.; Szebeni, J.; Rozsnyay, Z. Complement activation by polyethoxylated pharmaceutical surfactants: Cremophor-EL, Tween-80 and Tween-20. *Eur. J. Pharm. Sci.* **2012**, *45*, 492–498.
79. Benet, L.Z.; Broccatelli, F.; Oprea, T.I. BDDCS applied to over 900 drugs. *AAPS J.* **2011**, *13*, 519–547.
80. Raveendran, S.; Poulose, A.C.; Yoshida, Y.; Maekawa, T.; Kumar, D.S. Bacterial exopolysaccharide based nanoparticles for sustained drug delivery, cancer chemotherapy and bioimaging. *Carbohydr. Polym.* **2013**, *91*, 22–32.
81. Pinhassi, R.I.; Assaraf, Y.G.; Farber, S.; Stark, M.; Ickowicz, D.; Drori, S.; Domb, A.J.; Livney, Y.D. Arabinogalactan-folic acid-drug conjugate for targeted delivery and target-activated release of anticancer drugs to folate receptor-overexpressing cells. *Biomacromolecules* **2010**, *11*, 294–303.
82. Dosio, F.; Stella, B.; Arpicco, S.; Cattel, L. Macromolecules as taxane delivery systems. *Expert Opin. Drug Deliv.* **2011**, *8*, 33–55.
83. Ryu, J.H.; Koo, H.; Sun, I.C.; Yuk, S.H.; Choi, K.; Kim, K.; Kwon, I.C. Tumor-targeting multi-functional nanoparticles for theragnosis: New paradigm for cancer therapy. *Adv. Drug Deliv. Rev.* **2012**, *64*, 1447–1458.
84. Celli, J.P.; Spring, B.Q.; Rizvi, I.; Evans, C.L.; Samkoe, K.S.; Verma, S.; Pogue, B.W.; Hasan, T. Imaging and photodynamic therapy: Mechanisms, monitoring, and optimization. *Chem. Rev.* **2010**, *110*, 2795–2838.
85. Neri, R. Pharmacology and pharmacokinetics of flutamide. *Urology* **1989**, *34*, 19–21.
86. Bhattarai, N.; Gunn, J.; Zhang, M. Chitosan-based hydrogels for controlled, localized drug delivery. *Adv. Drug Deliv. Rev.* **2010**, *62*, 83–99.
87. Descalzo, A.B.; Martínez-Máñez, R.; Sancenon, F.; Hoffmann, K.; Rurack, K. The supramolecular chemistry of organic-inorganic hybrid materials. *Angew. Chem. Int. Ed. Engl.* **2006**, *45*, 5924–5948.
88. Lu, J.; Liong, M.; Li, Z.; Zink, J.I.; Tamanoi, F. Biocompatibility, biodistribution, and drug-delivery efficiency of mesoporous silica nanoparticles for cancer therapy in animals. *Small* **2010**, *6*, 1794–1805.
89. Jothy, S. CD44 and its partners in metastasis. *Clin. Exp. Metastasis* **2003**, *20*, 195–201.
90. Park, J.H.; Cho, H.J.; Yoon, H.Y.; Yoon, I.S.; Ko, S.H.; Shim, J.S.; Cho, J.H.; Park, J.H.; Kim, K.; Kwon, I.C.; *et al.* Hyaluronic acid derivative-coated nanohybrid liposomes for cancer imaging and drug delivery. *J. Control Release* **2014**, *174*, 98–108.
91. Tzircotis, G.; Thorne, R.F.; Isacke, C.M. Chemotaxis towards hyaluronan is dependent on CD44 expression and modulated by cell type variation in CD44-hyaluronan binding. *J. Cell Sci.* **2005**, *118*, 5119–5128.

92. Eftink, M.R.; Andy, M.L.; Bystrom, K.; Perlmutter, H.D.; Kristol, D.S. Cyclodextrin inclusion complexes: Studies of the variation in the size of alicyclic guests. *J. Am. Chem. Soc.* **1989**, *111*, 6765–6772.
93. Connor, E.E.; Mwamuka, J.; Gole, A.; Murphy, C.J.; Wyatt, M.D. Gold nanoparticles are taken up by human cells but do not cause acute cytotoxicity. *Small* **2005**, *1*, 325–327.
94. Nune, S.K.; Chanda, N.; Shukla, R.; Katti, K.; Kulkarni, R.R.; Thilakavathi, S.; Mekapothula, S.; Kannan, R.; Katti, K.V. Green nanotechnology from tea: Phytochemicals in tea as building blocks for production of biocompatible gold nanoparticles. *J. Mater. Chem.* **2009**, *19*, 2912–2920.
95. Alvarez, M.M.; Khoury, J.T.; Schaaff, T.G.; Shafigullin, M.N.; Vezmar, I.; Whetten, R.L. Optical absorption spectra of nanocrystal gold molecules. *J. Phys. Chem. B* **1997**, *101*, 3706–3712.
96. Varnavski, O.; Ramakrishna, G.; Kim, J.; Lee, D.; Goodson, T. Critical size for the observation of quantum confinement in optically excited gold clusters. *J. Am. Chem. Soc.* **2010**, *132*, 16–17.
97. Panigrahi, S.; Basu, S.; Praharaj, S.; Pande, S.; Jana, S.; Pal, A.; Ghosh, S.K.; Pal, T. Synthesis and size-selective catalysis by supported gold nanoparticles: Study on heterogeneous and homogeneous catalytic process. *J. Phys. Chem. C* **2007**, *111*, 4596–4605.
98. Citrin, P.; Wertheim, G. Photoemission from surface-atom core levels, surface densities of states, and metal-atom clusters: A unified picture. *Phys. Rev. B* **1983**, *27*, 3176–3200.
99. Kamel, S.; Ali, N.; Jahangir, K.; Shah, S.M.; El-Gendy, A.A. Pharmaceutical significance of cellulose: A review. *Express Polym. Lett.* **2008**, *2*, 758–778.
100. Torchilin, V. Multifunctional and stimuli-sensitive pharmaceutical nanocarriers. *Eur. J. Pharm. Biopharm.* **2009**, *71*, 431–444.
101. Sharma, R.A.; Gescher, A.J.; Steward, W.P. Curcumin: The story so far. *Eur. J. Cancer* **2005**, *41*, 1955–1968.
102. Grassi, G.; Marini, J.C. Ribozymes: Structure, function, and potential therapy for dominant genetic disorders. *Ann. Med.* **1996**, *28*, 499–510.
103. Grassi, G.; Dawson, P.; Guarnieri, G.; Kandolf, R.; Grassi, M. Therapeutic potential of hammerhead ribozymes in the treatment of hyper-proliferative diseases. *Curr. Pharm. Biotechnol.* **2004**, *5*, 369–386.
104. Grassi, G.; Schneider, A.; Engel, S.; Racchi, G.; Kandolf, R.; Kuhn, A. Hammerhead ribozymes targeted against cyclin E and E2F1 cooperate to down-regulate coronary smooth muscle cell proliferation. *J. Gene Med.* **2005**, *7*, 1223–1234.
105. Scaggiante, B.; Dapas, B.; Farra, R.; Grassi, M.; Pozzato, G.; Giansante, C.; Fiotti, N.; Grassi, G. Improving siRNA bio-distribution and minimizing side effects. *Curr. Drug Metab.* **2011**, *12*, 11–23.
106. Scaggiante, B.; Dapas, B.; Farra, R.; Tonon, F.; Abrami, M.; Grassi, M.; Musiani, F.; Zanconati, F.; Pozzato, G.; Grassi, G. Translation elongation. In *Translation and its Regulation in Cancer Biology and Medicine*; Parsyan, A., Ed.; Springer: Berlin/Heidelberg, Germany, 2014; pp. 241–265.
107. Dapas, B.; Farra, R.; Grassi, M.; Giansante, C.; Fiotti, N.; Uxa, L.; Rainaldi, G.; Mercatanti, A.; Colombatti, A.; Spessotto, P.; *et al.* Role of E2F1-cyclin E1-cyclin E2 circuit in human coronary smooth muscle cell proliferation and therapeutic potential of its downregulation by siRNAs. *Mol. Med.* **2009**, *15*, 297–306.
108. Farra, R.; Dapas, B.; Pozzato, G.; Scaggiante, B.; Agostini, F.; Zennaro, C.; Grassi, M.; Rosso, N.; Giansante, C.; Fiotti, N.; *et al.* Effects of E2F1-cyclin E1-E2 circuit down regulation in hepatocellular carcinoma cells. *Dig. Liver Dis.* **2011**, *43*, 1006–1014.

109. Werth, D.; Grassi, G.; Konjer, N.; Dapas, B.; Farra, R.; Giansante, C.; Kandolf, R.; Guarnieri, G.; Nordheim, A.; Heidenreich, O. Proliferation of human primary vascular smooth muscle cells depends on serum response factor. *Eur. J. Cell Biol.* **2010**, *89*, 216–224.
110. Farra, R.; Dapas, B.; Pozzato, G.; Giansante, C.; Heidenreich, O.; Uxa, L.; Zennaro, C.; Guarnieri, G.; Grassi, G. Serum response factor depletion affects the proliferation of the hepatocellular carcinoma cells HepG2 and JHH6. *Biochimie* **2010**, *92*, 455–463.
111. Grassi, M.; Cavallaro, G.; Scirè, S.; Scaggiante, B.; Daps, B.; Farra, R.; Baiz, D.; Giansante, C.; Guarnieri, G.; Perin, D.; *et al.* Current strategies to improve the efficacy and the delivery of nucleic acid based drugs. *Curr. Signal Transduct. Ther.* **2010**, *5*, 92–120.
112. Lang, C.; Sauter, M.; Szalay, G.; Racchi, G.; Grassi, G.; Rainaldi, G.; Mercatanti, A.; Lang, F.; Kandolf, R.; Klingel, K. Connective tissue growth factor: A crucial cytokine-mediating cardiac fibrosis in ongoing enterovirus myocarditis. *J. Mol. Med.* **2008**, *86*, 49–60.
113. Agostini, F.; Dapas, B.; Farra, R.; Grassi, M.; Racchi, G.; Klingel, K.; Kandolf, R.; Heidenreich, O.; Mercatahnti, A.; rainaldi, G.; *et al.* Potential applications of small interfering RNAs in the cardiovascular field. *Drug Future* **2006**, *31*, 513–525.
114. Huang, Y.; Hong, J.; Zheng, S.; Ding, Y.; Guo, S.; Zhang, H.; Zhang, X.; Du, Q.; Liang, Z. Elimination pathways of systemically delivered siRNA. *Mol. Ther.* **2011**, *19*, 381–385.
115. Layzer, J.M.; McCaffrey, A.P.; Tanner, A.K.; Huang, Z.; Kay, M.A.; Sullenger, B.A. *In vivo* activity of nuclease-resistant siRNAs. *RNA* **2004**, *10*, 766–771.
116. Jackson, A.L.; Linsley, P.S. Recognizing and avoiding siRNA off-target effects for target identification and therapeutic application. *Nat. Rev. Drug Discov.* **2010**, *9*, 57–67.
117. Wisse, E.; Jacobs, F.; Topal, B.; Frederik, P.; De, G.B. The size of endothelial fenestrae in human liver sinusoids: Implications for hepatocyte-directed gene transfer. *Gene Ther.* **2008**, *15*, 1193–1199.
118. Kaneda, Y. Gene therapy: A battle against biological barriers. *Curr. Mol. Med.* **2001**, *1*, 493–499.
119. Brown, M.D.; Schatzlein, A.G.; Uchegbu, I.F. Gene delivery with synthetic (non viral) carriers. *Int. J. Pharm.* **2001**, *229*, 1–21.
120. Rahman, M.A.; Amin, A.R.; Wang, X.; Zuckerman, J.E.; Choi, C.H.; Zhou, B.; Wang, D.; Nannapaneni, S.; Koenig, L.; Chen, Z.; *et al.* Systemic delivery of siRNA nanoparticles targeting RRM2 suppresses head and neck tumor growth. *J. Control. Release* **2012**, *159*, 384–392.
121. Davis, M.E.; Zuckerman, J.E.; Choi, C.H.; Seligson, D.; Tolcher, A.; Alabi, C.A.; Yen, Y.; Heidel, J.D.; Ribas, A. Evidence of RNAi in humans from systemically administered siRNA via targeted nanoparticles. *Nature* **2010**, *464*, 1067–1070.
122. Koping-Hoggard, M.; Tubulekas, I.; Guan, H.; Edwards, K.; Nilsson, M.; Varum, K.M.; Artursson, P. Chitosan as a nonviral gene delivery system. Structure-property relationships and characteristics compared with polyethylenimine *in vitro* and after lung administration *in vivo*. *Gene Ther.* **2001**, *8*, 1108–1121.
123. Lavertu, M.; Methot, S.; Tran-Khanh, N.; Buschmann, M.D. High efficiency gene transfer using chitosan/DNA nanoparticles with specific combinations of molecular weight and degree of deacetylation. *Biomaterials* **2006**, *27*, 4815–4824.
124. Thibault, M.; Nimesh, S.; Lavertu, M.; Buschmann, M.D. Intracellular trafficking and decondensation kinetics of chitosan-pDNA polyplexes. *Mol. Ther.* **2010**, *18*, 1787–1795.

125. Malmo, J.; Sorgard, H.; Varum, K.M.; Strand, S.P. siRNA delivery with chitosan nanoparticles: Molecular properties favoring efficient gene silencing. *J. Control Release* **2012**, *158*, 261–268.
126. Liao, Z.X.; Ho, Y.C.; Chen, H.L.; Peng, S.F.; Hsiao, C.W.; Sung, H.W. Enhancement of efficiencies of the cellular uptake and gene silencing of chitosan/siRNA complexes via the inclusion of a negatively charged poly(gamma-glutamic acid). *Biomaterials* **2010**, *31*, 8780–8788.
127. Shen, Y.; Wang, B.; Lu, Y.; Ouahab, A.; Li, Q.; Tu, J. A novel tumor-targeted delivery system with hydrophobized hyaluronic acid-spermine conjugates (HHSCs) for efficient receptor-mediated siRNA delivery. *Int. J. Pharm.* **2011**, *414*, 233–243.
128. Takei, Y.; Maruyama, A.; Kawano, S.; Nishimura, Y.; Asayama, S.; Nogawa, M.; Ikejima, K.; Hori, M.; Akaike, T.; Lemasters, J.J.; *et al.* Targeted gene transfer to sinusoidal endothelial cells and expression *in vivo*. *Transplant Proc.* **1999**, *31*, 790–791.
129. Choi, K.Y.; Silvestre, O.F.; Huang, X.; Hida, N.; Liu, G.; Ho, D.N.; Lee, S.; Lee, S.W.; Hong, J.I.; Chen, X. A nanoparticle formula for delivering siRNA or miRNAs to tumor cells in cell culture and *in vivo*. *Nat. Protoc.* **2014**, *9*, 1900–1915.
130. Naeye, B.; Raemdonck, K.; Remaut, K.; Sproat, B.; Demeester, J.; De Smedt, S.C. PEGylation of biodegradable dextran nanogels for siRNA delivery. *Eur. J. Pharm. Sci.* **2010**, *40*, 342–351.
131. Naeye, B.; Deschout, H.; Roding, M.; Rudemo, M.; Delanghe, J.; Devreese, K.; Demeester, J.; Braeckmans, K.; De Smedt, S.C.; Raemdonck, K. Hemocompatibility of siRNA loaded dextran nanogels. *Biomaterials* **2011**, *32*, 9120–9127.
132. Patnaik, S.; Arif, M.; Pathak, A.; Singh, N.; Gupta, K.C. PEI-alginate nanocomposites: Efficient non-viral vectors for nucleic acids. *Int. J. Pharm.* **2010**, *385*, 194–202.
133. Hu-Lieskovan, S.; Heidel, J.D.; Bartlett, D.W.; Davis, M.E.; Triche, T.J. Sequence-specific knockdown of EWS-FLI1 by targeted, nonviral delivery of small interfering RNA inhibits tumor growth in a murine model of metastatic Ewing's sarcoma. *Cancer Res.* **2005**, *65*, 8984–8992.
134. Bartlett, D.W.; Davis, M.E. Physicochemical and biological characterization of targeted, nucleic acid-containing nanoparticles. *Bioconjug. Chem.* **2007**, *18*, 456–468.
135. Bartlett, D.W.; Su, H.; Hildebrandt, I.J.; Weber, W.A.; Davis, M.E. Impact of tumor-specific targeting on the biodistribution and efficacy of siRNA nanoparticles measured by multimodality *in vivo* imaging. *Proc. Natl. Acad. Sci. USA* **2007**, *104*, 15549–15554.
136. Kulkarni, A.; Deng, W.; Hyun, S.H.; Thompson, D.H. Development of a low toxicity, effective pDNA vector based on noncovalent assembly of bioresponsive amino-beta-cyclodextrin:adamantane-poly(vinyl alcohol)-poly(ethylene glycol) transfection complexes. *Bioconjug Chem.* **2012**, *23*, 933–940.
137. Kulkarni, A.; DeFrees, K.; Hyun, S.H.; Thompson, D.H. Pendant polymer: amino-beta-cyclodextrin: siRNA guest: host nanoparticles as efficient vectors for gene silencing. *J. Am. Chem. Soc.* **2012**, *134*, 7596–7599.

Biofilm and Dental Biomaterials

Marit Øilo [1,*] and Vidar Bakken [2]

[1] Department of Clinical Dentistry, Faculty of Medicine and Dentistry, University of Bergen, Aarstadveien 19, Bergen NO-5009, Norway
[2] Department of Clinical Science, Faculty of Medicine and Dentistry, University of Bergen, Jonas Lies vei 65, The Laboratory Building, Bergen NO-5021, Norway; E-Mail: Vidar.Bakken@k2.uib.no

* Author to whom correspondence should be addressed; E-Mail: marit.oilo@iko.uib.no;

Academic Editor: Ihtesham ur Rehman

Abstract: All treatment involving the use of biomaterials in the body can affect the host in positive or negative ways. The microbiological environment in the oral cavity is affected by the composition and shape of the biomaterials used for oral restorations. This may impair the patients' oral health and sometimes their general health as well. Many factors determine the composition of the microbiota and the formation of biofilm in relation to biomaterials such as, surface roughness, surface energy and chemical composition, This paper aims to give an overview of the scientific literature regarding the association between the chemical, mechanical and physical properties of dental biomaterials and oral biofilm formation, with emphasis on current research and future perspectives.

Keywords: biofilm; biomaterials; dental materials; metals; polymers; ceramics

1. Introduction

Dental treatment involving biomaterials will alter the mechanical, physiological and chemical conditions in the oral cavity. The degree of change depends on the size and quality of the restoration inserted, which in turn, will affect the microbiology of the oral cavity. All surfaces in the oral cavity are covered by a pellicle of glycoproteins from saliva within seconds after cleaning [1,2]. Microorganisms are able to adhere to this layer and form biofilms organized as a consortium of bacteria, virus and fungi.

Most of the microorganisms present in the oral biofilm are harmless natural inhabitants, but some have the potential to cause damage to the mineralized tissues or infections in the soft tissues. Primary infections (caries) are responsible for less than half of the dental restorations produced annually [3]. Most of the restorations placed in general dental practices are replacements of old restorations, and the majority of these need replacement because of biofilm related secondary infections. By inserting foreign bodies such as dental restorations, new niches appear for the microorganisms. Such niches create areas promoting biofilm accumulation with a pathological potential. Increased numbers of microorganisms with a pathological ability will increase the risk of developing disease. Examples of such diseases are secondary caries along a restoration, fungal infections in relation to complete dentures or progression of periodontal disease on teeth supporting a partial denture.

However, patients with active caries, worn or broken fillings or badly fitting dentures, may reduce the pathological potential of the biofilm with new, optimally shaped dental restorations. The aim of using biomaterials for treating patients is to do more good than harm. The biomaterials' ability to inhibit biofilm formation is an important factor for clinical success [4]. There is a global interest in finding ways to reduce the biofilm formation on biomaterials in general. Biofilm infections of medical devices at other body sites than the oral cavity also constitute major problems [5].

2. Oral Biofilm

In the human oral cavity there are hundreds of different species of microorganisms, including bacteria, virus and fungi [6]. More than 700 unique bacterial species have been detected [7]. There can be more than 10^{11} microorganisms per mg of dental plaque [7,8]. These live in complex societies, usually organized in thin layers covering the oral surfaces—biofilm—for example as dental plaque on tooth surfaces [1,2]. Immediately after cleaning, the proteins from saliva will cover the tooth surfaces in a pellicle. Bacteria attach to this pellicle by microfilaments in their cell walls. When the bacteria increase in number, they will be able to communicate by secreting signal molecules and create a community [1,9]. The bacteria secrete proteins, polysaccharides, nucleic acids and other substances to the extracellular matrix, additionally containing proteins and nutrients from saliva. This matrix is the "glue" of the biofilm.

The microorganisms inside a matrix behave differently from bacteria floating freely in the saliva (planktonic growth). The biofilm community behaves as a unit in response to environmental changes rather than as single bacterial responses. The matrix protects the organisms inside from chemical treatment that could have been lethal for planktonic bacteria, for example antibacterial mouth rinses [10]. Complete removal of oral plaque is difficult due to limited access between teeth and in deep crevices on tooth surfaces. Dental restorations with pores, gaps and margins further complicate this. In healthy individuals the biofilm on teeth will function as a protective shield from foreign microorganisms and chemicals in the food, such as acids that can potentially dissolve dental enamel [11]. If, however, the biofilm is left untreated, an ecological shift might occur, favoring microorganisms that may have detrimental effects on teeth, surrounding tissues and the patient's oral and general health.

The thickness and structure of a biofilm will be affected by many factors, such as pH, nutrients, oxygen, time since last cleaning and the kind of surface to which it is attached [2,12–17]. The biofilm in the oral cavity will therefore differ in the different locations, such as on the cheek or in between teeth. A biofilm that is allowed to grow over days will have a different composition than a biofilm that is mechanically removed and renewed daily [2]. The bacteria adjust to the matrix and surrounding organisms by gene regulations. Biofilm formation is a big problem for all medical biomaterials in humid environments, such as artificial heart valves, artificial vocal cords and incubation tubes [5,18]. They often have the added complications that access for cleaning procedures is impossible. However, they are usually not in environments with such an abundance of microorganisms as in the oral cavity.

3. Oral Microorganisms and Oral Infections

Growth of pathogenic microorganisms can lead to disease if the host's immune response is not able to neutralize or destroy them [9]. The most common oral infections are pulpitis from untreated caries, gingivitis, and periodontitis [2]. The microorganisms in oral biofilm are mainly bacteria. However, adults usually have both fungal and viral species present in the oral microflora [6,7]. *Candida albicans* will normally be present only in small amounts in healthy adults because the healthy biofilm favor other microorganisms. The microflora develops gradually as the environment in the mouth changes. Microorganisms with and without known ability to provoke diseases will be present in the flora. As long as the majority of microorganisms are non-pathogenic, the host stays healthy, but if the conditions favor growth of pathogens, disease might develop [7].

Teeth are unique organs since they penetrate the oral mucosa and have one part secured in bone and one part exposed to the oral cavity [19]. The exposed part is continually colonized by the oral microorganisms. The periodontal pocket surrounding the tooth is designed as a protective barrier against the invasion of microorganisms to the mucosa and the alveolar bone. The cells in the pocket release exudate with antibacterial effect that hinders microorganisms from invading the pockets. Increased thickness of the biofilm will reduce the pH-, oxygen- and nutrient-levels in the bottom layers of the plaque (Figure 1) [2,9,12,17,20]. Food debris in the biofilm will ensure access to nutrients for the bacteria. This will be favorable for the species associated with dental caries (*Streptococcus mutans*, *Actinomyces spp.*, *Veillonella spp.*, *Lactobacillus spp.*) and gingival and periodontal inflammation (*Porphyromonas gingivalis*, *Treponema denticola*, *Tannerella forsythia etc.*).

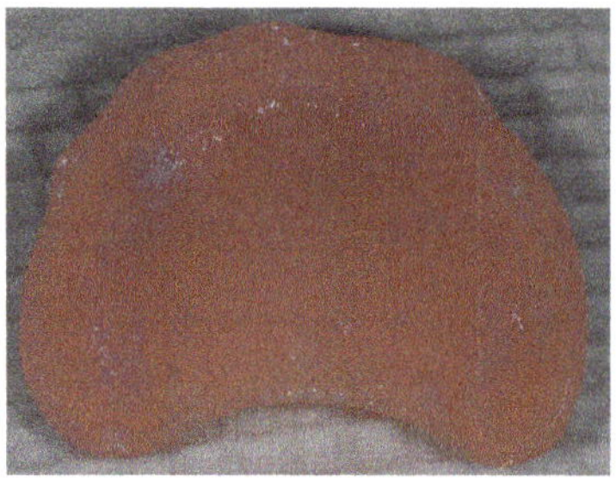

Figure 1. Biofilm on a denture belonging to a patient with poor oral hygiene routine. (photo: Marit Øilo).

Caries is in itself not an infection, but demineralization of dentin and enamel. The demineralization occurs as a consequence of acid production from bacteria in the biofilm when they process sugars in plaque [21]. The biofilm becomes acidic (pH < 5) and the hydroxyapatite in the enamel dissolves. When the lesion is very deep, the pulp of the tooth might be infected with oral bacteria and inflammation occurs, known as pulpitis [2]. Gingivitis occurs when the biofilm is too thick to be effectively washed away by the gingival exudate. The anaerobic bacteria are favored and intrude into the periodontal pocket. Proliferation of pathogenic bacteria in the pocket induces inflammatory response in the gingiva with swelling and increased bleeding which complicates the cleaning procedures. This response can in turn dissolve the alveolar bone supporting the teeth—periodontitis—which eventually will result in tooth loss [1,2,6,12].

Oral infections can affect the hosts' general health in many ways [8,22]. Firstly, it is unfavorable to have ongoing infections in general. Oral infections such as gingivitis can involve quite large areas, and is a constant burden for the hosts' immune system. This is, of course, most harmful for patients with other ongoing inflammations or diseases. Secondly, oral microbes may spread to other organs within the host either via the respiratory system or through the blood stream [1,2,8,20,22,23]. This is, again, most harmful for patients with other complications, such as artificial heart valves, or transplanted organs. Elderly people with pneumonia will often have oral microbes in their lungs, which may have been aspired from the oropharynx and caused the inflammation in the lungs [8,20,23]. The direct causality between oral infections and systemic disease has been difficult to prove. It is difficult to detect whether the oral microorganisms were present prior to the disease and were the direct cause of the infection or not. It is also possible that the oral microbes entered the loci after the primary inflammation occurred due to reduced effect in the immune response system. *Candida albicans* can spread directly from the oral cavity to the throat and stomach (Figure 2) [24,25]. General candida infections are extremely difficult to cure due to the high risk of re-colonization. Thirdly, individuals with poor general health will normally also have poor oral health, due to malfunctioning immune system and altered chemical and physical conditions in the body, such as reduced saliva flow, iron deficiency, malnutrition, medical treatment, *etc.* [26,27]. Sick and elderly have reduced ability to clean their teeth properly and are therefore more exposed to development of unfavorable biofilm and subsequent general infection [28].

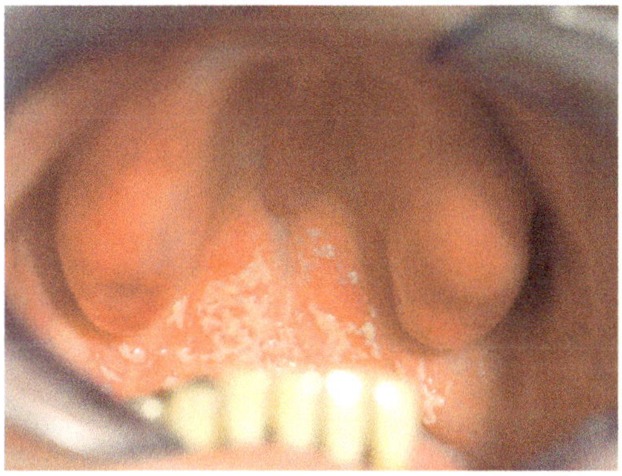

Figure 2. Patient with denture stomatitis. The infection has spread from the denture covered palate to the throat and stomach. The patient experienced no oral pain and was unaware of the situation (photo: Marit Øilo).

4. Biomaterials' Effect on Biofilm

Dental restorations affect the composition of the biofilm in many ways. There will always be steps, gaps or groves between tooth and restoration (Figure 3). These will complicate mechanical biofilm removal and alter the chemical balance in the biofilm in the region (Figure 4) [29,30]. Restorations differ from enamel with regard to surface roughness, surface energy and chemical composition [29–33].

Most adults have at least one dental restoration and the role of biofilm related infections related to the restoration as opposed to primary oral infections is not easily distinguished.

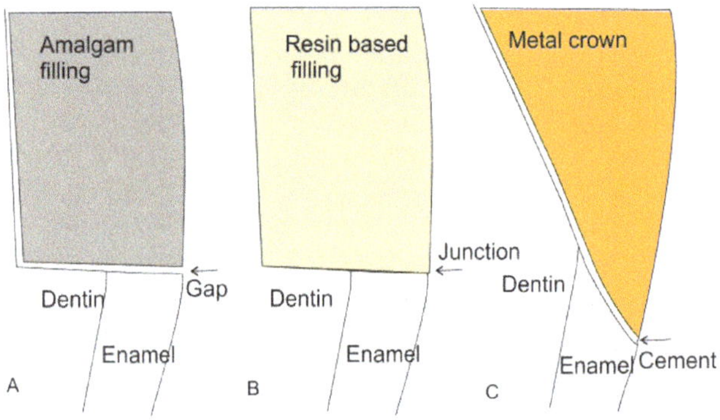

Figure 3. Different biomaterials create different types of junctions between tooth and restoration. The illustrations represent cross sections of the tooth-restoration border. (**A**) An amalgam filling is not directly attached to the tooth. The gap is a potential area for ingrowth of bacteria which causes secondary caries; (**B**) The resin based fillings are supposed to be adhesively attached to the tooth substance with no gaps in between. This requires that the filling is placed and cured with a special technique to avoid gaps caused by polymerization shrinkage; (**C**) A crown is fixed to the tooth by means of cement which fills the gap between tooth and restoration. This will always cause irregularities on the surface. The exposed cement is subjected to wash out and wear causing an increase of the groove over time.

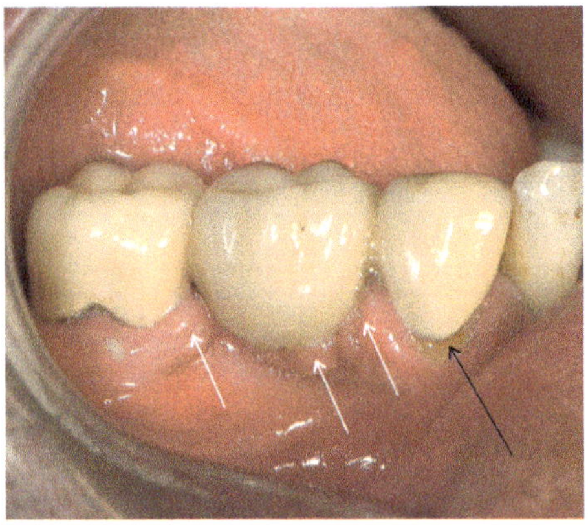

Figure 4. Secondary caries at the crown margin (black arrow) caused by accumulation of biofilm in the crevice between crown and tooth. Hyperplastic gingiva (white arrows) due to increased plaque accumulation in the connector area of the fixed partial denture. The connector area of the bridge has a bulky design which complicates interdental hygiene (photo: Marit Øilo).

Many different studies have been performed evaluating the effect on biofilm formation from surface qualities such as surface energy, roughness, topography and chemical composition of the restorative materials [34–36]. Both surface quality and chemical composition will affect the topography and surface energy and the different studies do not always disclose or differentiate properly between the different factors. Multiple factors are working simultaneously that all may affect the outcome. Comparison between different studies and among materials is not straight forward. Furthermore, the findings from *in vitro* and *in vivo* studies are not always in coherence with each other, which indicates that the oral biofilms *in vivo* are complicated and difficult to mimic in a laboratory setting. Most studies evaluate only a selection of pathogenic bacteria or only one surface variable [7,21]. As discussed above, it is not likely that one surface quality or single microorganisms are the cause of disease alone but rather a shift in the composition in the biofilm.

4.1. Surface Energy

Surfaces with a low surface energy usually display lower adherence to biofilms than similar surfaces with higher surface energy [35]. No effect of changes in surface energy was found in a study on surface nanoroughness, texture and chemistry [37]. Most dental materials, with the exception of ceramics, have a higher surface energy than enamel and have thus a greater risk of biofilm accumulation. Alteration in surface roughness will in most cases also alter the surface energy. It is therefore difficult to distinguish between the two factors. It seems that surface roughness plays a more important role than surface energy [36].

4.2. Surface Roughness

Increased surface roughness and complicated topography shows higher affinity to microbes than smoother surfaces and subsequently increased difficulty in complete removal of the biofilm by mechanical brushing [15,34,35,38]. The surface roughness seems to affect only the number of bacteria in the biofilm, not the species. A rough surface increases the surface area available for colonization compared to a smooth surface [36]. Furthermore, the crevices created by the roughness generate shelters for the bacteria so they have time for securing their attachment to the pellicle. On smooth surfaces, the brushing of tongue or cheek movement is sufficient for mechanical removal of the biofilm. A maximum surface roughness of $R_a = 0.2$ μm has been suggested as a threshold value for bacterial retention. Below this value, no further reductions were observed, while over this value, biofilm accumulation increased with increasing roughness [34]. Surface roughness can, however, be measured in many ways. R_a gives an arithmetic mean of the surface roughness, while the topography and maximum peak to valley can differ significantly among materials with similar R_a values. Nanoscale morphology has also been shown to influence the adhesion of proteins and subsequent formation of biofilm in *in vitro* models [39]. An absolute threshold value measured in R_a in probably not clinically relevant for all materials, but gives a general indication of the clinical acceptance level. Furthermore, the surface roughness of most materials will increase over time in the oral cavity.

4.3. Chemical Composition

The chemical composition of the dental material will further affect the bacterial adhesion since both proteins and microorganisms can chemically attach or attract to components in the material, by van der Waal forces, acid-base reactions or electrostatic interactions [40]. In most patients, there will be several different materials present in the mouth simultaneously with can interfere with the biofilm formation and the microbiota in general. The chemical interaction between material and microorganisms can lead to alterations in the surface properties over time.

4.4. Dental Restorations

The type of biomaterial in a dental restoration, can also affect the biofilm formation to a large extent [16]. Crowns and bridges are usually made of ceramics, metals or a combination of these. Ceramic materials have a smooth, polished surface which is easily cleaned. This is very good with regard to daily removal of the biofilm. Some ceramics have, however, very uneven crown margins from the machining procedures (Figure 5). A crown margin with many small defects will retain more plaque and bacteria than a smooth margin. Most metals can be polished to give very little retention for biofilms, although some alloys have a higher affinity to bacteria than others [41]. It seems that some bacteria are attracted to electrical charges in some alloys [16]. Biofilms on gold restorations, however, generally have low viability [41]. Some microbes are affected by the eluates from the metals. The mercury leaching in very small amounts from fresh amalgam restorations has, for instance, a bacteriostatic effect [41]. The border between the tooth and the restorations will always be problematic, since it is virtually impossible to make restorations with perfect adaptations all the way around [42,43]. The grooves or edges created by this will increase accumulation of plaque and complicate the mechanical removal of it afterwards.

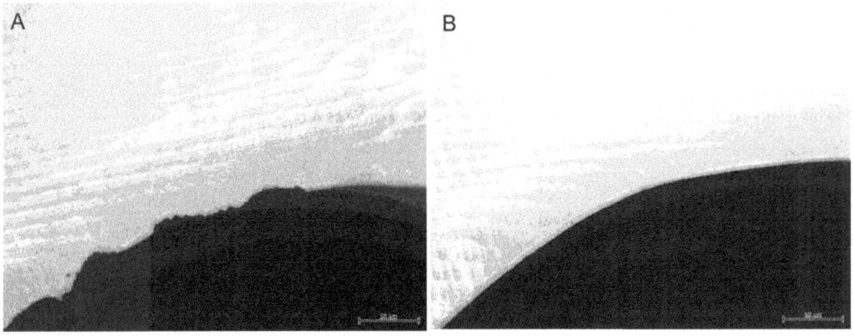

Figure 5. (**A**) A zirconia ceramic crown margin severely damaged by the machining procedure creating an uneven and biofilm-retaining margin between restoration and tooth; (**B**) A zirconia ceramic crown with a smooth undamaged crown margin that is easily cleaned. (photo: Marit Øilo).

Dental implants are most commonly made of titanium which are inserted into the alveolar bone to function as artificial roots [35]. They penetrate the oral mucosa and are therefore exposed to the oral microflora. Implants usually have a smooth polished surface in the areas that are exposed to the oral cavity which can be easily cleaned. They are, however, often made of several small components that are fixed together with tiny screws [44]. There will always be gaps and crevices between these parts, creating almost a "greenhouse" for bacterial growth. Pathogenic bacteria in such gaps can cause inflammation in the surrounding bone or inhibit osseointegration [25,35]. Modern implants have a very rough surface in the part that is to be osseointegrated in the alveolar bone to enhance early bone attachment [35]. If the bone reacts inadequately or does not grow to cover the rough area, this will be exposed to the oral cavity as well. This surface will easily be colonized by oral bacteria which then will be almost impossible to remove completely (Figure 6). If pathogenic organisms are allowed to settle in such an area, it will be extremely difficult to arrest inflammation (periimplantitis). Since the implant is in direct contact with bone, the inflammation will potentially be able to spread via the bone or the blood faster than similar inflammation around teeth [16,35]. The teeth are separated from the alveolar bone by fibrous tissue, which work as a barrier to bacterial ingrowth. Currently, research cannot fully explain the differences between periimplantitis and periodontitis [45]. The microorganisms surrounding an implant with healthy supporting tissues are similar to those surrounding a healthy tooth [46]. However, periimplantitis locations seem to have more bacterial species and fungal species than locations associated with periodontitis. This is possibly an effect of the titanium in the implant.

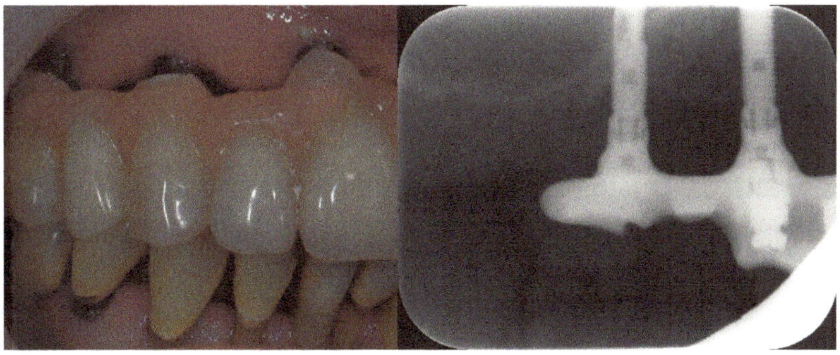

Figure 6. Implant in the upper jaw supporting a fixed denture. The rough area on the molar region implant is exposed due to previous periimplantitis (photo: Marit Øilo).

Polymers, such as resin based composites, glass ionomer and acrylics, have more pores and defects in the surface than metals, ceramics and enamel (Figure 7). Porosities will be filled with humidity and make perfect incubation chambers for certain microbes [14,24,47,48]. Biofilms on polymers develop quicker and will be more difficult to remove completely [41]. Acidic residues from the bacteria will roughen the surface, which will further complicate biofilm removal [15,16]. In order to achieve a surface roughness under the threshold value R_a = 0.2 μm on dental polymers special routines with stepwise polishing and finishing must be performed and repeated at regular intervals [38].

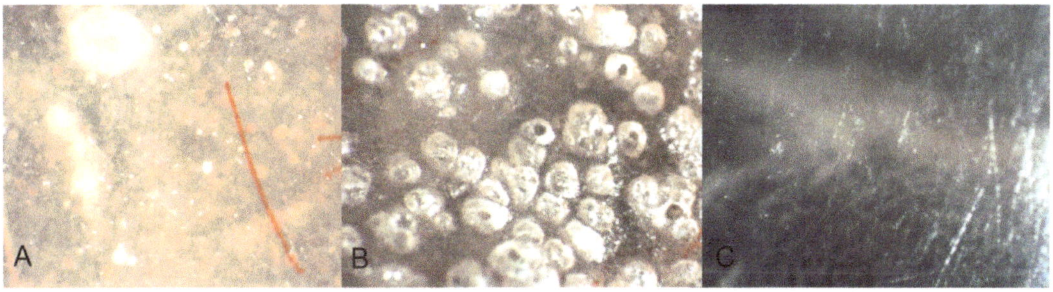

Figure 7. Different polymers have a different density and surface structure. The amount and size of pores and surface defects will affect the biofilm formation and composition. (**A**) polished heat cured acrylic; (**B**) unpolished cold cured acrylic; (**C**) polished polyamide (photo: Marit Øilo).

Removable dental prostheses, such as full or partial dentures or orthopedic appliances, will create new surfaces for biofilm formation, and thus increase the total amount of biofilm dramatically [30,49–51]. An increase of the total number of microorganisms alters the whole ecology of the oral cavity which can shift the balance between harmful and friendly microorganisms in the entire mouth [27]. The mucosa in direct contact with these appliances will be most affected, but the other structures in the mouth will be involved as well. Fungal species, such as *C. albicans*, can proliferate when the chemical and physical situations alter [24,52]. The oxygen level drops when large areas of the oral mucosa suddenly are covered by dentures. Organisms that demand oxygen (aerobic) will perish and make room for other species (anaerobic) [20,24,52–54]. Proliferations of *Candida* will cause an inflammatory response in the mucosa under the biofilm—denture stomatitis. The risk of re-colonization is great if the chemical and physical situation is not altered simultaneously with medical treatment [20,53–55]. The host's health and nutritional status are important factors for both development and the possibility to cure this infection [1,6].

Dentures rest directly on the oral mucosa. The stiff and hard acrylic dentures create wounds or irritations in the oral mucosa if they do not have an optimal fit. These irritated sites are easily colonized by fungi and bacteria causing inflammation and pain. These problems are either treated with adjustments in the acrylic, which very often results in a rougher surface or by lining the denture with a material to either increase fit or softness [56]. The liner materials available are cold cured acrylic, polyvinyl siloxanes, or acrylics containing plasticizers. All these are much more porous than heat cured acrylics and have increased problems with fungal infections [47,57,58]. The softer materials are most problematic but also most comfortable for the patients. Patients with poor general health are most exposed, firstly, because they usually have more problems with wearing dentures, and secondly, because they are more susceptible for infectious diseases [59]. Some soft denture materials, such as polyamides are probably less prone to biofilm formation than acrylics due to fewer pores [60].

5. Future Aspects and Current Research

Due to the many problems associated with biofilm formation on dental biomaterials, the challenge is to improve the materials in use. Reduced polymerization shrinkage of dental resin composites is believed to reduce the leakage of microorganisms into the gap and thereby reduce the rate of secondary caries, although the clinical relevance is difficult to detect [61]. Fluoride release from glass ionomer cements are meant to reduce the demineralization of the dentine and enamel on the tooth adjacent to the cement [62]. Fluoride release from other polymers have not been equally successful. Lately, there have been many attempts at altering the chemical composition in the materials with the intention of reducing biofilm formation [4,63,64]. Different strategies are used to achieve this, mainly reducing the initial attachment of bacteria by altering the surface properties or reducing the viability of the attached bacteria by chemical components [4]. Some dental materials include monomers with antibacterial eluates [65]. Antibacterial chemicals, such as chlorhexidine or silver nanoparticles, can be embedded in nanoparticles in resin, coatings, acrylics, sealers or cements [66,67]. The nanoparticles dissolve slowly from the material and destroy the bacteria in the biofilm attached to the surface of the material. More than 50 abstracts each year concerning antibacterial effects of dental restorative materials have been presented in the International Association of Dental Research meetings worldwide in recent years, illustrating the current interest for this clinical problem. There are several potential benefits from including slow releasing antibacterial particles in biomaterials. The number of bacteria in the biofilm can be reduced, the potential harm from biofilm formation can be reduced, and the oral environment be made healthier. The antibacterial effect can be used in specific localizations, such as in the layer between filling and tooth, where they are most effective and thus reduce the general effect on the patient [68,69]. Antibacterial liners can ensure that remaining bacteria in the caries lesion are killed and thereby arrest the development of the lesion without excessive removal of tooth substances [68,69]. Antibacterial sealers for endodontic treatment can probably reduce the number of visits and increase success rates of endodontic treatment [70]. Antibacterial surface modifications or coatings are investigated for a wide range of applications also outside medicine and dentistry [4].

There are several uncertainties regarding the long term effects regarding surface coatings or slow releasing particles which have not been fully investigated. Coatings will be worn, roughened and may not have sufficient mechanical properties over time. Gradual release of particles will eventually alter the mechanical properties of the material. Such release can be compared with corrosion. Corrosion will normally result in increased surface roughness and loss of substance and thereby increase the retention for bacteria. It is likely that the release of nanoparticles over a long time will reduce the fracture strength, surface hardness, and wear resistance of the material. An antimicrobial layer between tooth and restoration may reduce both initial and long term retention by altering the chemical adhesion between layers. These potential side effects are rarely discussed or evaluated in the studies available. The effect of any antibacterial chemical will always be dose-dependent. It will be difficult to ensure that the dose of the released particles is high enough to be effective, while not weakening the material. It is reasonable to assume that the effect will diminish over time. Additionally, there may be a serious risk of developing multi-resistant bacteria [71,72] or other adverse reactions to the released particles [73].

Author Contributions

The first author is the main author. The second author has contributed with knowledge in the microbiological section.

Conflicts of interest

The authors declare no conflict of interest.

References

1. Avila, M.; Ojcius, D.M.; Yilmaz, O. The oral microbiota: Living with a permanent guest. *DNA Cell Biol.* **2009**, *28*, 405–411.
2. Filoche, S.; Wong, L.; Sissons, C.H. Oral biofilms: Emerging concepts in microbial ecology. *J. Dent. Res.* **2010**, *89*, 8–18.
3. Mjör, I.A.; Moorhead, J.E.; Dahl, J.E. Reasons for replacement of restorations in permanent teeth in general dental practice. *Int. Dent. J.* **2000**, *50*, 361–366.
4. Hasan, J.; Crawford, R.J.; Ivanova, E.P. Antibacterial surfaces: The quest for a new generation of biomaterials. *Trends Biotechnol.* **2013**, *31*, 295–304.
5. Donlan, R.M. Biofilms and device-associated infections. *Emerg. Infect. Dis.* **2001**, *7*, 277–281.
6. Lamont, R.J. *Oral Microbiology and Immunology*; ASM Press: Washington, DC, USA, 2006.
7. Aas, J.A.; Paster, B.J.; Stokes, L.N.; Olsen, I.; Dewhirst, F.E. Defining the normal bacterial flora of the oral cavity. *J. Clin. Microbiol.* **2005**, *43*, 5721–5732.
8. Li, X.; Kolltveit, K.M.; Tronstad, L.; Olsen, I. Systemic diseases caused by oral infection. *Clin. Microbiol. Rev.* **2000**, *13*, 547–558.

9. Sbordone, L.; Bortolaia, C. Oral microbial biofilms and plaque-related diseases: Microbial communities and their role in the shift from oral health to disease. *Clin. Oral Investig.* **2003**, *7*, 181–188.
10. Marsh, P.D. Plaque as a biofilm: Pharmacological principles of drug delivery and action in the sub- and supragingival environment. *Oral Dis.* **2003**, *9*, 16–22.
11. Bartlett, D.W. The role of erosion in tooth wear: Aetiology, prevention and management. *Int. Dent. J.* **2005**, *55*, 277–284.
12. Von Fraunhofer, J.A.; Loewy, Z.G. Factors involved in microbial colonization of oral prostheses. *Gen. Dent.* **2009**, *57*, 136–143.
13. Sachdeo, A.; Haffajee, A.D.; Socransky, S.S. Biofilms in the edentulous oral cavity. *J. Prosthodont.* **2008**, *17*, 348–356.
14. Paranhos Hde, F.; da Silva, C.H.; Venezian, G.C.; Macedo, L.D.; de Souza, R.F. Distribution of biofilm on internal and external surfaces of upper complete dentures: The effect of hygiene instruction. *Gerodontology* **2007**, *24*, 162–168.
15. Morgan, T.D.; Wilson, M. The effects of surface roughness and type of denture acrylic on biofilm formation by streptococcus oralis in a constant depth film fermentor. *J. Appl. Microbiol.* **2001**, *91*, 47–53.
16. Busscher, H.J.; Rinastiti, M.; Siswomihardjo, W.; van der Mei, H.C. Biofilm formation on dental restorative and implant materials. *J. Dent. Res.* **2010**, *89*, 657–665.
17. Palmer, R.J., Jr. Supragingival and subgingival plaque: Paradigm of biofilms. *Compend. Contin. Educ. Dent.* **2010**, *31*, 104–106.
18. Francolini, I.; Donelli, G. Prevention and control of biofilm-based medical-device-related infections. *FEMS Immunol. Med. Microbiol.* **2010**, *59*, 227–238.
19. Leknes, K.N. The influence of anatomic and iatrogenic root surface characteristics on bacterial colonization and periodontal destruction: A review. *J. Periodontol.* **1997**, *68*, 507–516.
20. Coulthwaite, L.; Verran, J. Potential pathogenic aspects of denture plaque. *Br. J. Biomed. Sci.* **2007**, *64*, 180–189.
21. Nyvad, B.; Crielaard, W.; Mira, A.; Takahashi, N.; Beighton, D. Dental caries from a molecular microbiological perspective. *Caries Res.* **2013**, *47*, 89–102.
22. Debelian, G.J.; Olsen, I.; Tronstad, L. Systemic diseases caused by oral microorganisms. *Endod. Dent. Traumatol.* **1994**, *10*, 57–65.
23. Sumi, Y.; Kagami, H.; Ohtsuka, Y.; Kakinoki, Y.; Haruguchi, Y.; Miyamoto, H. High correlation between the bacterial species in denture plaque and pharyngeal microflora. *Gerodontology* **2003**, *20*, 84–87.
24. Seneviratne, C.J.; Jin, L.; Samaranayake, L.P. Biofilm lifestyle of candida: A mini review. *Oral. Dis.* **2008**, *14*, 582–590.
25. Ramage, G.; Martinez, J.P.; Lopez-Ribot, J.L. Candida biofilms on implanted biomaterials: A clinically significant problem. *FEMS Yeast Res.* **2006**, *6*, 979–986.
26. Marsh, P.D.; Martin, M.V. *Oral Microbiology*, 5th ed.; Elsevier: Edinburgh, Scotland, UK, 2009.
27. Marsh, P.D.; Percival, R.S.; Challacombe, S.J. The influence of denture-wearing and age on the oral microflora. *J. Dent. Res.* **1992**, *71*, 1374–1381.

28. Chiappelli, F.; Bauer, J.; Spackman, S.; Prolo, P.; Edgerton, M.; Armenian, C.; Dickmeyer, J.; Harper, S. Dental needs of the elderly in the 21st century. *Gen. Dent.* **2002**, *50*, 358–363.
29. Adamczyk, E.; Spiechowicz, E. Plaque accumulation on crowns made of various materials. *Int. J. Prosthodont.* **1990**, *3*, 285–291.
30. Addy, M.; Bates, J.F. Plaque accumulation following the wearing of different types of removable partial dentures. *J. Oral Rehabil.* **1979**, *6*, 111–117.
31. Chan, C.; Weber, H. Plaque retention on teeth restored with full-ceramic crowns: A comparative study. *J. Prosthet. Dent.* **1986**, *56*, 666–671.
32. Jameson, L.M.; Malone, W.F. Crown contours and gingival response. *J. Prosthet. Dent.* **1982**, *47*, 620–624.
33. Parkinson, C.F. Excessive crown contours facilitate endemic plaque niches. *J. Prosthet. Dent.* **1976**, *35*, 424–429.
34. Bollen, C.M.L.; Lambrechts, P.; Quirynen, M. Comparison of surface roughness of oral hard materials to the threshold surface roughness for bacterial plaque retention: A review of the literature. *Dent. Mater.* **1997**, *13*, 258–269.
35. Subramani, K.; Jung, R.E.; Molenberg, A.; Hammerle, C.H. Biofilm on dental implants: A review of the literature. *Int. J. Oral Maxillofac. Implants* **2009**, *24*, 616–626.
36. Teughels, W.; Van Assche, N.; Sliepen, I.; Quirynen, M. Effect of material characteristics and/or surface topography on biofilm development. *Clin. Oral Implants Res.* **2006**, *17*, 68–81.
37. Xing, R.; Lyngstadaas, S.P.; Ellingsen, J.E.; Taxt-Lamolle, S.; Haugen, H.J. The influence of surfacc nanoroughness, texture and chemistry of tizr implant abutment on oral biofilm accumulation. *Clin. Oral Implants Res.* **2015**, *26*, 649–656.
38. Aykent, F.; Yondem, I.; Ozyesil, A.G.; Gunal, S.K.; Avunduk, M.C.; Ozkan, S. Effect of different finishing techniques for restorative materials on surface roughness and bacterial adhesion. *J. Prosthet. Dent.* **2010**, *103*, 221–227.
39. Singh, A.V.; Vyas, V.; Patil, R.; Sharma, V.; Scopelliti, P.E.; Bongiorno, G.; Podestà, A.; Lenardi, C.; Gade, W.N.; Milani, P. Quantitative characterization of the influence of the nanoscale morphology of nanostructured surfaces on bacterial adhesion and biofilm formation. *PLoS One* **2011**, *6*, doi:10.1371/journal.pone.0025029.
40. Sakaguchi, R.; Powers, J. *Craig's Restorative Dental Materials*, 13th ed.; Elsevier: Philadelphia, PA, USA, 2012.
41. Auschill, T.M.; Arweiler, N.B.; Brecx, M.; Reich, E.; Sculean, A.; Netuschil, L. The effect of dental restorative materials on dental biofilm. *Eur. J. Oral Sci.* **2002**, *110*, 48–53.
42. Gotfredsen, K.; Carlsson, G.E.; Jokstad, A.; Arvidson Fyrberg, K.; Berge, M.; Bergendal, B.; Bergendal, T.; Ellingsen, J.E.; Gunne, J.; Hofgren, M.; *et al.* Implants and/or teeth: Consensus statements and recommendations. *J. Oral Rehabil.* **2008**, *35*, 2–8.
43. Felton, D.A.; Kanoy, B.E.; Bayne, S.C.; Wirthman, G.P. Effect of *in vivo* crown margin discrepancies on periodontal health. *J. Prosthet. Dent.* **1991**, *65*, 357–364.
44. Jansen, V.K.; Conrads, G.; Richter, E.J. Microbial leakage and marginal fit of the implant-abutment interface. *Int. J. Oral. Maxillofac. Implants* **1997**, *12*, 527–540.

45. Ong, C.T.; Ivanovski, S.; Needleman, I.G.; Retzepi, M.; Moles, D.R.; Tonetti, M.S.; Donos, N. Systematic review of implant outcomes in treated periodontitis subjects. *J. Clin. Periodontol.* **2008**, *35*, 438–462.
46. Heitz-Mayfield, L.J.; Lang, N.P. Comparative biology of chronic and aggressive periodontitis *vs.* Peri-implantitis. *Periodontology* **2000**, *53*, 167–181.
47. Boscato, N.; Radavelli, A.; Faccio, D.; Loguercio, A.D. Biofilm formation of candida albicans on the surface of a soft denture-lining material. *Gerodontology* **2009**, *26*, 210–213.
48. Rocha, E.P.; Luvizuto, E.R.; Sabotto, S.F. Biofilm formation and caries incidence with removable partial dentures. *Dent. Today* **2008**, *27*, 60, 62–63.
49. Zlataric, D.K.; Celebic, A.; Valentic-Peruzovic, M. The effect of removable partial dentures on periodontal health of abutment and non-abutment teeth. *J. Periodontol.* **2002**, *73*, 137–144.
50. Radford, D.R.; Challacombe, S.J.; Walter, J.D. Denture plaque and adherence of candida albicans to denture-base materials *in vivo* and *in vitro*. *Crit. Rev. Oral Biol. Med.* **1999**, *10*, 99–116.
51. McHenry, K.R.; Johansson, O.E.; Christersson, L.A. The effect of removable partial denture framework design on gingival inflammation: A clinical model. *J. Prosthet. Dent.* **1992**, *68*, 799–803.
52. Campos, M.S.; Marchini, L.; Bernardes, L.A.; Paulino, L.C.; Nobrega, F.G. Biofilm microbial communities of denture stomatitis. *Oral Microbiol. Immunol.* **2008**, *23*, 419–424.
53. Pereira-Cenci, T.; Del Bel Cury, A.A.; Crielaard, W.; Ten Cate, J.M. Development of candida-associated denture stomatitis: New insights. *J. Appl. Oral Sci.* **2008**, *16*, 86–94.
54. Ramage, G.; Tomsett, K.; Wickes, B.L.; Lopez-Ribot, J.L.; Redding, S.W. Denture stomatitis: A role for candida biofilms. *Oral Surg. Oral Med .Oral Pathol. Oral Radiol. Endod.* **2004**, *98*, 53–59.
55. Redding, S.; Bhatt, B.; Rawls, H.R.; Siegel, G.; Scott, K.; Lopez-Ribot, J. Inhibition of candida albicans biofilm formation on denture material. *Oral Surg. Oral Med. Oral Pathol. Oral Radiol. Endod.* **2009**, *107*, 669–672.
56. Rodrigues, S.; Shenoy, V.; Shetty, T. Resilient liners: A review. *J. Indian Prosthodont. Soc.* **2013**, *13*, 155–164.
57. Nikawa, H.; Jin, C.; Makihira, S.; Egusa, H.; Hamada, T.; Kumagai, H. Biofilm formation of candida albicans on the surfaces of deteriorated soft denture lining materials caused by denture cleansers *in vitro*. *J. Oral Rehabil.* **2003**, *30*, 243–250.
58. Pereira-Cenci, T.; da Silva, W.J.; Cenci, M.S.; Cury, A.A. Temporal changes of denture plaque microbiologic composition evaluated *in situ*. *Int. J. Prosthodont.* **2010**, *23*, 239–242.
59. Mutluay, M.M.; Oguz, S.; Floystrand, F.; Saxegaard, E.; Dogan, A.; Bek, B.; Ruyter, I.E. A prospective study on the clinical performance of polysiloxane soft liners: One-year results. *Dent. Mater. J.* **2008**, *27*, 440–447.
60. Hamanaka, I.; Takahashi, Y.; Shimizu, H. Mechanical properties of injection-molded thermoplastic denture base resins. *Acta Odontol. Scand.* **2011**, *69*, 75–79.
61. Sarrett, D.C. Clinical challenges and the relevance of materials testing for posterior composite restorations. *Dent. Mater.* **2005**, *21*, 9–20.
62. Wiegand, A.; Buchalla, W.; Attin, T. Review on fluoride-releasing restorative materials—fluoride release and uptake characteristics, antibacterial activity and influence on caries formation. *Dent. Mater. Off. Publ. Acad. Dent. Mater.* **2007**, *23*, 343–362.

63. Wang, Z.; Shen, Y.; Haapasalo, M. Dental materials with antibiofilm properties. *Dent. Mater.* **2014**, *30*, doi:10.1016/j.dental.2013.12.001.
64. Beyth, N.; Farah, S.; Domb, A.J.; Weiss, E.I. Antibacterial dental resin composites. *React. Funct. Polym.* **2014**, *75*, 81–88.
65. Weng, Y.; Howard, L.; Guo, X.; Chong, V.; Gregory, R.; Xie, D. A novel antibacterial resin composite for improved dental restoratives. *J. Mater. Sci. Mater. Med.* **2012**, *23*, 1553–1561.
66. Cheng, L.; Weir, M.D.; Xu, H.H.K.; Kraigsley, A.M.; Lin, N.J.; Lin-Gibson, S.; Zhou, X. Antibacterial and physical properties of calcium–phosphate and calcium–fluoride nanocomposites with chlorhexidine. *Dent. Mater.* **2012**, *28*, 573–583.
67. Kasraei, S.; Sami, L.; Hendi, S.; AliKhani, M.-Y.; Rezaei-Soufi, L.; Khamverdi, Z. Antibacterial properties of composite resins incorporating silver and zinc oxide nanoparticles on streptococcus mutans and lactobacillus. *Restor Dent Endod* **2014**, *39*, 109–114.
68. Elsaka, S.E. Antibacterial activity and adhesive properties of a chitosan-containing dental adhesive. *Quintessence Int.* **2012**, *43*, 603–613.
69. Polydorou, O.; Rogatti, P.; Bolek, R.; Wolkewitz, M.; Kümmerer, K.; Hellwig, E. Elution of monomers from three different bonding systems and their antibacterial effect. *Odontology* **2013**, *101*, 1–7.
70. Pizzo, G.; Giammanco, G.M.; Cumbo, E.; Nicolosi, G.; Gallina, G. *In vitro* antibacterial activity of endodontic sealers. *J. Dent.* **2006**, *34*, 35–40.
71. Wataha, J.C. Predicting clinical biological responses to dental materials. *Dent. Mater.* **2012**, *28*, 23–40.
72. Gnanadhas, D.P.; Marathe, S.A.; Chakravortty, D. Biocides—resistance, cross-resistance mechanisms and assessment. *Expert Opin. Investig. Drugs* **2013**, *22*, 191–206.
73. Ahamed, M.; AlSalhi, M.S.; Siddiqui, M.K.J. Silver nanoparticle applications and human health. *Clin. Chim. Acta* **2010**, *411*, 1841–1848.

Wear Characteristics of Metallic Biomaterials: A Review

Mohamed A. Hussein [1,2], Abdul Samad Mohammed [1,*] and Naser Al-Aqeeli [1]

[1] Department of Mechanical Engineering, King Fahd University of Petroleum & Minerals, Dhahran 31261, Saudi Arabia; E-Mails: mahussein@kfupm.edu.sa (M.A.H.); naqeeli@kfupm.edu.sa (N.A.-A.)

[2] Department of Mechanical Engineering, Kafrelsheikh University, Kafrelsheikh 33516, Egypt

* Author to whom correspondence should be addressed; E-Mail: samad@kfupm.edu.sa;

Academic Editor: Naozumi Teramoto

Abstract: Metals are extensively used in a variety of applications in the medical field for internal support and biological tissue replacements, such as joint replacements, dental roots, orthopedic fixation, and stents. The metals and alloys that are primarily used in biomedical applications are stainless steels, Co alloys, and Ti alloys. The service period of a metallic biomaterial is determined by its abrasion and wear resistance. A reduction in the wear resistance of the implant results in the release of incompatible metal ions into the body that loosen the implant. In addition, several reactions may occur because of the deposition of wear debris in tissue. Therefore, developing biomaterials with high wear resistance is critical to ensuring a long life for the biomaterial. The aim of this work is to review the current state of knowledge of the wear of metallic biomaterials and how wear is affected by the material properties and conditions in terms of the type of alloys developed and fabrication processes. We also present a brief evaluation of various experimental test techniques and wear characterization techniques that are used to determine the tribological performance of metallic biomaterials.

Keywords: metallic biomaterial; tribology; wear; friction

1. Introduction

For decades, metals have been used extensively in a variety of applications in the medical field. Specifically, metals are used for internal support and biological tissue replacements, such as joint replacements, dental roots, orthopedic fixation, and stents [1]. The metals and alloys that are primarily used in biomedical applications are stainless steels, Co alloys, and Ti alloys [2,3].

All biomaterials are required to satisfy various criteria, such as adequate strength, high resistance to corrosion, bioadhesion, biofunctionality, biocompatibility, high wear resistance, and low friction [4]. However, the various biomaterials that have been developed thus far do not satisfy all of the above requirements. Wear and corrosion have been reported to be the primary reasons for the failure of implant elements. Some of the applications of tribology in the biomedical field are as follows: wear of dentures [5,6], heart valves [7], plates and screws in bone fracture repair [8]; friction between garments and friction between garments and skin [9,10]; and lubrication of artificial heart pumps [11], pleural surfaces, and the pericardium. Wear is a major factor in controlling and determining the long-term clinical performance of a metallic biomaterial.

Hence, the objective of this work is to address the current state of knowledge of the wear of metallic biomaterials and how wear is affected by material properties and conditions in terms of the type of alloys developed, fabrication processes, experimental test techniques, and characterization methodologies.

2. Desired Properties of Biomaterials

A biomaterial should satisfy the criteria given below.

- **Mechanical properties:** Stress shielding can be prevented by matching the modulus of elasticity of biomaterials to that of bone, which varies from 4 to 30 GPa [12,13]. Additionally, the material should have a low modulus combined with high strength to prolong the service period of the implant and prevent loosening, thereby preventing the need for revision surgery.
- **Biocompatibility:** The developed material should be compatible with living systems and not cause any bodily harm, which includes all of the negative effects a material can have on the components of a biological system (bone, extra- and intracellular tissues, and ionic composition of plasma) [11–13].
- **High wear resistance:** The material should have a high wear resistance and exhibit a low friction coefficient when sliding against body tissues. An increase in the friction coefficient or a decrease in the wear resistance can cause the implant to loosen [14,15]. Moreover, the wear debris generated can cause inflammation that is destructive to the bone supporting the implant.
- **High corrosion resistance:** An implant that is made of a biomaterial with a low corrosion resistance can release metal ions into the body, which in turn produces toxic reactions [16].
- **Osseointegration:** Osseointegration was first defined as “a direct structural and functional connection between ordered, living bone and the surface of a load-carrying implant” [17]. The roughness, chemistry, and topography of the surface play a major role in good osseointegration [18]. Implant loosening results from the non-integration of the implant surface into the adjacent bone [19]. Few researchers mention that osseiontegration is undesirable due to the risk of not being able to remove the implant after use [20]. However, a few of them have also demonstrated that the implant could be removed safely [20]. Thus osseointegration is a desirable

property for a biomaterial in some applications such as in implant where it is to be made sure that the implant will integrate properly with the bone and other tissues [21].

- **Non-toxic:** The material should be neither genotoxic (which can alter the DNA of the genome) nor cytotoxic (causes damage to individual cells).
- **Long fatigue life:** The material should exhibit a high resistance to failure by fatigue to prevent implant failure and stress shielding from fatigue fracture. The failure of implants by fatigue has been reported for hip prostheses [22].

3. Types of Biomaterials

The materials that are used to build biomedical devices (orthopedic, dental, bone cements, *etc.*) can be classified into metallic materials, ceramics, polymers, and composites. Metallic materials within these four categories, despite some shortcomings, such as the release of metallic ions and wear debris, are widely used due to their high strength, toughness, and good biocompatibility.

3.1. Metallic Alloys for Biomaterials

The high reliability of metallic biomaterials, in terms of their mechanical performance, has resulted in their use "mainly for the fabrication of medical devices for the replacement of hard tissue such as artificial hip joints, bone plates, and dental implants" [2]. Multiple types of materials and alloys have been investigated in the medical field for their various properties and characteristics [1]. Different alloy systems have been developed for use in the medical field, including stainless steels, Co alloys, and Ti alloys. Tables 1–3 summarize the chemical composition of alloys that are registered in the ASTM Standard and have been developed for biomedical applications [23]. A brief description of each material is given below.

Table 1. Titanium alloys for biomedical applications [23].

Alloy	Microstructure
1. Pure Ti	(ASTM F67-89)
2. Ti-6Al-4V ELI (ASTM F136-84, F620-87)	α+β type
3. Ti-6Al-4V (ASTM F1108-88)	α+β type
4. Ti-6Al-7Nb (ASTM F1295-92, ISO5832-11)	α+β type(Swiss)
5. Ti-5Al-2.5Fe (ISO5832-10)	α+β type (Germany)
6. Ti-5Al-3Mo-4Zr	α+β type (Japan)
7. Ti-15Sn-4Nb-2Ta-0.2Pd	α+β type (Japan)
8. Ti-15Zr-4Nb-2Ta-0.2Pd	α+β type (Japan)
9. Ti-13Nb-13Zr (ASTM F1713-96)	near β type (U.S.A.), Low modulus
10. Ti-12Mo-6Zr-2Fe (ASTM F1813-97	β type (U.S.A.), Low modulus
11. Ti-15Mo	β type (U.S.A.), Low modulus
12. Ti-16Nb-10Hf	β type (U.S.A.), Low modulus
13. Ti-15Mo-5Zr-3Al	β type (Japan), Low modulus
14. Ti-15Mo-2.8Nb-0.2Si-0.26O	β type (U.S.A.), Low modulus
15. Ti-35Nb-7Zr-5Ta	β type (U.S.A.), Low modulus
16. Ti-29Nb-13Ta-4.6Zr	β type (Japan), Low modulus
17. Ti-40Ta, Ti-50Ta	β type (U.S.A), High corrosion resistance

Table 2. Chemical compositions of stainless steels registered in ASTM standard for biomedical applications [23].

ASTM designation	Alloy	Cr	Ni	Mo	N	Mn	C	P	S	Si	Cu	Fe
(F138-92)	Bar and Wire											
	Grade 1	17.00–19.00	13.00–15.50	2.00–3.00	−0.1	−2.0	−0.08	−0.025	−0.01	−0.75	−0.5	balance
	Grade 2	17.00–19.00	13.00–15.50	2.00–3.00	−0.1	−2.0	−0.03	−0.025	−0.01	−0.75	−0.5	balance
(F139-96)	18Cr-14Ni-2.5Mo	17.00–19.00	13.00–15.00	2.25–3.00	−0.1	−2.0	−0.03	−0.025	−0.01	−0.75	−0.5	balance
(F621-92)	Sheet and Strip	Same chemical composition as specified in Specification F138, grade 1 and 2										
(F1314-95)	Forgings Nitrogen strengthened 22Cr-12.5Ni-5Mn-2.5Mo	20.5–23.5	11.5–13.5	2.0–3.0	0.2–0.4	4.0–6.0	−0.03	−0.025	−0.01	−0.75	−0.5	balance
					(0.10 < Nb < 0.30, 0.10 < V < 0.30)							
(F1586-95)	Bar and Wire Nitrogen strengthened 21Cr-10Ni-3Mn-2.5Mo	19.5–22.0	9.0–11.0	2.0–3.0	0.25–0.5	2.00–4.25	−0.08	−0.025	−0.01	−0.75	−0.25	balance
						0.25 < Nb < 0.80						

Table 3. Chemical compositions of Co alloys registered in ASTM standard for biomedical applications [23].

ASTM designation	Alloy	Cr	Mo	Ni	W	Fe	Ti	C	Si	P	S	Mn	Co
(F75-92)	Co-Cr-Mo Cast alloy	27.0–30.0	5.0–7.0	−1.0		−0.75		−0.35	−1.0			−1.0	balance
(F90-96)	Co-20Cr-15W-10Ni Wrought alloy	19.0–21.0		9.0–11.0	14.0–16.0	−3.0		0.05–0.15	−0.4	−0.03	−0.03	1–2	balance
(F562-95)	Co-35Ni-20Cr-10Mo Wrought alloy	19.0–21.0	9–10.5	33.0–37		−1.0	−1.0	0.025	−0.15	−0.015	−0.01	−0.15	balance
							(B < 0.0015)						
(F563-95)	Co-Ni-Cr-Mo-W-Fe Wrought alloy	18-22	3–4	15-25	3-4	4-6	0.5-3.5	0.05	0.5		0.01	1.0	balance
(F799-96)	Co-28Cr-6Mo forgings	26.0-30.0	5–7	−1.0		−0.75		−0.35	-1.0			−1.0	balance
(F1058-91)	Co-Cr-Ni-Mo-Fe Wrought alloy												
	Grade 1	19.0–21.0	6.0–8.0	14.0–16.0		balance		0.15	−1.2	−0.015	−0.015	1.5–2.5	39.0–41
							(Be < 0.01)						
	Grade 2	18.5–21.5	6.5–7.5	15.0–18.0		balance		0.15	−1.2	−0.015	−0.015	1.0–2.0	39.0–42
							(Be < 0.001)						
(F1537-94)	Co-28Cr-6Mo Wrought alloy	26.0–30.0	5.0–7.0	−1.0		−0.75		0.35	−1.0			1.0	balance
							(N < 0.25)						

3.1.1. Ti Alloys

The high biocompatibility of Ti and Ti alloys has resulted in their preferential use over other alloy systems in the medical and dentistry fields [24–26]. The primary characteristics of Ti alloys that have resulted in their being one of the main choices in the biomedical field include good mechanical properties, excellent corrosion behavior because of a TiO_2 solid oxide layer, good biocompatibility, a relatively low Young's modulus, light weight, and non-magnetic behavior. The aforementioned characteristics make Ti and Ti alloys the preferred choices for implantation. However, Ti alloys exhibit poor tribological properties [27] because of "low resistance to plastic shearing, low work hardening, and low protection exerted by surface oxides" [27].

3.1.2. Stainless Steels

The austenitic stainless steel SUS 316L is the only reported stainless steel that is used in the biomedical field. However, a few researchers have found that the Ni contained in this alloy causes allergic reactions [2]. Moreover, pitting, crevice, and stress corrosion have been reported for implants fabricated from SUS 316L [23]. To prevent Ni allergic reactions, an austenitic stainless steel with high nitrogen content has been developed. Therefore, the new research trend is to develop Ni-free stainless steels.

3.1.3. Co Alloys

The wear resistance of Co alloys is higher than that of both Ti alloys and stainless steel alloys [23]. In artificial hip joints, the head of the joint is subjected to wear. Thus, hip joints have been fabricated from Co alloys, such as Co-Cr-Mo alloys, which exhibit high strength and ductility. Dispersing carbide in Co alloys has been reported to increase the resistance to wear of these alloys [2]. Furthermore, the transformation of the metastable γ phase to the ε martensitic phase (via a deformation-induced transformation) has been found to improve the wear resistance of Co alloys [2]. Compared to cast Co-Cr alloys, wrought Co-Cr alloys can be used for implant devices with high strength requirements. However, the Ni content in wrought Co-Cr alloys causes allergic reactions [23].

Some of the mechanical properties of metallic biomaterials are compared in Table 4. Examples of the metallic alloys used in biomedical applications, their advantages and disadvantages are summarized in Table 5.

As observed in Table 4, the Young's Modulus of Co-Cr alloys and stainless steel is found to be 10 × that of the bone, which may cause stress shielding. However, the Young's modulus of titanium and its alloys is approximately 0.5× that of stainless steel, and hence the risk of stress shielding is less in titanium and its alloys compared to that of Co-Cr alloys and stainless steel.

Table 4. Comparison of mechanical properties of metallic biomaterials with bone [28].

Material	Young's Modulus, E (GPa)	Yield Strength, (MPa)	Tensile Strength (MPa)	Fatigue Limit, (MPa)
Stainless steel	190	221–1213	586–1351	241–820
Co-Cr alloys	210–253	448–1606	655–1896	207–950
Titanium (Ti)	110	485	760	300
Ti-6Al-4V	116	896–1034	965–1103	620
Cortical bone	15–30	30–70	70–150	

Table 5. Comparison of metallic biomaterials used in the human body.

Metals and alloys	Selected examples	Advantages	Disadvantages	Principal applications [29]
Titanium-based Alloys	CP-Ti, Ti-Al-V, Ti-Al-Nb, Ti- 13Nb-13Zr, Ti-Mo-Zr-Fe	High biocompatibility [24–26]. Low Young's modulus excellent corrosion resistance, low density	Poor tribological properties [27], Toxic effect of Al and V on long term	Bone and joint replacement, fracture fixation, dental implants, pacemaker encapsulation
Cobalt and Cr alloys	Co-Cr-Mo, Cr-Ni-Cr-Mo	High wear resistance [20]	Allergy consideration with Ni, Cr and Co [2] much higher modulus than bone	Bone and joint replacement, dental implants, dental restorations, heart valves
Stainless steels	316L stainless steel	High wear resistance [23]	Allergy consideration with Ni, Cr and Co [2] much higher modulus than bone	Fracture fixation, stents, surgical instruments
Others	Ni-Ti	Low Young's modulus	Ni cause allergy [2]	Bone plates, stents, orthodontic wires
	Platinum and Pt-Ir	High corrosion resistant under extreme voltage potential and charge transfer conditions [30]		Electrodes
	Hg-Ag-Sn amalgam	Easy *in situ* formability to a desired shape susceptible to corrosion in the oral environment [30]	Concerns related to Hg toxicity [30]	Dental restorations

3.1.4. Limitations of Current Metallic Biomaterials

The presence of elements such as Ni, Cr, and Co in both stainless steel and Co-Cr alloys has toxic effects [31]. Ni toxicity leads to dermatitis. The long-term existence of Al and V ions in Ti alloys has been found to cause Alzheimer's disease, osteomalacia, and neuropathy in the long term [32]. The presence of Co has also been reported to have carcinogenic effects [33]. Recently, it is reported in [34] that stainless steels and Co-Cr alloys usually contain some harmful elements, such as Ni, Co, and Cr. In addition, 6Al-4V alloy is composed of cytotoxic elements like Al and V, which may cause severe problems once released inside the human body.

A high friction coefficient and wear debris formation can produce an inflammatory reaction, leading to the loosening of implants due to osteolysis [35]. A high modulus of elasticity leads to stress shielding, which causes implant failure. Figure 1 summarizes the reported causes of implant failure.

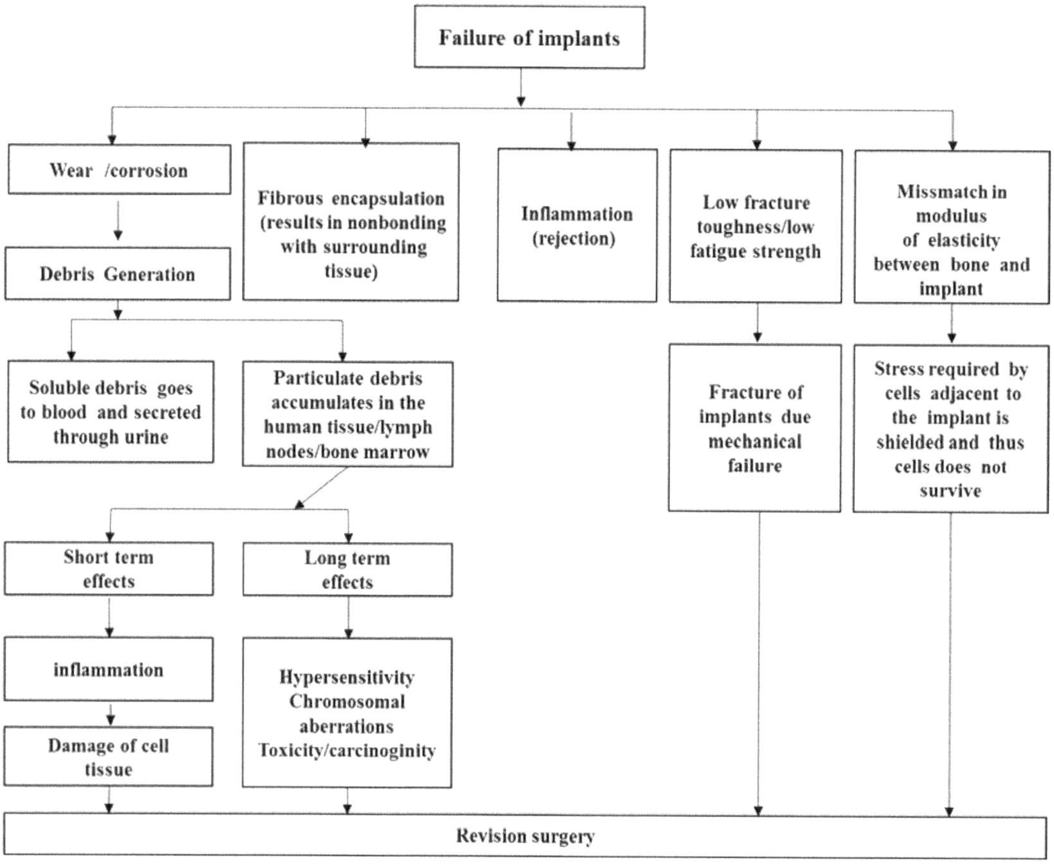

Figure 1. Reported causes of implant failure [18].

4. Wear of Metallic Biomaterials

The aforementioned discussion illustrates that the wear resistance of the biomaterial plays a significant role in the proper functioning of the material. Hence, the wear resistance of a biomaterial is clinically important. Several studies have been performed to investigate the tribological properties of developed biomaterials. We first summarize various wear test configurations that have been used by various researchers, followed by a description of the different techniques that are used to characterize wear and a review of various results for the wear and friction that have been obtained by different researchers.

4.1. Wear Testing Methods

Given the aforementioned limitations, especially in terms of the tribological properties, it is critical to characterize the wear and friction of developed biomaterials using a suitable test methodology. The methods that are most commonly used in the literature to study the tribological behavior of metallic biomaterials are the block-on-disc [36,37], ball-on-disc [38–41], and pin-on-disc [42], as shown in Figure 2. The temperature of the tests was selected to be 37 ± 0.1 °C to simulate real [37,38] or ambient [36] conditions. The wear tests were conducted in an environment of simulated body fluids (Ringer's solution) [36] or fluids containing NaCl and phosphate-buffered solutions (PBS) [26]. However, in a few studies, tests were also conducted under dry sliding conditions [37,39]. Table 6 summarizes the advantages and disadvantages of various wear test configurations.

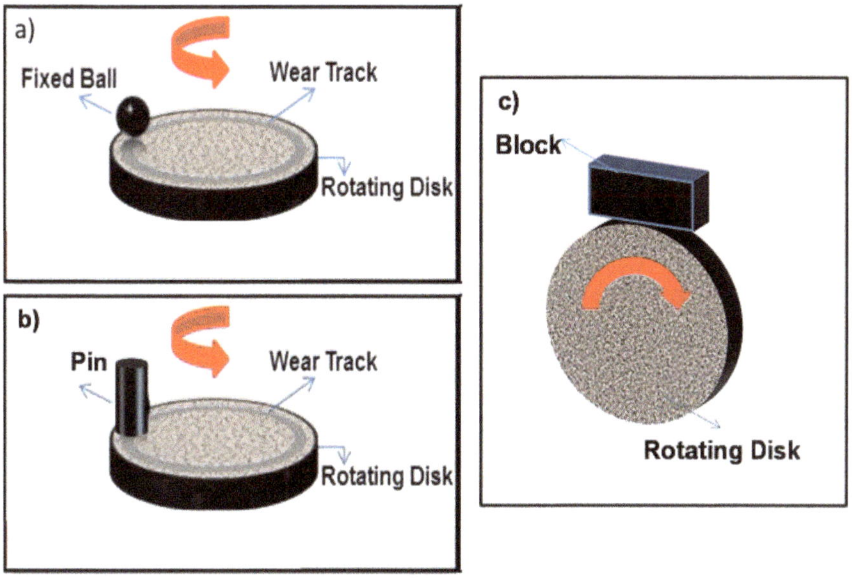

Figure 2. Schematic of (**a**) ball-on-disc, (**b**) pin-on-disc, and (**c**) block-on-disc wear test configurations.

Table 6. Advantages and disadvantages of various wear test configurations [43].

Test	Advantages	Disadvantages
Pin-on-Disk	After run-in, surface pressure remains constant. Easy to determine wear volume and wear rate. The model closely simulates a linear friction bearing.	Difficult to align pin. If the pin does not stand perfectly vertical on the plate, edge contact results. A very long run-in time is therefore necessary. The front edge of the pin can skim off lubricant. This makes a defined lubrication state impossible.
Ball-on-Disk	High surface pressures are possible. The ball skims off lubricant less than a pin does. The model is similar to a linear friction bearing and a radial friction bearing.	Very small contact ratio: The contact surface of the ball is small compared to the sliding track on the disk. The contact area is enlarged by wear. Difficult to determine the wear volume of the ball.
Block-on-disc	The model is capable of simulating a variety of harsh field conditions, e.g., high temperature, high speed, and high loading pressure.	

4.2. Characterization Techniques for the Wear of Biomaterials

Different characterization techniques have been used to evaluate tribological behavior. Chemical analyses, such as energy-dispersive X-ray spectroscopy (EDXS/EDS), have been used for selected regions to determine the composition. A scanning electron microscope/ light microscope (SEM/LM) has been used to determine the wear mechanism or to perform a post-test examination [30]. A 3D Talysurf instrument has been used to measure the 3D surface roughness [38]. X-ray photoelectron spectroscopy (XPS) has been used to study the surface chemistry. The volume loss in the wear track has been measured using an optical profilometer/3D-profilometer. An atomic force microscope (AFM) has also recently been used [44] to study different tribological phenomena, such as friction, surface roughness, scratching, and adhesion. Moreover, interesting mechanical properties, such as the modulus of elasticity and hardness, could be measured using AFM with a depth-sensing indentation system.

4.3. Wear Performance of Different Biomaterials

Table 7 lists various developed materials and their fabrication techniques, the types of wear tests, and the parameters used to characterize wear, along with a summary of the results obtained by various researchers.

Table 7. Wear studies of metallic biomaterials.

First author, year	Material & fabrication processes	Experimental test techniques & parameters	Main results
Cvijovic'-Alagic *et al.* [36]	Ti-13Nb-13Zr Ti-6Al-4V Arc melting	A **block-on-disc** tribometer was used to conduct wear and friction tests in a simulated body fluid (Ringer's solution). **Temperature**: ambient **Normal load**: 20–60 N **Sliding speed**: 0.26–1.0 m/s	The Ti-6Al-4V alloy showed a higher wear resistance than the Ti-13Nb-13Zr alloy. Abrasion was the primary wear mechanism.
Stefano Gialanella *et al.* [37]	NiTi Commercial alloy	A **block-on-disc** was used to measure dry sliding wear. A profilometer was used to quantify wear. **Sliding speed**: 0.837 ms^{-1} **Sliding distance**: 1004 m **Loads**: 50 to 200 N	A NiTi/WC-Co coupling exhibited a high wear rate. **Wear mechanism**: a transition from delamination wear to a regime featuring a mixture of delamination and oxidation wear.
K.S. Suresh *et al.* [38]	Ti-13Nb-13Zr Equal channel angular pressing (ECAE)	A tribometer was used as a lubricity fretting test system for texture and wear behavior; fretting wear and3D surface texture measurements were performed. **Normal loads**: 6 N **Frequency**: 20 Hz **Temperature**: 37 ± 0.1 °C	The grain size and the texture of material affected the wear of the surface. There was no difference in the friction coefficient between the ECAE processed and as-received samples.
Li-juan Xu *et al.* [39]	β-type Ti-15Mo-xNb arc-melting vacuum-pressure casting system	A **ball-on-disc** was used for dry wear tests. **Normal load**: 1 N and 2 N **Test-disc rate**: 100 r/min	The lowest friction coefficient was obtained for a Ti-15Mo-5Nb alloy under a 1-N load. Adhesion was the primary wear mechanism.
M. Fellah *et al.* [41]	Ti-6Al-7Nb and AISI 316L stainless steel	**Ball-on–disc** and sphere-on-plane **Load**: 3 N, 6 N and 10 N **Sliding speed**: 1 mm/s, 15 mm/s and 25 mm/s	The same mechanisms of wear and friction were found for all of the tested samples.
S.J. Li *et al.* [42]	Ti-Nb-Ta-Zr and Ti-6Al-4V induction skull melting method	**Reciprocal pin-on-disc** in a 0.9% NaCl solution **Reciprocating velocity**: 45 rpm **Sliding distance**: 30 km	The wear resistance of Ti-29Nb-13Ta-4.6Zr was enhanced by incorporating Nb_2O_5 oxide particles into the diffusion-hardened surface of the alloy.

Table 7. *Cont.*

First author, year	Material & fabrication processes	Experimental test techniques & parameters	Main results
Animesh Choubey *et al.* [45]	CP Titanium, Ti-6Al-4V, Ti-5Al-2.5Fe, Ti-13Nb-13Zr and Co-28Cr-6Mo	**Ball on flat fretting** wear tester: Hanks' balanced salt solution **Normal load**: 10 N for 10,000 cycles **Frequency**: 10 Hz	The primary wear mechanisms of Ti alloys were tribomechanical abrasion, transfer layer formation and cracking.
A. Iwabuchi *et al.* [46]	Co-29Cr-6Mo alloy and Ti-6Al-4V	**Fretting apparatus and a reciprocating** sliding tribometer: Quasi-body fluid, Hanks's solution. **Normal load**: 5 N; **frequency**: 10 Hz **Temperature** in the solutions: 37 ± 2 °C	Co alloy exhibited good wear resistance; Ti alloy exhibited good fretting resistance.
X. Luo *et al.* [47]	ASTM F1537 Co-Cr alloy	**Pin-on-disc** tribometer **Load**: 20 N **Rotation speed**: 60 (rpm)	The tribocorrosion properties of the Co-Cr alloy were enhanced by a layer of the S-phase.
Akihiko Chiba *et al.* [48]	Co-Cr-Mo forged	**Pin-on-disc** **Load:** 9.8 N 24 rpm	Forged CoCr exhibited a lower wear loss than a cast CoCr alloy.
S. M. T. Chan *et al.* [49]	(CoCr), stainless steel (SS)	**Pin-on-disc** **sliding speed**: 0.5 mm/s, 5 mm wear track radius **Normal load**: 1.8 N	
Alfons Fischer *et al.* [50]	AISI 316L CoCr29Mo6	**Pin-on-disc** for dry sliding wear tests **Load**: 5 N **Relative velocity**: 0.1 m/s **Ambient temperature:** 25 °C	Ni-free high-nitrogen steel and LC-CoCrMo alloy exhibited higher wear resistance and dry friction than Ni-containing austenitic steels.
A. Igual Muñoz *et al.* [40]	Co-Cr-Mo Low and high carbon	**Tribocorrosion techniques** **Load**: 1.2 N; **frequency**: 1 Hz **Temperature**: 37 ± 0.1 °C Simulated body fluids [NaCl and phosphate-buffered solutions (PBS) with and without albumin]	LC CoCrMo had a higher wear resistance in NaCl and PBS albumin than HC. No differences were observed for the alloys in the other solutions.
M. Alvarez-Vera, *et al.* [51]	Co-Cr alloy with boron additions (0, 0.3, 0.6 and 1 B wt%) by casting method	three-axial hip joint simulator	Wear resistance as the boron increased.
L. Mohan, *et al.* [52]	Commercial Ti-13Nb-3Zr alloy oxygen implanted	Reciprocating type wear tester normal forces: 3, 5 and 10 N. The stroke length: 10 mm and an alumina ball of 6 mm diameter was used as the counter surface	The implanted samples display a lower friction coefficient as compared to the substrate one.
H. Attar *et al.* [53]	Commercially pure titanium (CP-Ti) parts produced using selective laser melting (SLM) and casting	a pin-on-disc at room temperature A stainless steel disc of 45 mm diameter loads: (15 N, 20 N,25 N and 30 N) sliding speed: 0.5 m/s for 15 min.	SLM CP-Ti showed better wear resistance compared to casting as a result of fine grains and higher microhardness.

Li *et al.* [42] studied the effects of the Nb content, surface modification, the material of the counterface, and heat treatment on the wear characteristics of Ti-Nb-Ta-Zr and Ti-6Al-4V (TAV1) alloys. The authors found that increasing the Nb content improved the wear resistance. Heat treatment enhanced the resistance to wear of Ti-29Nb-13Ta-4.6Zr (TNZT1) because of the formation of oxide particles from Nb_2O_5. The material of the counterface was reported to have a significant effect on the wear loss. Although no wear was observed for the sliding of polyethylene (UHMWPE) and a pig bone on the oxidized surface of Ti-29Nb-13Ta-4.6Zr, the wear loss from the sliding of these materials on stainless steel was higher than on both the TNZT1 and TAV1 alloys. Figure 3 compares the morphology for different alloys sliding against different counterfaces.

Choubey *et al.* [45] investigated the tribological characteristics of commercially pure (CP) Ti, Ti-6Al-4V, and Ti-5Al-2.5Fe in Hanks's solution (a simulated body fluid solution). The observed wear mechanisms were tribo-mechanical abrasion, transfer layer formation, and cracking. The predominant wear mechanism was tribo-mechanical wear. Figure 4 compares the steady-state coefficient of friction (COF) for the studied materials. The COF of Co-28Cr-6Mo under fretting was 0.4, whereas a superior (lower) value of 0.3 was observed for a Ti-5Al-2.5 Fe/steel couple.

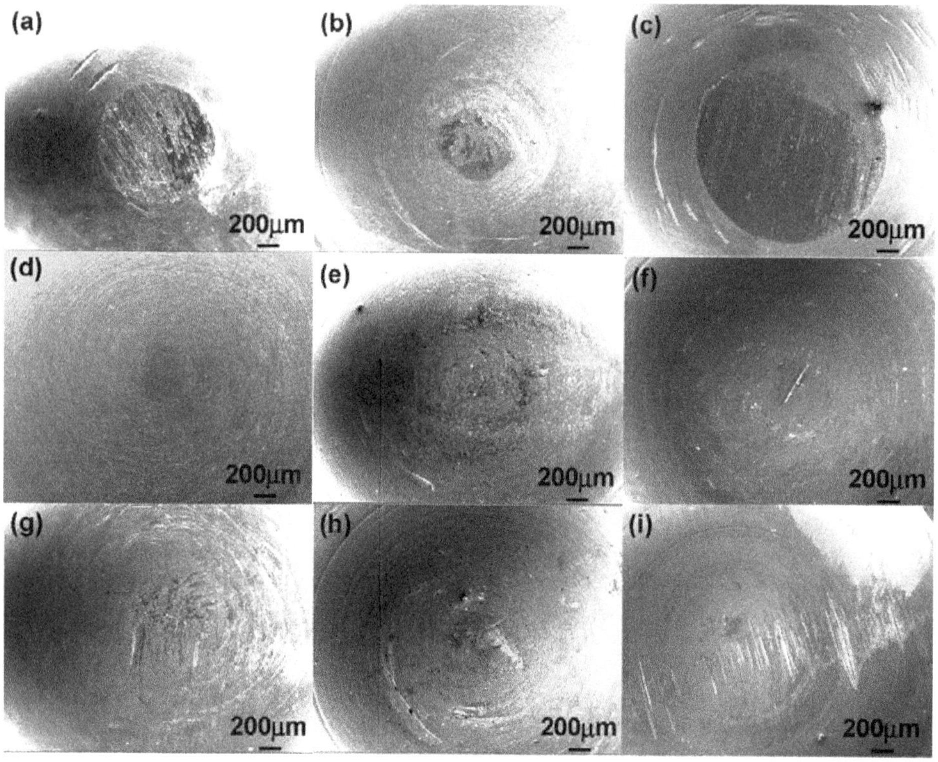

Figure 3. Wear morphology of (**a**) TNZT1, (**b**) TNZT3, and (**c**) TAV1 sliding on a stainless steel plate; (**d**) TNZT1, (**e**) TNZT3, and (**f**) TAV1 sliding on UHMWPE; (**g**) TNZT1, (**h**) TNZT3, and (**i**) TAV1 sliding on a pig bone in 0.9% NaCl [36].

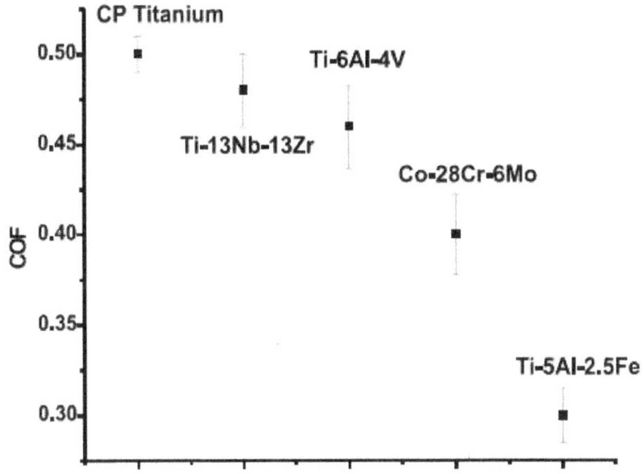

Figure 4. Steady-state friction coefficient for the studied materials [45].

Cvijovic'-Alagic *et al.* [36] compared the tribological behavior of Ti-13Nb-13Zr and Ti-6Al-4V ELI under different heat treatment conditions. The results showed that the wear resistance of the Ti-6Al-4V ELI alloy was superior to that of the Ti-13Nb-13Zr alloy under all of the heat treatment conditions. The martensitic microstructure of the WQ Ti-6Al-4V ELI alloy increased the hardness of the alloy, resulting in superior resistance to plastic deformation during the wear test. The wear mechanism observed for this alloy was predominantly abrasion. The lower hardness of the Ti-13Nb-13Zr alloy resulted in a higher amount of wear loss than for the Ti-6Al-4V alloy.

Suresh *et al.* [38] studied the wear behavior of an ultra-fine grained (UFG) Ti-13Nb-Zr alloy. The samples were processed by equal channel angular extrusion (ECAE). The effect of the surface roughness on the wear behavior was also studied. The authors concluded that both the grain size and surface texture affected the mechanical properties and wear of the as-received alloy. However, there was no significant difference between the average friction coefficient of the ECAE-processed alloy and that of the as-received sample. In addition, no major change was observed in the fretting wear after ECAE processing for both samples. This result was attributed to the absence of a significant increase in the hardness of the samples after ECAE. Abrasion was observed to be the predominant wear mechanism. A few researchers [54,55] reported that the UFG nanostructured materials increase the hardness and lead to enhancement in the wear resistance and resistance to wear debris formation. The UFG materials obtained by the severe plastic deformation (SPD) process exhibit lower friction coefficients and better wear resistance [56]. UFG Ti samples in two processing states—equal channel angular pressing (ECAP) for eight passes and ECAP + further cold rolling for 75% strain—were studied and the results showed that the UFG structure results in a lower adhesion component and consequently lowers friction coefficient, which in turn improved the wear property [56].

Xu *et al.* [39] investigated the wear resistance of Ti-15Mo-xNb (x = 5%, 10%, 15%, and 20%) alloys under dry conditions. The friction coefficient was observed to increase with the Nb content. Adhesive wear was observed to be the primary wear mechanism.

Fellah *et al.* [41] examined the wear behavior of Ti-6Al-7Nb and AISI 316L stainless steel alloys at different sliding speeds and loads. Figure 5 compares the mean values of the friction coefficient for both alloy systems under different conditions. The same friction and wear mechanisms were observed for the tested samples. The change in the wear rate with sliding speed was not significant for the Ti-6Al-7Nb alloy. An increase in the friction coefficient with the sliding speed was observed for both alloys. The wear mechanisms that were observed at high speed were plastic deformation and adhesive wear.

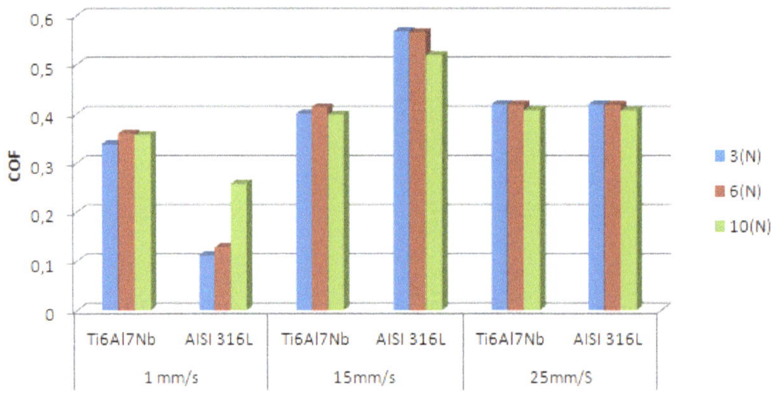

Figure 5. Mean friction coefficients of AISI 316L and Ti-6Al-7Nb [41].

Iwabuchi *et al.* [46] conducted a fretting wear test and a reciprocating sliding wear test on a Co-29Cr-6Mo alloy, a Ti-6Al-4V alloy, and a SUS304 stainless steel alloy in Hanks's solution. The results showed the good fretting resistance of the Ti-6Al-4V alloy relative to that of the standard SUS304 stainless steel alloy, whereas the corresponding resistance to sliding wear was poor because of abrasion. The Co-29Cr-6Mo alloy exhibited good wear resistance, and the synergy effect was considerably stronger in the fretting test than in the sliding wear test. Figure 6 compares the mean value of the friction coefficient in pure water to that in Hanks's solution. The figure clearly shows higher mean values for the friction coefficient in the fretting test than the sliding test for both liquids.

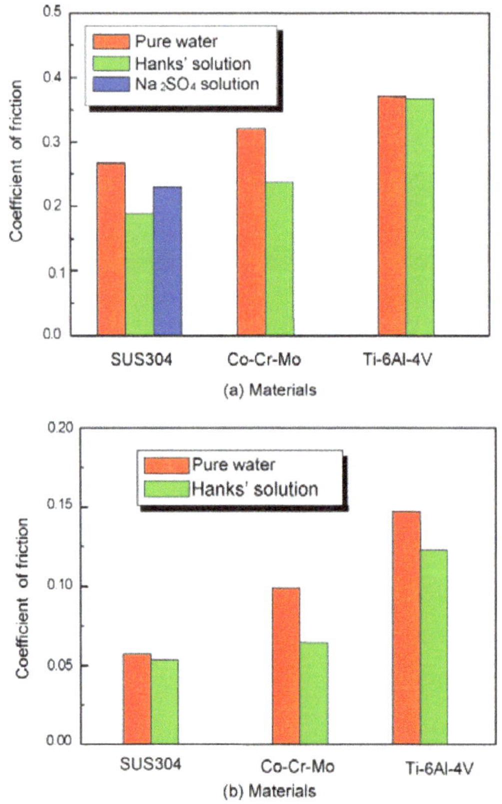

Figure 6. Mean friction coefficient against an Al_2O_3 ball in (**a**) fretting and (**b**) sliding [46].

Chiba *et al.* [48] studied the wear characteristics of a forged Co-Cr-Mo alloy and a cast CoCr alloy with high carbon contents. The results showed that the wear resistance of the forged CoCr alloy was higher than that of the cast CoCr alloy. The higher wear loss of the cast CoCr alloy resulted from the precipitation of carbide in this alloy. Figure 7 is a comparison of the SEM micrographs of the wear scars on both surfaces. The number of grooves and scratches in the cast CoCr alloy was higher than that in the forged CoCr alloy. The measured friction coefficient of the forged alloy was higher than that of the cast alloy.

Muñoz *et al.* [40] used tribo-corrosion to compare the wear of CoCrMo alloys in two forms—low carbon (LC) and high carbon (HC)—in four different simulated body fluids. The results showed that the wear behavior of these alloys depended on the surrounding environment. There was no difference in the wear resistance between the LC and HC alloys in both NaCl and PBS without albumin. The LC alloy had a higher wear resistance than the HC alloy in NaCl and PBS with albumin. The difference in the alloy behavior under different solutions was attributed to the enhancement of corrosion from sliding and chemical effects at the surface that affected both the mechanical wear and third-body behavior. Figure 8 compares the wear tracks for both the LC and HC alloys under different solution conditions. The optical microscope images clearly show scratches over the scar length, illustrating that abrasion was the predominant wear mechanism. The wear debris accumulated around the track for both alloys in the NaCl solution and was higher than in the PBS solution. Interestingly, the presence of albumin in the solutions reduced the wear debris for both alloys.

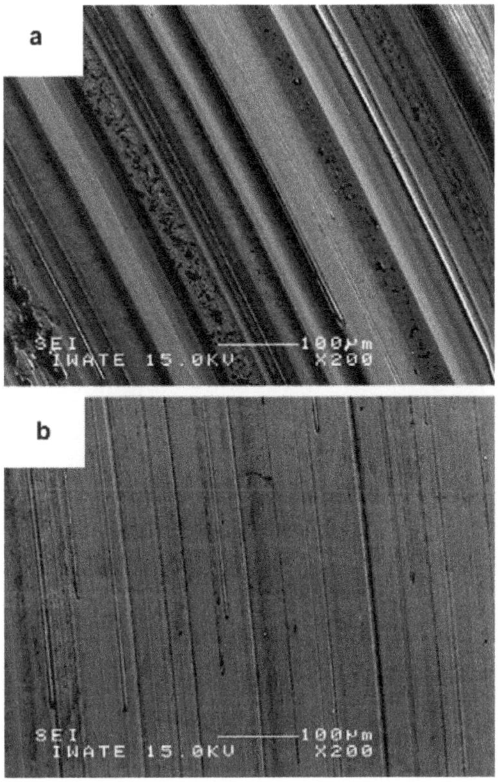

Figure 7. SEM micrographs of wear scars of (**a**) cast CoCr alloy and (**b**) forged CoCr alloy [48].

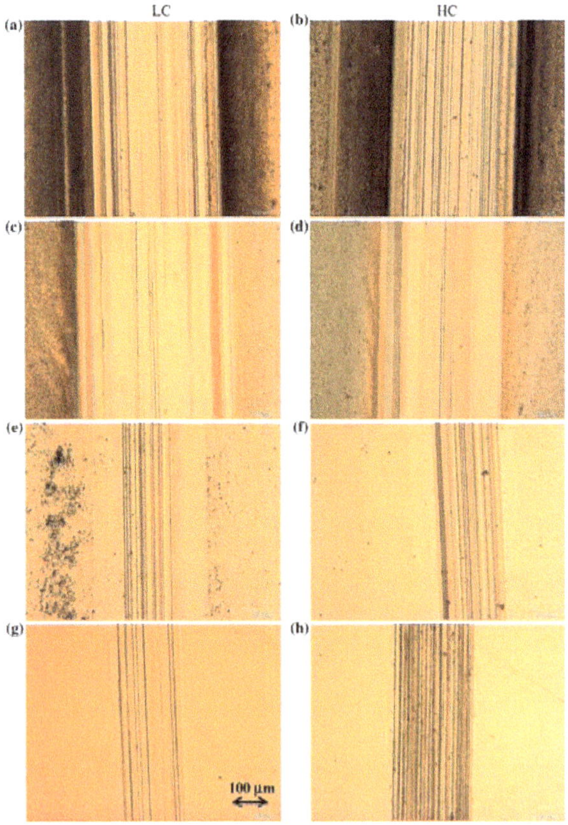

Figure 8. Optical microscopy images of wear trace LC and HC alloys during sliding in (**a**) and (**b**) NaCl; (**c**) and (**d**) NaCl with albumin; (**e**) and (**f**) PBS; and (**g**) and (**h**) PBS with albumin [40].

Fischer *et al*. [50] examined the wear characteristics of CrNiMo steels, low-carbon Co-base alloy CoCr29Mo, CrNiMnMoN steel, and Ni-free CrMnMoN steel alloys in sliding wear. They evaluated the wear characteristics in sliding wear using a pin-on-disk tribometer and torsional fatigue tests followed by electron microscopy. The results showed that Ni-free CrMnMo exhibited superior wear behavior compared to CrNiMo steels. This improvement was found to result from planar slip because of the formation of nano-sized wear particles in the CrMnMo alloy containing C and N. However, the CrNiMo containing Ni exhibited wavy-slip. The wear particles that formed were either nano- or micron-sized and had a higher wear rate than that of Ni-free high-nitrogen steel and the LC CoCrMo alloy. This difference in the wear rate was attributed to the tribo-chemical reactions of the nano-sized wear particles, which resulted in surface nano-fatigue. In contrast, the micron-sized wear particles produced abrasion, micro-fatigue, and micro-ploughing that inhibited tribo-chemical reactions.

By decreasing the length-scale of the contact, the hardness and yield stress will increase [57–59]. Therefore, it is important to study the tribological and mechanical properties of the biomaterials at the relevant scale.

Beake *et al*. [60], studied the nano-scratch and nano-fretting of Ti6Al4V, 316L stainless steel and CoCr alloy. The nano-scratch and nano-fretting tests were performed with a commercial nanomechanical test system. Tests were performed at 25 °C using a 3.7 mm sphero-conical diamond indenter. The results showed that the CoCr alloy possesses better wear resistance over a wide range of experimental conditions compared to Ti6Al4V, and 316L stainless steel alloys exhibited decreasing wear resistance with an increase in the fretting load.

Sun *et al.* [61], investigated the Micro-abrasion mechanisms of cast CoCrMo in simulated body fluids. A modified Phoenix Tribology TE/66 micro-abrasion tester was semi immersed in a liquid tank. Moreover, the nanoindentation and nano-scratch test were performed to study the nano-scale material deformation occurring during micro abrasion process. The results showed that the abrasive wear rate and wear mechanisms of the CoCrMo are dependent on the nature of the third body abrasives, their entrainment into the contact, and the presence of the proteins. The interaction between the specimen and the abrasive was affected by the presence of protein due to its influence on the solution viscosity. The existence of protein at lower abrasive volume fractions acts as a boundary lubricant and reduces the wear loss; however, at a high volume fraction of abrasive, the existence of protein enhanced the wear loss.

This study is important for *in vivo* wear corrosion study of the alloys in the hip joint as the hard particles sizes are small and less in quantity.

4.4. Techniques to Improve Wear Resistance of Metallic Biomaterials

In addition to the various processing techniques and compositional changes that enhance the wear resistance of metallic biomaterials, the following surface modification techniques for improving the wear resistance of the biomaterials have also been reported in the literature.

- **Ion implantation (physical deposition)** is considered a simple technique for significantly modifying the physical and/or chemical properties of the near surface of a material in which suitable ions are embedded into the surface of a material from a beam of ionized particles. This technique has been reported to improve the wear performance of Ti6Al4V and Co28Cr6Mo alloys [62].
- **Nitriding** (a thermo-chemical surface treatment) has been used to increase the resistance of a Ti6Al4V alloy to dry wear [63,64].
- **Carburization and boriding** techniques are used to enhance surface hardness, which in turn improves the wear resistance.
- **Plasma spray coating** has also been used to enhance the wear resistance of few biomaterials.

5. Summary

An extensive literature survey has shown that the wear resistance of Ti and Co alloys has been improved under different conditions in extensive studies. We summarize the results of these studies for Ti- and Co-based metallic biomaterials below.

Ti alloys

- In general, adding Nb to Ti alloys enhances the wear resistance of these alloys and slightly increases the friction coefficient primarily because of the increase in the hardness of the alloy. The heat treatment of these alloys has been observed to further increase the wear resistance because of the formation of Nb_2O_5 particles.
- Abrasive wear has been observed to be the predominant wear mechanism.
- Hence, surface treatments and coating are necessary to enhance the resistance of alloys to wear and friction.

Co alloys

- Forged CoCr alloys exhibit higher wear resistance than cast CoCr alloys. However, the friction coefficient of the forged CoCr alloys has been observed to be higher than that of the cast alloys.
- The wear behavior of LC and HC CoCrMo alloys depends on the surrounding environment.
- Ni-free CrMnMo exhibits improved wear behavior compared to CrNiMo steels.

It is important to study the tribological and mechanical properties of the biomaterials at the relevant scale.

Acknowledgments

The authors would like to acknowledge the support provided by King Abdulaziz City for Science and Technology (KACST) through the Science & Technology Unit at King Fahd University of Petroleum & Minerals (KFUPM) for funding this work through project No.12-ADV3036-04 as part of the National Science, Technology and Innovation Plan and King Fahd University of Petroleum & Minerals (KFUPM) through Project No. IN131020.

Author Contributions

Mohammed A. Hussein and Abdul Samad, Mohammed conceived the idea for this paper. Mohammed A. Hussein was instrumental in collecting and putting all the literature together in the form of this review. Abdul Samad, Mohammed contributed by guiding and structuring the complete document into a comprehensive review paper and checking for the language coherency. Naser Al-Aqeeli contributed by helping in editing the final document. All the authors contributed to the proof reading of the final document.

Conflicts of Interest

The authors declare no conflict of interest.

References

1. Park, J.; Lakes, R.S. *Biomaterials an Introduction*, 3rd ed.; Springer: Berlin/Heidelberg, Germany, 2007.
2. Niinomi, M. Metallic biomaterials. *J. Artif. Organs* **2008**, *11*, 105–110.
3. Karanjai, M.; Sundaresan, G.V.N.; Rao, T.R.; Mohan, R.; Kashyap, B.P. Development of titanium based biocomposite by powder metallurgy processing with *in situ* forming of Ca–P phases. *Mat. Sci. Eng. A* **2007**, *447*, 19–26.
4. Patel, N.R.; Gohil, P.P. A review on biomaterials: Scope. Application s & human anatomy significance. *Int. J. Emerg. Technol. Adv. Eng.* **2012**, *2*, 91–101.
5. Litonjua, L.A.; Andreana, S.; Bush, P.J.; Cohen, R.E. Tooth wear: Attrition, erosion, and abrasion. *Quintessence Int.* **2003**, *34*, 435–446.
6. Turssi, C.P.; Purquerio, B.; Serra, M.C. Wear of dental resin composites: Insights into underlying processes and assessment methods—A review. *J. Biomed. Mater. Res.* **2003**, *65B*, 280–285.
7. Reul, H.; Schmitz, C.; Pfaff, E.M.; Hohlstein, C.; Schmidt, P.A.; Rau, G.; Arru, P. *In-vitro* assessment of the wear development mechanism and stabilization of wear in the Edwards MIRA/Sorin Bicarbon mechanical heart valve orifice ring. *J. Heart Valve D* **2002**, *11*, 409–418.

8. Shahgaldi, B.F.; Compson, J. Wear and corrosion of sliding counterparts of stainless-steel hip screw-plates. *Injury* **2000**, *31*, 85–92.
9. Dowson, D. Tribology and the skin structure. In *Bioengineering of the Skin: Methods and Instrumentation, Part III: General Aspects*; Berardesca, E., Elsner, P., Wilhelm, K.P., Maibach, H.I., Eds.; CRC Press: Boca Raton, FL, USA; 1995; p. 159.
10. Sivamani, R.K.; Goodman, J.; Gitis, N.V.; Maibach, H.I. Coefficient of friction: Tribological studies in man—An overview. *Skin Res. Technol.* **2003**, *9*, 227–234.
11. Walowit, J.A. The analysis, design, and testing of a blood lubricated hydrodynamic journal bearing. *ASAIO J.* **1997**, *43*, M556.
12. Lawrence, K.J. Anisotropy of Young's modulus of bone. *Nature* **1980**, *283*, 106–107.
13. Black, J.; Hastings, G.W. *Handbook of Biomaterials Properties*; Chapman and Hall: London, UK, 1998.
14. Alvarado, J.; Maldonado, R.; Marxuach, J.; Otero, R. Biomechanics of hip and knee prostheses. *Appl. Eng. Mechan. Med. GED–Univ. Puerto Rico Mayaguez* **2003**, *6*, 22.
15. Ramsden, J.J. The design and manufacture of biomedical surfaces. *CIRP Ann.-Manuf. Technol.* **2007**, *56*, 687–711.
16. Hallab, N.J.; Anderson, S.; Stafford, T.; Glant, T.; Jacobs, J.J. Lymphocyte responses in patients with total hip arthroplasty. *J. Orthop. Res.* **2005**, *23*, 384–391.
17. Branemark, P.I. Osseointegration and its experimental background. *J. Pros. Dent.* **1983**, *50*, 399–410.
18. Geetha, M.; Singh, A.K.; Asokamani, R.; Gogia, A.K. Ti based biomaterials, the ultimate choice for orthopaedic implants—A review. *Progr. Mater. Sci.* **2009**, *54*, 397–425.
19. Viceconti, M.; Muccini, R.; Bernakiewicz, M.; Baleani, M.; Cristofolini, L. Large-sliding contact elements accurately predict levels of bone–implant micromotion relevant to osseointegration. *J. Biomech.* **2000**, *33*, 1611–1618.
20. Wennerberg, A.; Albrektsson, T.; Jimbo, R. *Implant Surfaces and Their Biological and Clinical Impact*, 1st ed.; Springer-Verlag: Berlin/Heidelberg, Germany, 2015; p. 168.
21. Barfeie, A.; Wilson, J.; Rees, J. Implant surface characteristics and their effect on osseointegration. *Br. Dental J.* **2015**, *218*, 1–9.
22. Teoh, S.H. Fatigue of biomaterials: A review. *Int. J. Fatigue* **2000**, *22*, 825–837.
23. Niinomi, M. Recent metallic materials for biomedical applications. *Metall. Mater. Trans. A* **2002**, *33*, 477–486.
24. Ribeiro, A.L.R.; Fla, R.C.J.; Cardoso, F.; Belon, R.; Vaz, F.F.L.G. Mechanical, physical, and chemical characterization of Ti–35Nb–5Zr and Ti–35Nb–10Zr casting alloys. *J. Mater. Sci.: Mater. Med.* **2009**, *20*, 1629–1636.
25. Niespodziana, K.; Jurczk, K.; Jurczk, M. The synthesis of titanium alloys for biomedical applications. *Rev. Adv. Mater. Sci.* **2008**, *18*, 236–240.
26. Henriques, V.A.R.; Galvani, E.T.; Petroni, S.L.G.; Paula, M.S.M.; Lemos, T.G. Production of Ti–13Nb–13Zr alloy for surgical implants by powder metallurgy. *J. Mater. Sci.* **2010**, *45*, 5844–5850.
27. Ohidul Alam, M.; Haseeb, A.S.M.A. Response of Ti–6Al–4V and Ti–24Al–11Nb alloys to dry sliding wear against hardened steel. *Tribol. Int.* **2002**, *35*, 357–362.
28. Ratner, J.B.B.D.; Hoffman, A.S.; Shoen, F.J.; Lemons, J.E. *Biomaterials Science: An Introduction to Materials in Medicine*; Academic Press: Waltham, MA, USA, 1996; pp. 37–50.

29. Williams, D. *An Introduction to Medical and Dental Materials, Concise Encyclopedia of Medical & Dental Materials*; Williams, D., Ed.; Pergamon Press: Oxford, UK; MIT Press: Cambridge, MA, USA, 1990; pp. xvii–xx.
30. Pilliar, R.M. *Metallic Biomaterials*; Springer Science + Business Media: New York, NY, USA, 2009.
31. Yoshimitsu, O.; Emiko, G. Comparison of metal release from various metallic biomaterials *in vitro*. *Biomaterials* **2005**, *26*, 11–21.
32. Nag, S.; Banerjee, R.; Fraser, H.L. Microstructural evolution and strengthening mechanisms in Ti-Nb-Zr-Ta, Ti-Mo-Zr-Fe and Ti-15Mo biocompatible alloys. *Mater. Sci. Eng. C* **2005**, *25*, 357–362.
33. McGregor, D.B.; Baan, R.A.; Partensky, C.; Rice, J.M.; Wibourn, J.D. Evaluation of the carcinogenic risks to humans associated with surgical implants and other foreign bodies-a report of an IARC monographs programmed meeting. *Eur. J. Cancer* **2000**, *36*, 307–313.
34. Yang, Y.; Yang, C.; Zhao, H.; Qu, S.; Li, X.; Li, Y. New developments of Ti-based alloys for biomedical applications. *Materials* **2014**, *7*, 1709–1800.
35. Liang, P.G.; Ferguson, A.; Hodge, E.S. Tissue reaction in rabbit muscle exposed to metallic implants. *J. Biomed. Mater. Res.* **1967**, *1*, 135–149.
36. Cvijovic, A.I.; Cvijovic, Z.; Mitrovic, S.; Rakin, M.; Veljovic, D.; Babic, M. Tribological behaviour of orthopaedic Ti-13Nb-13Zr and Ti-6Al-4V alloys. *Tribol. Lett.* **2010**, *40*, 59–70.
37. Gialanella, S.; Ischia, G.; Straffelini, G. Phase composition and wear behavior of NiTi alloys. *J. Mater. Sci.* **2008**, *43*, 1701–1710.
38. Suresh, K.S.; Geetha, M.; Richard, C.; Landoulsi, J.; Ramasawmy, H.; Suwas, S.; Asokamani, R. Effect of equal channel angular extrusion on wear and corrosion behavior of the orthopedic Ti-13Nb-13Zr alloy in simulated body fluid. *Mater. Sci. Eng. C* **2012**, *32*, 763–771.
39. Xu, L.; Xiao, S.; Tian, J.; Chen, Y. Microstructure, mechanical properties and dry wear resistance of β-type Ti–15Mo–xNb alloys for biomedical applications. *Trans. Nonferrous Met. Soc. China* **2013**, *23*, 692–698.
40. Muñoz, A.I. Effect of the environment on wear ranking and corrosion of biomedical CoCrMo alloys. *J. Mater. Sci. Mater. Med.* **2011**, *22*, 437–450.
41. Fellah, M.; Labaïz, M.; Assala, O.; Iost, A. Comparative Tribological study of biomaterials AISI 316L and Ti-6Al-7Nb. *TMS* **2014**, 237–246.
42. Li, S.J.; Yang, R.; Li, S.; Hao, Y.L.; Cui, Y.Y.; Niinomi, M.; Guo, Z.X. Wear characteristics of Ti-Nb-Ta-Zr and Ti-6Al-4V alloys for biomedical applications. *Wear* **2004**, *257*, 869–876.
43. Bhushan, B. *Introduction to Tribology*, 2nd ed.; John Wiley & Sons, Ltd.: New York, NY, USA, 2013; p. 621.
44. Bhushan, B. Nanotribology, nanomechanics and nanomaterials characterization. *Philos. Trans. R. Soc. A* **2008**, *366*, 1351–1381.
45. Choubey, A.; Basu, B.; Balasubramaniam, R. Tribological behaviour of Ti-based alloys in simulated body fluid solution at fretting contacts. *Trends Biomater. Artif. Organs* **2005**, *18*, 141–147.
46. Iwabuchi, A.; Lee, J.W.; Uchidate, M. Synergistic effect of fretting wear and sliding wear of Co-alloy and Ti-alloy in Hankssolution. *Wear* **2007**, *263*, 492–500.
47. Luo, X.; Li, X.; Sun, Y.; Dong, H. Tribocorrosion behavior of S-phase surface engineered medical grade Co-Cr alloy. *Wear* **2013**, *302*, 1615–1623.

48. Chiba, A. Pin-on-disk wear behavior in a like-on-like configuration in a biological environment of high carbon cast and low carbon forged Co-29Cr-6Mo alloys. *Acta Mater.* **2007**, *55*, 1309–1318.
49. Chan, S.M.T.; Neu, C.P.; Komvopoulos, K.; Reddi, A.H.; Di Cesare, P. Friction and wear of hemiarthroplasty biomaterials in reciprocating sliding contact with articular cartilage. *J. Tribol.* **2011**, *133*, 1–7.
50. Fischer, A.; Weiß, S.; Wimmer, M.A. The tribological difference between biomedical steels and CoCrMo-alloys. *J. Mech. Behav. Biomed. Mater.* **2012**, *9*, 50–62.
51. Alvarez-Vera, M.; Ortega-Saenz, J.A.; Hernandez-Rodríguez, M.A.L. A study of the wear performance in a hip simulator of a metal-metal Co-Cr alloy with different boron additions. *Wear* **2013**, *301*, 175–181.
52. Mohan, L.; Anandan, C. Wear and corrosion behavior of oxygen implanted biomedical titanium alloy Ti-13Nb-13Zr. *Appl. Surf. Sci.* **2013**, *282*, 281–290.
53. Attar, H.; Prashanth, G.; Chaubey, A.K.; Calin, M.; Zhang, S.L.C.; Eckert, J. Comparison of wear properties of commercially pure titanium prepared by selective laser melting and casting processes. *Mater. Lett.* **2015**, *142*, 38–41.
54. Rack, H.J.; Qazi, J.I. Titanium alloys for biomedical applications. *Mater. Sci. Eng. C* **2006**, *26*, 1269–1277.
55. Latysh, V.; Krallics, G.; Alexandrov, I.; Fodor, A. Application of bulk nanostructured materials in medicine. *Curr. Appl. Phys.* **2006**, *6*, 262–266.
56. Stolyarov, V.V.; Shuster, L.S.; Migranov, M.S.; Valiev, R.Z.; Zhu, Y.T. Reduction of friction coefficient of ultrafine-grained CP titanium. *Mater. Sci. Eng.: A* **2004**, *371*, 313–317.
57. Spary, I.D.; Bushby, A.J.; Jennet, N.M. On the indentation size effect in spherical indentation. *Philos. Mag.* **2006**, *86*, 5581–5593.
58. Hou, X.D.; Bushby, A.J.; Jennett, N.M. Study of the interaction between indentation size effect and Hall–Petch effect with spherical indenters on annealed polycrystalline copper. *J. Phys. D: Appl. Phys.* **2008**, *41*, 074006–074007.
59. Hou, X.D.; Jennett, N.M. Application of a modified slip-distance theory to the indentation of single-crystal and polycrystalline copper to model the interactions between indentation size and structure size effects. *Acta Mater.* **2012**, *60*, 4128–4135.
60. Beake, B.D.; Liskiewicz, T.W. Comparison of nano-fretting and nano-scratch tests on biomedical materials. *Tribol. Int.* **2013**, *63*, 123–131.
61. Sun, D.; Wharton, J.A.; Wood, R.J.K. Micro-abrasion mechanisms of cast CoCrMo in simulated body fluids. *Wear* **2009**, *267*, 1845–1855.
62. Díaz, C.; Lutz, J.; Mändl, S.; García, J.A.; Martínez, R.; Rodríguez, R.J. Improved bio-tribology of biomedical alloys by ion implantation techniques. *Nucl. Instrum. Methods Phys. Res.* **2009**, *267*, 1630–1633.
63. Manhabosco, T.M.; Tamborim, S.M.; Dos Santos, C.B.; Müller, I.L. Tribological, electrochemical and tribo-electrochemical characterization of bare and nitrided Ti6Al4V in simulated body fluid solution. *Corros. Sci.* **2011**, *53*, 1786–1793.
64. Sathish, S.; Geetha, M.; Pandey, N.D.; Richard, C.; Asokamani, R. Studies on the corrosion and wear behavior of the laser nitrided biomedical titanium and its alloys. *Mater. Sci. Eng. C* **2010**, *30*, 376–382.

Combined Characterization of the Time Response of Impression Materials via Traditional and FTIR Measurements

Giacomo Derchi [1,2], Enrico Manca [2], Amirreza Shayganpour [3,4], Antonio Barone [1], Alberto Diaspro [3,4] and Marco Salerno [3,*]

[1] Department of Surgical, Medical, Molecular and Critical-Area Pathology, University of Pisa, via Paradisa 2, I-56124 Pisa, Italy; E-Mails: gnolo78@gmail.com (G.D.); barosurg@gmail.com (A.B.)

[2] Tuscan Stomatological Institute, via Aurelia 335, Lido di Camaiore, I-55041 Lucca, Italy; E-Mail: info@enricomanca.it

[3] Nanophysics Department, Istituto Italiano di Tecnologia, via Morego 30, I-16163 Genova, Italy; E-Mails: amirreza.shayganpour@iit.it (A.S.); alberto.diaspro@iit.it (A.D.)

[4] University of Genova, viale Causa 13, I-16145 Genova, Italy

* Author to whom correspondence should be addressed; E-Mail: marco.salerno@iit.it;

Academic Editor: Ihtesham ur Rehman

Abstract: We investigated the temporal response of four dental impression materials, namely three siloxanes (Imprint 4, Flexitime, Aquasil) and one polyether (Impregum). The null hypothesis was that the nominal working times are confirmed by instrumental laboratory tests. We also aimed to identify alternative techniques with strong physical-chemical background for the assessment of temporal response. Traditional characterization was carried out by shark fin test device and durometer at both ambient and body temperature. Additionally, Fourier-transform infrared spectroscopy was performed at room temperature. From shark fin height and Shore hardness *versus* time the working time and the setting time of the materials were evaluated, respectively. These were in reasonable agreement with the nominal values, except for Impregum, which showed longer working time. Spectroscopy confirmed the different character of the two types of materials, and provided for Imprint 4 and Aquasil an independent evaluation of both evolution times, consistent with the results of the other techniques. Shark fin test and durometer measurements showed deviations in setting time, low sensitivity to temperature for

Flexitime, and longer working time at higher temperature for Impregum. Deviations of working time appear in operating conditions from what specified by the manufacturers. Fourier-transform infrared spectroscopy can provide insight in the correlation between material properties and their composition and structure.

Keywords: dental impression materials; polyether; siloxane; working time; setting time; Shore hardness; shark fin test; Fourier-transform infrared spectroscopy

---

1. Introduction

Today several types of dental impression materials exist, including polysulphides, based on traditional covalent bonding between polymer networks, and alternative polymers with networking based on Hydrogen bonding, for example the two polysaccharides of agar and alginate [1]. However, the most common impression materials used are still polyethers (PEs) and addition silicones (ASs) [2]. The PEs are based on moderately low molecular weight PE, silica filler and plasticizer, which on mixing give rise to a rubber by cationic polymerization. ASs are based on a main chain containing Silicon (Si), geometrically similar to the Carbon (C) chain of resins. Materials from both classes may exhibit excellent dimensional stability and wettability, which means minimal voids [1–3].

The major operating parameters of impression materials correlate with the time elapsed from mixing the two components that start the reticulation. The working time t_W can be defined as the time during which there is no significant change in the rheological and mechanical properties of the material, such that re-arrangement of the material is still possible within this time; the setting time t_S is the time for the network of reticulated polymer to take place, converting the viscous liquid to a stable stretch-resistant solid. While $t_W < t_S$ by definition, the ideal material would exhibit comparatively long t_W and short t_S, both on the scale of minutes, which would allow for both confident operation by the dentist and short waiting time for both dentist and patient. However, the two time response requirements counteract to each other and obviously a trade-off must be accepted.

The characteristic times of impression materials have been measured traditionally by shark fin test (SFT) apparatus [4] and hand portable durometers based on Shore hardness scale [5]. These simple techniques are inexpensive, easy to use, and allow for a fast measurement in real time during the setting, making it possible to directly access the material kinetics. More accurate physical techniques that allow for elastic measurements, for example nanoindentation—via either traditional indenters [6] of atomic force microscopy [7]—or bending in dynamic mechanical analysis [8–10] or compression/tension in universal testing machine [1,11,12], are more complex and require 10–30 min for each measurement, due to specimen preparation and loading. On the other hand, durometer and SFT measurements suffer from a poor scientific background. The only advanced instruments usable so far for both accurate and fast analysis of impression materials are viscosimeters or rheometers [13]. However, alternative means with similar performance and non-invasive character may also be found in Raman or infra-red (IR) spectroscopy, which today are routine techniques for the investigation of polymers through their molecular vibrations [14]. IR in particular has found widespread application thanks to its convenient implementation as Fourier-transformed IR (FTIR) [15].

In this work, in addition to SFT and durometer we also used FTIR to study the time response of dental impression materials for the first time to our knowledge. Application of advanced nanoscience and nanotechnology practices in dentistry is an expanding reality [16–18], for example the measurement of surface roughness and stiffness via atomic force microscopy has become a new standard technique with unsurpassed sensitivity and resolution [19,20]. In this streamline, FTIR should make it possible to explore the connections between composition and functionality of impression materials.

The null hypothesis which we aimed to check here by means of the experimental techniques described above is that the working times specified by the manufacturers are in agreement with the results of instrumental tests done in the laboratory.

2. Results

In Figures 1–3, the colors of the data points for the different materials have been assigned to be as close as possible to the actual material colors. In Figure 1 one representative set of FTIR spectra for all the four materials at the shortest possible time from mixing (~15 s) is shown, across the wavenumber range where major features emerge. Clearly, the spectra of all three ASs are similar, whereas the spectrum of the PE material is quite different. Thus, FTIR first allows us to group the materials into classes with similar chemistry.

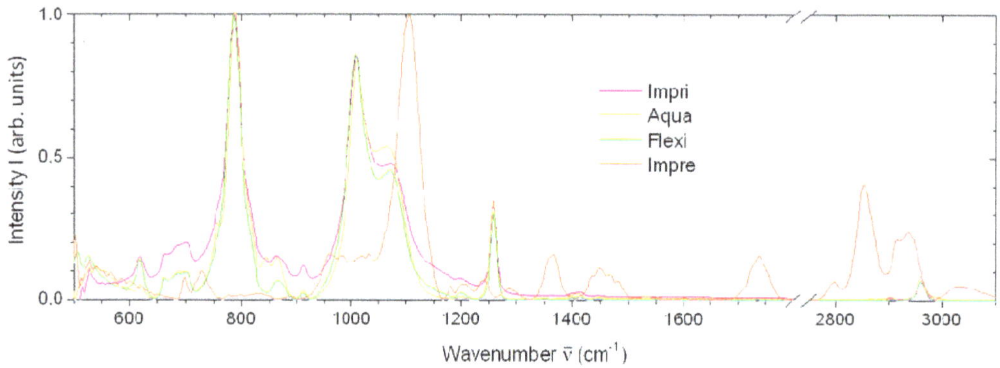

Figure 1. Fourier-transformed IR (FTIR) spectra of all the four materials investigated, at the shortest possible acquisition time since mixing (t ≈ 15 s).

Next, we try to assign some prominent features in Figure 1. The Impri spectrum (magenta line) shows the strongest peaks at ~789, ~1011 and ~1259 cm^{-1}. The first is assigned to the Si–C bond, and was used as the reference peak for normalization. The peak at ~1011 cm^{-1} is assigned to the Si–O bond. On the long wavenumber side of the 1011 cm^{-1} peak a shoulder due to a secondary peak at around 1068 cm^{-1} also appears, which is assigned to Si–O–Si asymmetric stretching. In fact, the whole 1000–1100 cm^{-1} broad band is distinctive of siloxane Si–O–Si bonds [21]. The peak at 1259 cm^{-1} can be assigned to a Si–CH_3 symmetric bending [21]. Additional features of Impri spectrum arise probably due to silica nanoparticle fillers, which are typically used for fine tuning of either rheological or stiffness properties of the fluid and set material, respectively.

For the other AS materials only minor differences appear with respect to Impri in the fingerprint region. For example, in Aqua a shoulder appears at ~758 cm^{-1}, tentatively assigned to C–O [22] or, together with bands at 708, 716, 728 and 758 cm^{-1}, to the CO_3^{2-} in-plane bending [23]. Possible assignment to a styrene component can be excluded, as that should also appear in another sharp peak at 704 cm^{-1} which is absent [24]. Additionally, the 912 cm^{-1} peak splits into two peaks at 843 and 865 cm^{-1}. The spectrum of Flexi also looks similar to both above, but closer to Impri, with no above peak shoulder and splitting.

For the spectrum of Impre, the highest peak appears at ~1105 cm^{-1}, which is the C–O stretching typical of ether C–O–C linkage in the 1000–1300 cm^{-1} range [21]. The second highest peak is at 2853 cm^{-1}, due to the CH stretch. Peaks at 698 and 730 cm^{-1} could be due to =CH bending. The band around 995 cm^{-1} could be the crystalline band due to symmetric stretching of C–O–C [25]. The bands at 2939 and 2922 cm^{-1} could be due to methylene (CH_2) stretching [25]. Multiple peaks also appear within the 1250–900 cm^{-1} band, which could be due to cyclic rather than aromatic ether [21]. The peaks at 1364, 1738, 2795 and 2930 cm^{-1} are compatible with those observed in [25]. Finally, the bands at 1460 and 3030 cm^{-1} could be CH_2 scissoring and =C–H stretching [21].

The difficulty in assigning all the peaks is due to the presence of additional components mixed in our materials, including nanoparticles, color pigments, and possible retarders or accelerators to appropriately tune the time response. However, identification of those peaks that change intensity over the setting time scale is of major interest here, as discussed at the end of this section.

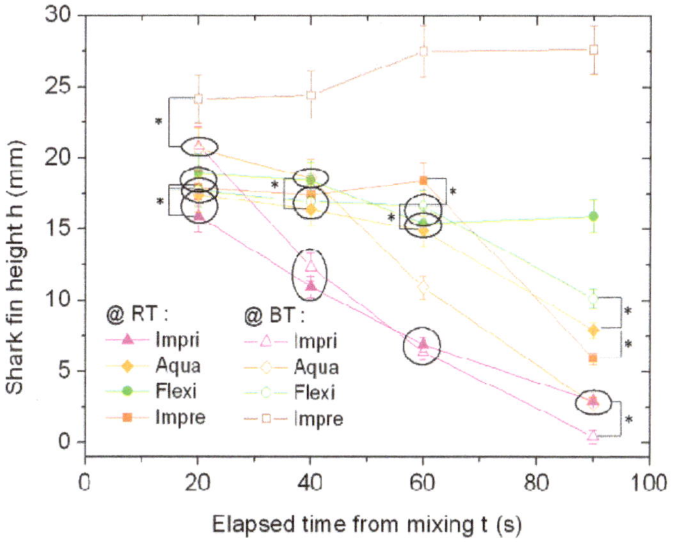

Figure 2. Shark fin test (SFT) measurements for the four materials made at room temperature (RT) (filled symbols) and body temperature (BT) (void symbols).

Characterization of the temporal response of the impression materials is straightforward when using the SFT and Shore hardness techniques. In Figure 2 the SFT heights taken for the four materials at different times are shown. Each datapoint represents the mean of several measurements ($N \geq 4$). The standard error was approximately 7% for most datapoints, and it has been represented in Figure 2 as error bars corresponding to ±1 σ (standard deviation). In Figure 2 also the most significant results of analysis of variance (ANOVA) with Bonferroni pair comparison tests have been represented: the datapoints enclosed by circles were not significantly different among each other ($p > 0.05$), whereas for pairs of datapoints statistically different ($p < 0.05$) a line connecting the points and signed by a single star (*) is displayed. All the other pairs without either enclosing circle (no difference) or line with star (difference at 0.05 level) must be considered being even more significantly different

($p < 0.01$), which would require a double star (**) as the graphic description, not placed on the graphic for the sake of clarity. The measurements have been repeated with device at room temperature (RT) and at body temperature (BT) at the time of pouring the mixed material into the mold. At BT (void symbols in Figure 2) the three ASs have similar behavior, since the shark fin height starts at the shortest time (t = 20 s) from similar values (18–21 mm) and decreases in all cases, even if in a milder way for Flexi and faster than all for Impri. On the contrary, Impre shows an anomalous behavior, since the fin height does not decrease with time in the considered range. At RT (filled symbols in Figure 2) instead, the Impre fin height decreases with time showing similar trend to Impri and Aqua (the former still being the fastest one in decrease, even if less fast than for BT). On the contrary, Flexi stays quite constant in fin height during the considered time, similar to Impre at BT. Thus, Flexi at RT does not crosslink in one minute, whereas it starts to change at BT: this probably occurs due to a special component, described by the manufacturer as "advance termasense" formulation [26]. Obviously, an increase in temperature activates an increase of viscosity for Flexi and slows down the increase of viscosity for Impre. Additionally, Impre values at BT are shifted to higher values than for the corresponding datapoints of the same material at RT, at all times, and also *versus* all the other materials, which means a significant decrease in viscosity for Impre with the increase in temperature.

The above datapoints in Figure 2 have been used to extract t_w values, taken as the intermediate time at which the shark fin height has decreased to half its initial value. It is assumed here that the SFT height correlates mostly with the fluidity of the material before setting. The results of this analysis are listed in Table 1.

Table 1. For each cell, the top row value is t_w and the bottom row one is t_s, where available. All numerical values are means ($N \geq 4$) in seconds. The column of dentist's perception is a qualitative score, higher score means best, + shows only a slight preference. §: for all "heavy body" formulations.

Materials	Methods of Assessment							
	Manufacturer Specifications (BT)	Dentists' Perception (BT)	Durometer			SFT		FTIR (RT)
			Literature [27] §	RT	BT	RT	BT	
Impri	60	1	-	-	-	55	46	68
(light-bodied)	95	-	225	223	98	-	-	353
Aqua	70	2	-	-	-	87	62	57
(regular set)	300	-	300	319	115	-	-	284
Flexi	180	2+	-	-	-	>>90	>90	>600
(medium flow)	360	-	350	332	251	-	-	>>600
Impre	60	3	-	-	-	83	>>90	>41
(light-bodied)	240	-	-	330	117	-	-	>185

Similarly, we assumed that the Shore hardness increasing over the time (see Figure 3) is instead a measurement of the solidification of the material, *i.e.*, its setting. Therefore, in that case the time up to one half of the final hardness has been used to obtain t_s values (see Table 1).

Qualitatively, the hardness *versus* time profiles in Figure 3 are much more finely sampled than the SFT ones, since the respective measurements are faster and could be made all on the same specimen, in just another small region close to the previous ones. Similar to Figure 2, also in Figure 3 the error bars represents ±1 σ (the standard error being here approximately 12%). However, given the large number of datapoints for these measurements, only the most significant values (*i.e.*, time at half of the maximum final hardness and final hardness) have been compared with ANOVA (results not shown in the plot, for the sake of clarity). The hardness curves at BT (void symbols) have similar intermediate time ($p > 0.05$) and are all shifted towards shorter times with respect to the corresponding curves at RT ($p < 0.01$) but for Flexi, which at BT shows similar t_S ($p > 0.05$) than the other three materials at RT. Thus, heat accelerates material setting, and only Flexi seems relatively insensitive to this effect, as its BT curve is also shifted to shorter times than the respective RT curve ($p < 0.01$), yet to a much lower extent (approximately −20% only in t instead of the −65% for the other materials). For Impri, an additional comparison between spiral tip and manual mixing points out that in the latter case setting is only slightly slower ($p < 0.05$), likely due to less uniform mixing.

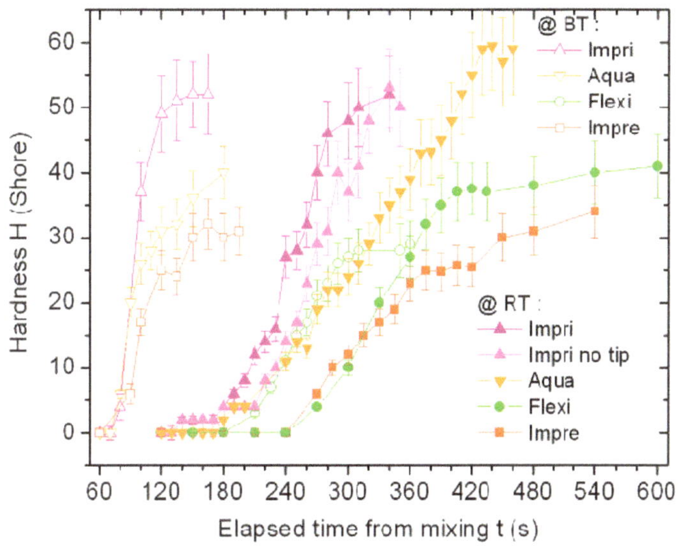

Figure 3. Shore hardness measurements for the four materials made at RT (filled symbols) and BT (void symbols).

The final hardness values of Impri are similar between BT and RT datasets ($p > 0.05$), while Aqua at RT reached a much higher value than at BT ($p < 0.01$) and higher than all other materials at both temperatures (mostly $p < 0.01$, only $p < 0.05$ for Impri at both RT and BT): thus, heating made Aqua to set slower yet to higher extent. The maximum final hardness at RT was higher than that at BT also for Flexi ($p < 0.05$), whereas it was not for Impre and Impri ($p > 0.05$).

Finally, from the FTIR spectra of the four materials as in Figure 1 but at different times since mixing, we sought the peaks whose intensities were changing over the time, to describe the materials setting. Candidate peaks are plotted in the left column of Figure 4 (Figure 4a,c,e,g). From the change in peak intensity, where available, we calculated discrete degree of setting (DS) values (right column of Figure 4, *i.e.*, Figure 4b,d,f,h) and fitted these datapoints *versus* time with a sigmoid. This curve has been used to find the times at which DS increased up to 10% and 90% of its final maximum value, defined by us as the t_W and t_S, respectively. Thus, we assumed that the FTIR information described the full materials physics and chemistry, and used its initial deviation from the starting value (+10% in units of final maximum) and its approach to the final value (−10% with respect to maximum) as markers for the respective characteristic times. Again, the numerical results appear in Table 1.

Deviations between the significant peaks of different spectra ($N \geq 2$, details see Material and methods) were all below 10%, so similar absolute errors (±10%) are expected in the resulting t_w and t_s values obtained with this method.

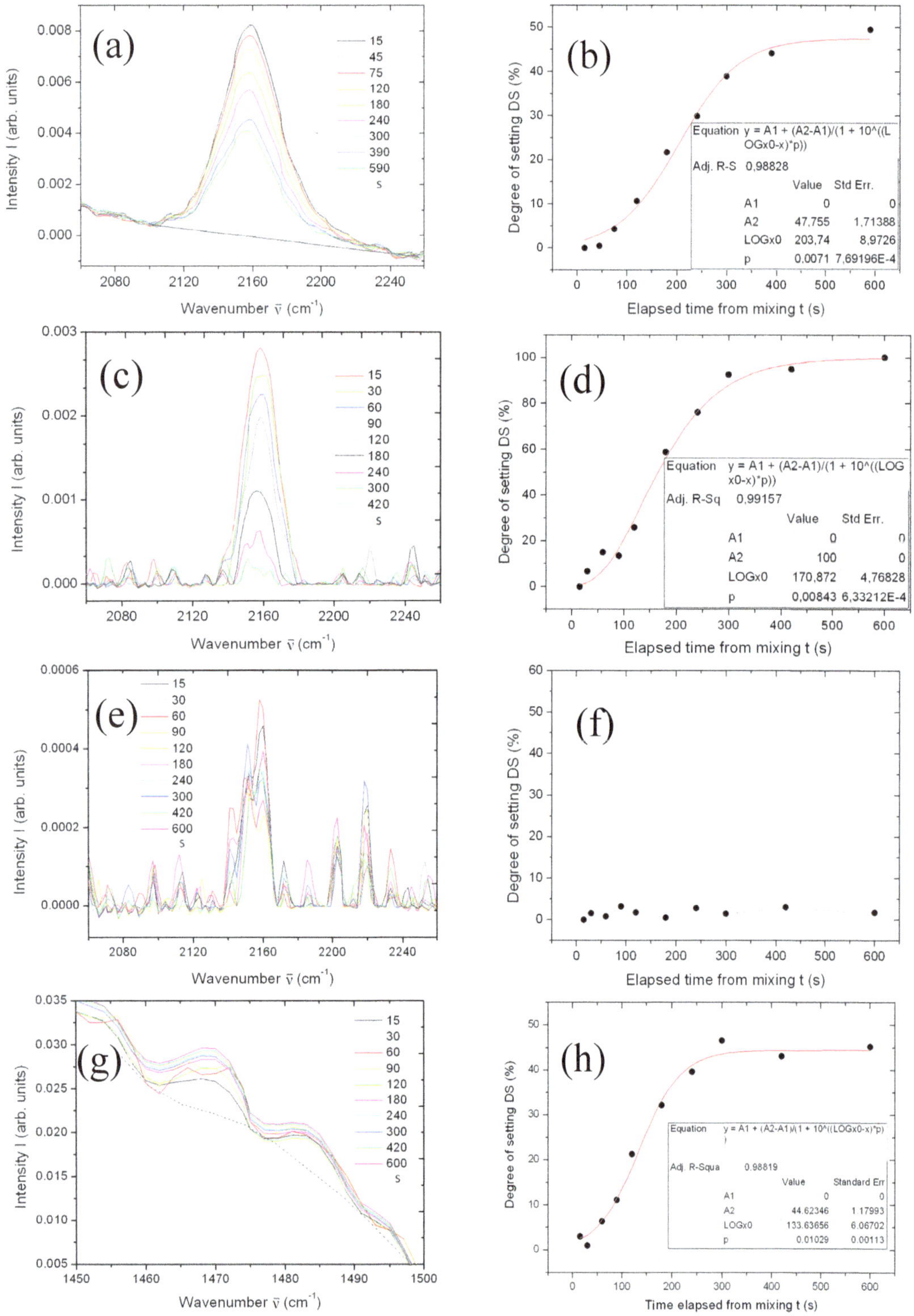

Figure 4. (**a**), (**c**), (**e**) and (**g**): regions with the peaks selected as representative of the expected setting process of Impri, Aqua, Flexi, and Impre over the time. (**b**), (**d**), (**f**) and (**h**): DS calculated from the respective peak signal, *versus* time, and—where possible—sigmoidal fit to the change.

3. Discussion

To our knowledge, FTIR has been used so far on dental impression materials only to assess the amount of OH groups present on the surface and characterize the wetting properties [28]. Here a new use is attempted to assess the time response during setting. With respect to the SFT and Shore hardness datasets, the FTIR ones—where available—are more detailed in the transition (*i.e.*, a better fit can be made), and account for the chemical changes at molecular level which comprise both the viscous and elastic properties of the material, so can provide both t_W and t_S.

FTIR is commonly used to assess the setting of dental restorative resins, based on the chain network change in double C bond during polymerization [8]. The chemistry of dental impression materials is obviously different. As the Si chain in ASs replaces the C chain in resins, bonds to Si have to be considered to evaluate the ASs reticulation. In the simplest picture, we expect the change in viscosity assessing the working time to be described by the same setting reaction, despite the complexity of reactions engineered in the materials formulation with the goal of tuning t_W and t_S to opposite requirements.

Since Si does not form a stable chain with itself [1], stability is obtained in silicones by interpolation of O, forming polysiloxanes. The pending Si bonds along the chain can be saturated by H, but the Si–H bond is extremely reactive and is usually substituted by an alkyl group like methyl (CH_3), making the Si–C bond inert. If such a chain is terminated with hydroxyl groups (OH) the molecule is called a silanol, and this is called a condensation silicone, after the polymerization reaction used for its manufacturing [1] via elimination of small molecules like H_2O. The dental impression materials are usually ASs, thanks to their better biocompatibility and higher stability, given by a variety of factors preventing them from curing. In ASs chain termination group is a vinyl (–CH=CH_2), so one has poly-vinyl-siloxane or vinyl-poly-siloxane (VPS), a moderately low molecular weight silicone that contains silane groups [2].

The main IR absorptions of silicones, appearing at ~770, 1000–1100 and ~1260 cm^{-1}, allow for detection down to 1% level sensitivity [29]. For the time progression of silicones setting one peak of Si–O should be found to be increasing, concurrently with a decreasing Si–H peak. Indeed, for Impri and Aqua we found a significant decrease of the Si–H peak at 2158 cm^{-1} over time, and used this peak according to Equation (1). Unfortunately, for Flexi the 2158 cm^{-1} peak was less clearly defined and did not change with time (see Figure 4e,f). Additionally, we failed to identify other FTIR peaks that exhibited a clear trend in time. One possible explanation (see Table 1) would be that the characteristic times of change are much longer than those considered here. Another possibility is that the reaction chemistry of Flexi is not associated with FTIR-active vibrations, and alternative spectroscopic techniques such as Raman scattering should be used. We rather think that for Flexi the setting process in more complex than for Impri and Aqua and can not be described by a single peak change, but a chain of reactions is involved which is hard to reconstruct without detailed information from the manufacturer.

For Impre a single peak clearly changing with time (increasing) was identified at 3032 cm^{-1} (data not shown). However, when assessment of DS was attempted according to Equation (2), the obtained datapoints provide values of 36 and 184 s for t_W and t_S, respectively, quite far from the SFT and durometer tests. We speculate that the 3032 cm^{-1} peak is due to an initiation reaction. In fact, it is expected that the aziridine ring opening is promoted by the action of an aromatic sulfonate ester [30].

Alternatively, after the assumption that PE is cured via cationic polymerization of aziridine rings [31], we tried to use for Impre the peak at 1240 cm^{-1} (data not shown). In this case the ring opening should be accompanied by decrease in C–N amine bonds stretching signal, to appear in the 1080–1340 cm^{-1} range. However, the 1240 cm^{-1} peak showed no change up to the times considered (10 min at RT). Instead, similar to the observation in [32] for PEEK, a weak band at 1470 cm^{-1} was observed to increase, which could be due to sulfonation starting the crosslinking (see Figure 4g). This band and that one at 1483 cm^{-1} were tentatively used as the signal and reference, according to Equation (2), to obtain the DS datapoint in Figure 4h. From the resulting best fit curve, again very short time values as for the 3032 cm^{-1} peak were found, namely 41 and 185 s. We assume that these values are just a lower limit for characteristic times, related to the initiation only of the PE crosslinking via sulfonation [1].

In Table 1 also the qualitative opinions of the two dentists participating in this study have been included for the working time, which is the parameter of highest practical interest. According to the scores describing their clinical experience, they agree that the best material in workability is Impre, followed by Aqua and Flexi with a slight preference only for one dentist for the latter. This is confirmed by the laboratory tests presented here, even if in contrast with the manufacturer specifications, which show a t_W for Impre as short as for the other three ASs.

In commenting the numerical data summarized in Table 1, one can first observe that, as expected, the intermediate time obtained by the SFT trends (representing t_W) is significantly lower than that obtained by the durometer measurements (representing t_S). Obviously, the former measurement is related to the properties of the material as a viscous paste, whereas the second is related to the properties of the material as a solid, after setting into the final mold.

With respect to the nominal values from the manufacturers, concerning the t_W obtained from SFT: Impri, Aqua, and Flexi are in good agreement; Impre instead shows longer t_W, even at BT (where it is actually even longer than at RT). Concerning the t_S obtained from the durometer: Impri is in agreement with the nominal value at BT (but is much longer at RT, in agreement with the literature); Aqua is in agreement with the nominal value at RT (but is much shorter at BT); Flexi is in agreement at RT (but shorter at BT); Impre has RT and BT t_S around the nominal value (but the RT one is closer).

Concerning the FTIR results, for Aqua both t_W and t_S are in agreement with the other techniques and the nominal values, whereas for Impri the t_W is in agreement with the nominal value but the t_S seem to be much longer.

4. Experimental Section

The materials were selected and provided by the dentists involved in the study. They were three materials from the AS class, namely Imprint 4 (3M-ESPE, St. Paul, MN, USA) [27], Flexitime (Heraeus Kulzer, Hanau, Germany) [26] and Aquasil (Dentsply, Konstanz, Germany) [33], and one from the PE class, namely Impregum (3M-ESPE) [34]. In the following they have been named shortly Impri, Flexi, Aqua and Impre, respectively. The materials were dispensed with a gun for dual-compartment cartridges equipped with ~7 cm long tips with spiral-shaped baffle, providing automatic mixing of the base and catalyst in the required volume proportions (1:2 for Impre and 1:1 for all the other materials).

FTIR was performed on a Vertex 70 (Bruker Optics, Billerica, MA, USA) in attenuated total reflection. The spectra were acquired on the whole range of 400–4000 cm^{-1}, with an aperture size of

3 mm and a resolution of 4 cm^{-1}. For minimization of noise effects, five scans were averaged per spectrum and apodized by Blackman-Harris three-term correction function. After acquisition of the transmission signal, the spectra were converted into absorbance in arbitrary units and the baseline was calculated with 3rd order polynomial and removed. Finally, the spectra were normalized to the highest peak intensity. The number of measured spectra at each time was $N = 2$ for Impri and Aqua, $N = 3$ for Impre and $N = 4$ for Flexi.

For SFT measurements, both an original SFT device of stainless steel by 3M-ESPE (lent on purpose) and a home-made SFT device of aluminum were used, which demonstrated equivalent results within the experimental error. Before each measurement the SFT device was either kept at room temperature (RT, 22 ± 2 °C) or heated at body temperature (BT, 37 ± 1 °C) for 10 min in a thermostated water bath. In the latter case, on removal from the bath the device was quickly dried by wiping with cotton cloth and blowing with compressed air (total time ~10 s). The device was loaded with ~6 mL of mixed material, in ~5 s, and this time was summed up with extra rest time, before letting the SFT piston equipped with V-shaped slit gently touch the polymer surface and start sinking under ~147 g load. The slit had 1 and 2 mm arch length for the tests made at RT and BT, respectively. The total time before sinking start was taken as the independent variable for the measured shark fin height. Time values of 20, 40, 60 and 90 s were used for all the materials. After 10 min the SFT device was disassembled, the excess material flown out of the split ring was cut off, and the shark fin height was measured with a caliper. Whereas the instrumental error was ±0.02 mm, a measuring accuracy of ±0.1 mm was assumed due to operator uncertainty in assignment of fin limits. All measurements were repeated at least four times ($N \geq 4$).

The durometer measurements were taken with a Affri 3001 Shore A portable hardness tester (BAMR, Cape Town, South Africa). The sensing tip of this device protrudes from the reference end-of-scale head level of ~1.5 mm. A metal mold was used for the material to be indented, of ~7 cm length, ~0.5 cm width, and ~4 mm depth. The mold depth was high enough that, on full penetration in the material, the durometer tip was still far from touching the mold bottom. The mold width was small enough that the durometer head was sitting on the top sides of the mold as a reference level. The mold length was high enough for each measurement to be taken at a position 2–3 mm apart along the specimen, not to be affected by the previous one. The 100 end-limit of A Shore hardness scale, according to manufacturer's calibration, corresponds to ~3.8 kg load. Assuming a spherical tip apex of ~0.5 mm diameter, this load would mean a maximum stress of ~190 MPa. The measurements were repeated at 10 s intervals after the first 1 min since mold loading, to avoid the initial full sinking in the liquid material, which would spoil the tip and affect the subsequent measurements. Same as for SFT, the materials were always coming from storage in the ambient, but the metal mold was either at RT as well or heated up to BT by means of a hotplate. All measurements were repeated four times ($N = 4$).

To try to extract both t_w and t_s from the FTIR spectra, a quantity called degree of setting (DS) has been defined as follows. Assuming that a peak decreasing with time due to setting is identified, its intensity I_{sign}—taken as the signal—is normalized to the intensity of a reference peak from the same spectrum, I_{ref}, and the ratio I_{sign}/I_{ref} is rescaled between each time instant t and the respective constant value from the initial spectrum at time t_0:

$$DS = 100[1 - (I_{sign}/I_{ref})(t)/(I_{sign}/I_{ref})(t_0)] \quad (1)$$

Equation (1) is similar to that used for resin composites to assess the degree of conversion resulting from polymerization, where I_{sign} is usually the intensity of the peak at 1637 cm^{-1} and I_{ref} that at 1580 cm^{-1}, related to stretching of the aliphatic and aromatic C=C bonds, respectively [8]. If, on the contrary, the peak used for I_{sign} is increasing over the time, the following formula should be used to have a DS increasing with time:

$$DS = 100[-1 + (I_{sign}/I_{ref})(t)/(I_{sign}/I_{ref})(t_0)] \quad (2)$$

In addition to the above quantitative evaluations of the characteristic times of the materials, we also asked the two dentists involved in the study to provide a qualitative evaluation of the material performances, limited to the working time. This opinion was based on a number of clinical treatments $N \geq 10$ for each considered material, in several different conditions of treated patients.

5. Conclusions

As a method for assessing the working time of dental impression materials, the SFT alone is commonly used. Yet, this empirical technique does not give insight in the physics and chemistry of setting and has long being criticized in the literature. Complementary techniques based on stronger scientific foundations should also be used for this task. We have studied the behavior of four dental impression materials based not only on SFT and durometer but also on FTIR measurements, probing the molecular bond vibrations. The materials investigated were one PE (Impre) and three ASs (Impri, Flexi, Aqua), which were studied *versus* the time since mixing at intervals as short as 15–20 s. This allowed us to extract the characteristic working and setting time of the polymers, after appropriate definition. The null hypothesis that laboratory measurements would agree with the nominal time values from the manufacturers was not confirmed. There exists an overall ambiguity on the temperature of supposed assessment, if BT or RT, whose experimental results are closest to the nominal values. Additionally, the working time of Impre turned out to be longer than the nominal value. For Flexi and Impre no single time-evolving peaks associated with their setting reaction could be identified from FTIR. This may be a limitation intrinsic in the technique, which in the future should be complemented with additional information on the material formulations or with techniques such as thermogravimetric analysis or differential scanning calorimetry.

Acknowledgments

Dr. Sanjay Thorat is gratefully acknowledged for helpful discussion about FTIR measurements.

Author Contributions

G.D. and E.M. selected and provided the dental materials, performed the SFT measurements, and gave the clinical opinions on the materials included in Table 1. G.D., E.M. and M.S. designed the experiment. A.S. and M.S. made the FTIR and the durometer measurements. M.S. drafted the manuscript. All the authors discussed the results and critically revised the manuscript.

Conflicts of Interest

The authors declare no conflict of interest.

References

1. Darvell, B.W. *Materials Science for Dentistry*, 9th ed.; Woodhead Publishing: Cambridge, UK, 2009.
2. Farah, J.W.; Powers, J.M. Elastomeric impression materials. *Dent. Advis.* **2003**, *20*, 87–89.
3. Rubel, B.S. Impression materials: A comparative review of impression materials most commonly used in restorative dentistry. *Dent. Clin. N. Am.* **2007**, *51*, 629–642.
4. Balkenhol, M.; Wo, B.; Kanehira, M.; Finger, W.J. Shark fin test and impression quality: A correlation analysis. *J. Dent.* **2006**, *35*, 409–415.
5. Farr, J.C.; Hoffmann, H.; Maurer, A.; Osswald, P.; Wanek, E. *Shore Hardness Increase of Tray Materials during Setting*; Poster Session No. 674; The International Association for Dental Research: Alexandria, VA, USA, 2012.
6. Salerno, M.; Derchi, G.; Thorat, S.; Ceseracciu, L.; Ruffilli, R.; Barone, A.C. Surface morphology and mechanical properties of new-generation flowable resin composites for dental restoration. *Dent. Mater.* **2011**, *27*, 1221–1228.
7. Salerno, M.; Patra, N.; Diaspro, A. Atomic force microscopy nanoindentation of a dental restorative midifill composite. *Dent. Mater.* **2012**, *28*, 197–203.
8. Thorat, S.; Patra, N.; Ruffilli, R.; Diaspro, A.; Salerno, M. Preparation and characterization of a BisGMA-resin dental restorative composites with glass, silica and titania fillers. *Dent. Mater. J.* **2012**, *31*, 635–644.
9. Thorat, S.; Diaspro, A.; Salerno, M. Effect of alumina reinforcing fillers in BisGMA-based resin composites for dental applications. *Adv. Mater. Lett.* **2013**, *4*, 15–21.
10. Thorat, S.B.; Diaspro, A.; Salerno, M. *In vitro* investigation of coupling-agent-free dental restorative composite based on nano-porous alumina fillers. *J. Dent.* **2014**, *42*, 279–286.
11. Instron Brochure, Dental Materials and Implants–Mechanical Testing Instruments. Available online: http://www.instron.com (accessed on 30 April 2015).
12. Mazzoni, A.; Marchesi, G.; Cadenaro, M.; Mazzotti, G.; Di Lenarda, R.; Ferrari, M.; Breschi, L. Push-out stress for fibre posts luted using different adhesive strategies. *Eur. J. Oral Sci.* **2009**, *117*, 447–453.
13. Lawson, N.C.; Cakir, D.; Ramp, L. Flow profile of regular and fast-setting elastomeric impression materials using a shark fin testing device. *J. Esthet. Restor. Dent.* **2011**, *23*, 171–176.
14. Siesler, H.W. Vibrational spectroscopy of polymers. *Int. J. Polym. Anal. Charact.* **2011**, *16*, 519–541.
15. Bhargava, R.; Wang, S.; Koenig, J.L. FTIR microspectroscopy of polymeric systems. *Adv. Polym. Sci.* **2003**, *163*, 137–191.
16. Sharma, S.; Cross, S.E.; Hsueh, C.; Wali, R.P.; Stieg, A.Z.; Gimzewski, J.K. Nanocharacterization in dentistry. *Int. J. Mol. Sci.* **2010**, *11*, 2523–2545.
17. Uskoković, V.; Bertassoni, L.E. Nanotechnology in dental sciences: Moving towards a finer way of doing dentistry. *Materials* **2010**, *3*, 1674–1691.

18. Salerno, M.; Diaspro, A. Dentistry on the bridge to nanoscience and nanotechnology. *Front. Bioeng. Biotechnol.* **2015**, doi:10.3389/fmats.2015.00019.
19. Salerno, M.; Patra, N.; Thorat, S.; Derchi, G.; Diaspro, A. Combined effect of polishing on surface morphology and elastic properties of a commercial dental restorative resin composite. *Sci. Adv. Mater.* **2012**, *4*, 126–134.
20. Salerno, M.; Giacomelli, L.; Derchi, G.; Patra, N.; Diaspro, A. Atomic force microscopy *in vitro* study of surface roughness and fractal character of a dental restoration composite after air-polishing. *Biomed. Eng. Online* **2010**, *9*, 1–11.
21. Stuart, B. *Infrared Spectroscopy: Fundamentals and Applications*; Wiley: Hoboken, NJ, USA, 2004.
22. Kebukawa, Y.; Nakashima, S.; Otsuka, T.; Nakamura-Messenger, K.; Zolensky, M.E. Rapid contamination during storage of carbonaceous chondrites prepared for micro FTIR measurements. *Meteorit. Planet. Sci.* **2009**, *44*, 545–557.
23. Frost, R.L. Raman spectroscopic study of the magnesium carbonate mineral-hydromagnesite $Mg_5[(CO_3)_4(OH)_2]\cdot 4H_2O$. *J. Raman Spectrosc.* **2011**, *42*, 1690–1694.
24. Learner, T. *Analysis of Modern Paints*; Research in Conservation Technical Report Series; Getty Institute: Los Angeles, CA, USA, 2005.
25. Wen, Z.; Shen, D.; Zhao, H. The rheo-optical FTIR study of polyester-polyether segmented copolymer. *Chin. J. Polym. Sci.* **1988**, *6*, 325–331.
26. Heraeus Kulzer-Flexitime-Confidently Flexible and Precise. Available online: http://heraeus-kulzer-us.com/en/ourproducts/dentistry_1/impressionmaterials/Flexitime.aspx (accessed on 30 April 2015).
27. 3M-ESPE. *Imprint 4—VPS Impression Material*; Technical Data Sheet; 3M-ESPE: St. Paul, MN, USA, 2013.
28. Ozden, N.; Ayhan, H.; Erkut, S.; Can, G.; Pişkin, E. Coating of silicone-based impression materials in a glow-discharge system by acrylic acid plasma. *Dent. Mater.* **1977**, *13*, 174–178.
29. Andriot, M.; Chao, S.H.; Colas, A.; Gray, S.; de Buyl, F.; Degroot, J.V.; Dupont, A.; Easton, T.; Garaud, J.L.; Gerlach, E.; *et al. Silicones in Industrial Applications*; Dow Corning Technical Report; Dow Corning: Midland, MI, USA, 2007.
30. Bradna, P. *Impression Materials*; Institute of Medical Biochemistry, Charles University and Institute of Dental Research in Prague: Prague, Cszech Republic, 2012.
31. Stipho, S.; Maiolo, C.; Sharma, S.; Perry, R.; Kugel, G. Flow of Impression materials using REALITY-method with 1.0 mm slit-size. *J. Dent. Res. A* **2006**, *85*, 0413.
32. Al Lafi, A.G. FTIR spectroscopic analysis of ion irradiated poly (ether ether ketone). *Polym. Degrad. Stab.* **2014**, *105*, 122–133.
33. Dentsply, Aquasil Ultra-Precisely. Available online: http://aquasilultra.com/ (accessed on 30 April 2015).
34. 3M-ESPE. *Impregum-Polyether Impression Materials*; 3M-ESPE: St. Paul, MN, USA, 2010.

12

Methods for Biomimetic Mineralisation of Human Enamel: A Systematic Review

Chris Ying Cao [1], May Lei Mei [1], Quan-li Li [2], Edward Chin Man Lo [1] and Chun Hung Chu [1,*]

[1] Faculty of Dentistry, the University of Hong Kong, Hong Kong, China;
E-Mails: caoying0713@gmail.com (C.Y.C.); modelbaby1981@gmail.com (M.L.M.);
hrdplcm@hku.hk (E.C.M.L.)

[2] Stomatological Hospital and College, Anhui Medical University,
Key Lab. of Oral Diseases Research of Anhui Province, Hefei 230032, China;
E-Mail: u3001279@connect.hku.hk

* Author to whom correspondence should be addressed; E-Mail: chchu@hku.hk;

Academic Editor: Greta Patzke

Abstract: Biomimetic mineralisation is an alternative restorative methodology that imitates the natural process of mineralisation. We aimed to systematically review the laboratory methods on the biomimetic mineralisation of demineralised enamel. A search in the PubMed, ScienceDirect, and ISI Web of Science databases was performed. Clinical trials, reviews, non-English articles, animal teeth, non-tooth substrates, and irrelevant studies were excluded. After screening the titles and abstracts of initially searched articles, 20 papers remained for full-text analysis. Eight articles were identified from the references of the remaining papers. A total of 28 studies were included in this systematic review. We found that protein or protein analogues were used to mimic the function of natural protein in 23 studies. Bioactive components inspired by mussel, an agarose hydrogel model, a glycerine-enriched gelatine technique, and ethylenediaminetetraacetic acid, were also used for biomimetic mineralisation of enamel. These laboratory studies reported success in the biomimetic mineralisation of enamel. Potential further research on the biomimetic mineralisation of enamel was discussed.

Keywords: biomimetic; mineralisation; enamel; proteins

---

1. Introduction

Unlike the other mineralised tissues of the human body (*i.e.*, bone, dentine, and cementum), mature enamel is acellular and has more than 95% mineral content and less than 1% organic material, thus, making it the hardest mineralised tissue in the human body [1]. The crystals found in enamel are nanorod-like hydroxyapatite (HAP) crystals with hierarchical levels of the enamel microstructure from the nanoscale to microscale [2]. The highly organised hierarchical microstructure (prism) provides dental enamel its unique strength and anti-abrasive properties [3]. During enamel biomineralisation, ameloblasts secrete proteins, such as amelogenin and ameloblastin, to mediate the crystallites to form the well-organised prism pattern [4]. Although the initial formation of enamel apatite occurs under a protein-rich gel-like matrix environment, the biological macromolecules are almost degraded or removed gradually. This makes mature enamel a non-living tissue. Damaged enamel is permanent and does not remodel. As the outermost layer covering the tooth, enamel is susceptible to dental caries and erosive wear. Traditional treatments replace the defective enamel with dental restorative materials, such as amalgam, composite resin, and ceramics, however, these restorative materials are very different from natural enamel in chemical components, crystal structures, and physicochemical properties. They cannot fit well with enamel at the interface of the enamel, and the restorative materials are at risk of marginal leakage, which leads to hypersensitivity and secondary caries. Therefore, it is of great interest for scientists in different fields to design an alternative strategy to repair defective enamel.

Enamel is relatively stable in a healthy oral environment with the presence of saliva. The demineralisation and remineralisation of enamel are continuously taking place at the tooth-pellicle-plaque-saliva interface [5]. It is well established that dental caries is a dynamic disease process caused by the imbalance between demineralisation and remineralisation [6]. This concept of dental caries provides the scope for remineralising early enamel carious lesions. Recently, various remineralising strategies have been developed to restore early enamel lesions and prevent enamel demineralisation, such as fluoride [7,8], surfactants [9], electrolytic deposition [10], hydrothermal methods [1,10], and hydrogen peroxide [11]. However, these methods are performed under stringent conditions, such as extremely low acidity or high electric field environments, which cannot be translated to clinical applications. Although surfactants, such as reverse micelles, mimic the natural biomineralisation process of enamel by modifying the HAP nanorods to self-assemble into an enamel-like structure, the biocompatibility limits their clinical application [12]. Dental enamel is produced by ameloblasts. Enamel formation is a protein-mediated biomineralisation process that takes place within an extracellular matrix [4,13,14]. Amelogenin is the major component of this matrix and it is essential for proper enamel biomineralisation [15]. Biomimetic mineralisation is a methodology that imitates the natural process of mineralisation [16]. It is generally accepted that the biomimetic synthesis of enamel-like apatite structures under a physiological condition is an alternative restorative pathway. This paper is a systematic review of the methods that successfully achieved the biomimetic mineralisation of demineralised human enamel.

2. Results

The initial search from three electronic databases identified 2291 potential articles (294 in PubMed, 1567 in ScienceDirect, and 430 in ISI Web of Science), 2271 were excluded after the initial screening of titles and abstracts. After retrieving the full text of the remaining 20 articles, eight articles were found from the bibliographies of the included articles. Therefore, 28 articles were included in the systematic review (Figure 1). All 28 articles were *in vitro* studies published (including those Epub ahead of print) between 1997 and 2014.

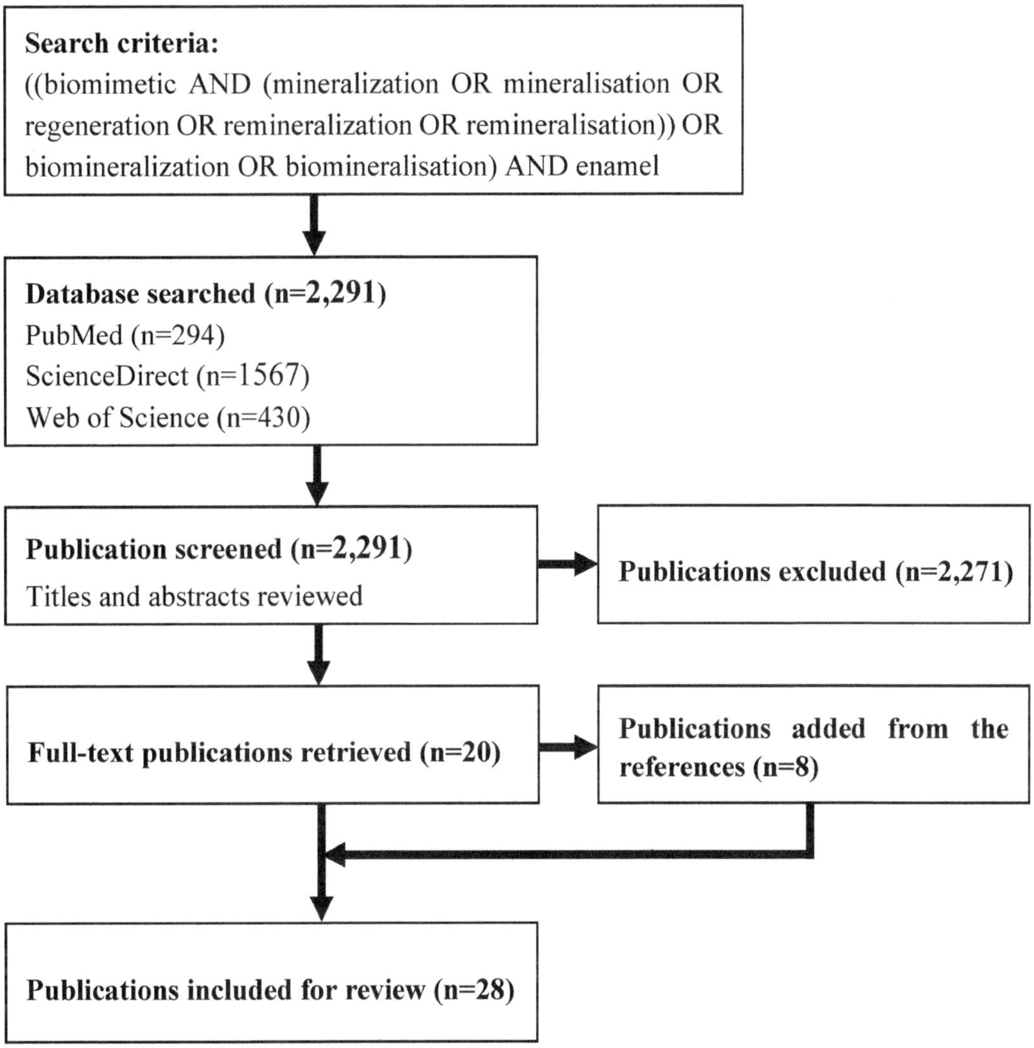

Figure 1. Flowchart of the search strategy.

Table 1 summarised the different methods of biomimetic mineralisation of enamel used in the included articles. Protein or protein analogues were used to mimic the function of natural proteins for the biomimetic mineralisation of enamel in 23 studies. Two studies used bioactive components inspired by mussel as a template to facilitate the hydroxyapatite nucleation on enamel [17,18]. A study constructed an agarose hydrogel model for the biomimetic mineralisation of enamel [19]. A study developed a glycerine-enriched gelatine technique to form fluorapatite layers on enamel [20]. A study used ethylenediaminetetraacetic acid (EDTA) as the mediating agent to reconstruct fluoride hydroxyapatite on enamel [21].

Mineral ions were provided by different remineralising mediums in the included articles. The most commonly used remineralising medium was artificial saliva. In addition, simulated body fluid (SBF), calcium phosphate remineralising solution/hydrogel, apatite nanocrystals slurry, bioactive glass slurry, and human saliva were also used. Fluoride at different concentrations was added in the remineralising mediums in 15 studies. The highest concentration of fluoride was 1100 ppm and the lowest concentration was 0.45 ppm.

Different demineralising agents were used to treat enamel surface before biomimetic mineralisation. Phosphoric acid at concentration from 30% to 37% was used in 17 studies. In addition, a demineralising solution [22–26], citric acid [27], nitric acid [28–30], acidified gelatine gel [31], and a bi-layer demineralisation protocol of methylcellulose gel buffered with a lactic acid layer [32] were also used. In fact, the different demineralization agents are used to create two different models of enamel defects. The strong acid (phosphoric acid or nitric acid) is used to create an erosive lesion, while the demineralising solutions (or some gels) usually with pH of ~4.6 are used to generate artificial caries. These studies can regenerate enamel-like crystals on the enamel surface, and the main findings can be found in Table 1.

The functions of the proteins or protein analogues are summarised in Table 2. Amelogenin was added into a calcium phosphate solution containing fluoride [28,29] or mixed with hydrogel containing calcium and phosphate [26,33,34] to promote the orderly enamel-like structure formation. In one study, an enamel matrix derivative was mixed with calcium chloride agarose hydrogel to guide the continuous growth of crystals, prevent the crystal fusion of premature crystals, and control the nucleation and growth of the enamel prism-like crystals [35]. A polyamidoamine (PAMAM) dendrimer solution was applied to the etched enamel surface to provide nucleation sites and a mineralisation template for nanorod-like HAP formation with high uniformity [30,36,37]. Peptide was the most commonly used protein analogue. The sequence of the different peptides used in the included articles was summarised in Table 2. Only one study added peptide into a calcium phosphate solution containing fluoride to induce the formation of stable amorphous calcium phosphate (ACP) nano-precursors and prevent them from aggregation and precipitation [38]. In the other seven studies, a peptide solution was applied to the etched enamel surface to bind with the HAP surface, capture calcium and phosphate ions from artificial saliva, and promote HAP with a small average crystalline size [27,31,39–43]. Glutamic acid (Glu) was added into SBF to induce nano-apatite particle assembly and aggregation on the etched enamel surface, resulting in the regeneration of the enamel-like structure [44]. Polyacrylic acid (PAA) was added to bioactive glass and applied to the etched enamel surface to promote the oriented bundle formation, release Ca and P ions to form an ACP layer, and act as a reservoir of ions available for remineralisation [32]. The nano-complexes of the phosphorylated chitosan-ACP solution were applied to the etched enamel surface to act as the aggregates of prenucleation clusters, to provide stable ACP, and to adsorb on the enamel surface [25]. Etched enamel was incubated with a casein phosphopeptide-amorphous calcium phosphate (CPP-ACP) solution [22], coated with CPP-ACP paste [23], or brushed with CPP-ACP paste [24] to act as a calcium and phosphate reservoir and promote enamel subsurface lesion remineralisation.

Table 1. Summary of the *in vitro* studies on biomimetic mineralisation on human enamel.

Authors, Year [Ref.]	Surface Treatment	Biomineralisation Method	Sources of Ca and P	Main Finding
Reynolds, 1997 [22]	Demineralising solution (4d)	CPP-ACP	$CaCl_2$ solution, K_3PO_4 buffer	The CPP-stabilised CaP solution can remineralise subsurface enamel lesions.
Busch, 2004 [20]	30% PA (30s)	Glycerin-enriched gelatin 620 ppm F	$CaCl_2$ solution, PO_4 gelatin	Ordered enamel-like mineral was induced on human teeth.
Kirkham *et al.*, 2007 [31]	Acidified gelatine gel (6w)	Peptide	pH cycling	Peptide treatment increased remineralisation and inhibited demineralisation.
Kumar *et al.*, 2008 [23]	Demineralising solution (4d)	CPP-ACP 1100 ppm F	pH cycling	CPP-ACP remineralised initial enamel lesions.
Fan *et al.*, 2009 [28]	3% HNO_3 solution (50s)	Amelogenin 0.45 ppm F	Calcification solution	The synthetic nanorod crystals formed on etched enamel in the presence of amelogenin.
Xie *et al.*, 2011 [21]	37% PA (30s)	EDTA 950 ppm F	Ca-EDTA complex-phosphate solution	EDTA induced the assembly of hexagonal prism-like FHAP microcrystals.
Jayarajan *et al.*, 2011 [24]	Demineralising solution (5h)	CPP-ACP, CPP-ACPF 900 ppm	Artificial saliva	CPP-ACPF showed more amount of remineralisation than CPP-ACP.
Fan *et al.*, 2011 [29]	5% HNO_3 solution (30s)	Amelogenin 1–10 ppm F	Remineralisation solution	Densely packed arrays of FHAP nanorods were observed.
Hsu *et al.*, 2011 [39]	35% PA (30s)	Peptide	SBF	The nanomechanical properties of the acid-demineralised enamel were greatly improved.
Hsu *et al.*, 2011 [42]	35% PA (30s)	Peptide	SBF	Peptide promoted the uniform deposition of nano-crystalline CaP over enamel surfaces.
Li *et al.*, 2011 [44]	37% PA (10s)	Glu-apatite nanoparticles	SBF	Nanoparticles turned into rod-like crystals on enamel surface.
Zhou *et al.*, 2012 [17]	37% PA (2m)	Polydopamine 1 ppm F	Calcification solution	No significant difference was observed in the remineralisation of enamel, except in dentine.
Fan *et al.*, 2012 [26]	Demineralising solution (3d)	Amelogenin 5 ppm F	Artificial saliva	Application of the amelogenin-hydrogel significantly improved enamel hardness.
Chung *et al.*, 2012 [40]	34% PA (15s)	Peptide	Artificial saliva	Peptide promoted the uniform deposition of apatites with small crystalline size.
Li *et al.*, 2013 [18]	37% PA (15s)	Nacre water-soluble matrix	SBF	Water-soluble matrix facilitated the formation of HAP nanorods.
Chung *et al.*, 2013 [27]	1M citric acid (2m)	Peptide	Artificial saliva	Peptide promoted the formation of nano-HAP crystals.

Table 1. *Cont.*

Authors, Year [Ref.]	Surface Treatment	Biomineralisation Method	Sources of Ca and P	Main Finding
Chen *et al.*, 2013 [30]	3% HNO_3 solution (50s)	PAMAM 1 ppm F	Calcification solution	PAMAM could induce the formation of HAP crystals on demineralised enamel.
Ruan *et al.*, 2013 [33]	30% PA (30s)	Amelogenin 136 ppm F	Amelogenin-CaP hydrogel, artificial saliva	The enamel-like layer was formed on natural enamel.
Wu *et al.*, 2013 [36]	37% PA (45s)	PAMAM	Artificial saliva	PAMAM could induce nanorod-like HAP remineralisation on acid-etched enamel.
Chung *et al.*, 2013 [41]	34% PA (30s)	Peptide	Artificial saliva	Peptide attracted ions from artificial saliva to form HAP crystals and fill enamel caries.
Chung *et al.*, 2013 [43]	34% PA (30s)	Peptide	Artificial saliva	Peptide promoted the deposition of small-grain HAP crystals.
Cao *et al.*, 2014 [19]	37% PA (60s)	Agarose hydrogel 500 ppm F	$CaCl_2$ hydrogel, Phosphate solution	Enamel prism-like tissue was generated on etched enamel surface.
Zhang *et al.*, 2014 [25]	Demineralising solution (7d)	Phosphorylated chitosan-ACP	Pchi-ACP solution	Remineralising rate of Pchi-ACP treatment was higher than that of fluoride treatment.
Ruan *et al.*, 2014 [34]	30% PA (30s)	Amelogenin 0.45 ppm F	Amelogenin-CaP hydrogel, artificial saliva	Enamel-like organized apatite crystals can be formed in amelogenin-chitosan system.
Milly *et al.*, 2014 [32]	8% methylcellulose gel, lactic acid (14d)	Polyacrylic acid, Bioactive glass	Bioactive glass slurry	BAG and PAA-BAG treatments enhanced remineralisation of enamel white spot lesions.
Cao *et al.*, 2014 [35]	37% PA (60s)	Enamel matrix derivative 500 ppm F	EMD-$CaCl_2$ hydrogel, Phosphate solution	EMD facilitated enamel prism-like tissue formation on demineralised human enamel.
Chen *et al.*, 2014 [37]	37% PA (45s)	PAMAM	Artificial saliva	PAMAM could produce an enamel prism-like structure on acid-etched enamel.
Li *et al.*, 2014 [38]	37% PA (60s)	Oligopeptide 1 ppm F	Metastable calcium phosphate solution	Oligopeptide induced the formation of HAP crystals on etched enamel.

BAG—Bioactive glass; CPP—Casein phosphopeptide; CPP-ACP—Casein phosphopeptide-amorphous calcium phosphate; CPP-ACPF—Casein phosphopeptide-amorphous calcium phosphate fluoride; EDTA—Ethylenediamine tetraacetic acid; EMD—Enamel matrix derivative; F—fluoride; FHAP—Fluoridated hydroxyapatite; Glu—glutamic acid; HAP—Hydroxyapatite; PA—Phosphoric acid; Pchi—Phosphorylated chitosan; PAA—Polyacrylic acid; PAMAM—Poly (amido amine) dendrimer; PVP—poly(vinylpyrrolidone); SBF—Simulated body fluid.

Table 2. Protein or its analogues and their functions in biomimetic mineralisation of human enamel.

Protein or Its Analogues	Function of Protein or Protein Analogues	Approach [Ref.]
Amelogenin	• Promote the oriented bundle formation • Mediate the orderly enamel-like structure formation • Control the organised growth of apatite crystals • Form transitional complexes with ACP • Promote the hydrolysis of ACP to stable HAP • Stabilise the Ca-P clusters and guide their arrangement into linear chains	Amel—Ca/P/F solution: [28,29] Amel—Ca/P hydrogel: [26,33,34]
EMD	• Affect the formation of enamel prism-like tissue • Guide the continuous growth of crystals • Prevent the crystal fusion of premature crystals • Control crystal morphology and subsequent elongation • Control the nucleation and growth of the crystals	EMD—Ca agarose hydrogel: [35]
PAMAM dendrimers: PAMAM-COOH [30] ALN-PAMAM-COOH [36] PAMAM-PO_3H_2 [37]	• Absorb onto the etched enamel surface • Provide nucleation sites and mineralisation template for HAP • Regulate the growth of crystals • Induce nanorod-like HAP with high uniformity • Induce *in situ* remineralisation of HAP on etched enamel	PAMAM—etched enamel: [30,36,37]
Peptide: 3NSS [27,41,43] QQRFEWEFEQQ [31] $C_{18}H_{35}O$-TKREEVD [38] 8DSS [39,42] 3DSS [40]	• Form 3D biomimetic scaffolds capable of nucleating HAP • Capture Ca and P ions from artificial saliva • Limit Ca and P ions to depart from the demineralised enamel • Possess a high affinity to bind with the HAP surface • Promote HAP with a small average crystalline size • Induce formation of stable ACP nano-precursors and prevent them from aggregation and precipitation	Peptide—etched enamel: [27,31,39–43] Peptide—Ca/P/F solution: [38]
Glutamic acid	• Induce the nano apatite assembly on enamel • Provide the active sites to control enamel formation • Interact with a (001) surface of the nano apatite • Induce the oriented aggregation of nano apatite	Glutamic acid—SBF: [44]
Polyacrylic acid	• Sequester calcium and phosphate ions • Form amorphous calcium phosphate nano-precursors • Reduce the abrasiveness of bioactive glass particles	Polyacrylic acid—bioactive glass: [32]
Phosphorylated chitosan	• Bind calcium ions to form nucleating sites • Stabilise ACP to form the nano-complexes of Pchi-ACP • Adsorb onto the surfaces of HAP crystals • Inhibit spontaneous precipitation of calcium phosphate	Pchi—$CaPO_4$ solution: [25]
Casein phosphopeptide	• Stabilise calcium, phosphate, and hydroxide ions • Prevent spontaneous precipitation of calcium phosphate • Promote enamel subsurface lesion remineralisation	CPP—$CaPO_4$ solution: [22] CPP—ACP Paste: [23,24]

ACP—Amorphous Calcium Phosphate; ALN—Alendronate; Amel—Amelogenin; Ca—Calcium; CPP—Casein Phosphopeptide; DSS—Aspartic acid-Serine-Serine; EMD—Enamel Matrix Derivative; F—Fluoride; HAP—Hydroxyapatite; NSS—Asparagine-Serine-Serine; P—Phosphorus; PAMAM—Poly (Amido Amine) Dendrimer; Pchi—Phosphorylated Chitosan; QQRFEWEFEQQ—Glutamine-Glutamine-Arginine-Phenylalanine-Glutamic acid-Tryptophan-Glutamic acid-Phenylalanine-Glutamic acid-Glutamine-Glutamine; SBF—Simulated Body Fluid; TKREEVD—Threonine-Lysine-Arginine-Glutamic acid-Glutamic acid-Valine-Aspartic acid.

3. Discussion

Enamel biomineralisation is a highly regulated process involving precise genetic control, as well as protein-protein interactions, protein-mineral interactions, and interactions involving the cell membrane. Ameloblast secretes enamel matrix proteins in the extracellular space between ameloblasts and dentine to control the initiation, orientation, nucleation, and growth of HAP crystals [4]. Amelogenin is the main secretory product of ameloblasts, making up more than 90% of the organic component in enamel [45]. The assembly of amelogenin has been shown to be crucial for the proper development of enamel crystallites. However, mature enamel is unable to heal and repair itself after demineralisation of the surface and subsequent cavitation. Biomimetic strategies for enamel regeneration may have the potential to repair enamel surface lesions. It is, therefore, important to study the structures and functions of amelogenin for the biomimetic mineralisation of enamel.

Iijima and Moradian-Oldak used the natural amelogenin *in vitro* to control calcium and phosphate crystallisation, and they found the growth of nanorod-like apatite crystals with a similar habit, size, and orientation to natural enamel [46]. Although this study was carried out through a cation selective membrane system as a model of tooth enamel formation, it highlighted the potential of amelogenin in the design and development of enamel-like biomaterials. Fan *et al.* used amelogenin with a modified biomimetic deposition method to remineralise the surface of etched enamel, and they were able to form mineral layers containing organised needle-like fluoridated hydroxyapatite crystals [28]. A follow-up study by Fan *et al.* reported the conditions of supersaturation degree, fluoride, and amelogenin concentration on the formation of densely packed fluoridated HAP on etched enamel *in vitro* [29]. An amelogenin-containing chitosan hydrogel was developed for superficial enamel reconstruction. The mechanism of the enamel reconstruction is through amelogenin supramolecular assembly, stabilising Ca-P clusters, and guiding their arrangement into linear chains [33,34]. However, it was difficult to extract and purify natural proteins. Thus, researchers fabricated protein analogues that could play the roles of the proteins in the biomineralisation process. Peptide/oligopeptide was fabricated to promote the formation of HAP crystals. A peptide with numerous repetitive nucleotide sequences of Aspartic-Serine-Serine (DSS), based on the sequence of dentin phosphoprotein had a high affinity to calcium phosphate compounds and promoted the formation of HAP crystals [39,40,42]. Triplet-repeating Asparagine-serine-serine (3NSS) peptide, a derivative of the DSS peptide, was designed by replacing the COOH group in Aspartic with the $CONH_2$ group. The authors declared that enamel remineralisation with this 3NSS peptide showed a higher degree of recovery compared to DSS peptide. This might due to the difference between the ionic-attraction ability of a COOH group and $CONH_2$ group [27,41,43]. Kirkham *et al.* designed a self-assembling peptide (P_{11}-4, Ace-Gln-Gln-Arg-Phe-Glu-Trp-Glu-Phe-Glu-Gln-Gln-NH_2), which could provide a biomimetic scaffold for HAP nucleation [31]. Recently, Li *et al.* fabricated an oligopeptide amphiphile ($C_{18}H_{35}O$-Thr-Lys-Arg-Glu-Glu-Val-Asp), which consisted of a derivative of stearic acid and a hydrophilic C-terminal amino residue of amelogenin [38]. In general, the peptides were shown to increase the remineralisation of enamel. PAMAM dendrimers have also been used as "artificial proteins" to evaluate their binding capacity and the surface charge with enamel crystals [12,47]. PAMAM-COOH acted as the organic template on the demineralized enamel surface to induce the formation of HAP crystals with the same structure, orientation and mineral phase of the natural enamel [30]. Due to the specific adsorption on HAP, alendronate was conjugated to PAMAM-COOH and induced *in situ* remineralisation of HAP on acid-etched enamel [36]. PAMAM-PO_3H_2 was designed with a similar dimensional scale and peripheral functions to that of amelogenin in enamel [37]. These dendrimers formed nanospheres and had a possible role in guiding crystal growth. In addition to the proteins and protein analogues, there are a number of other available biopolymers that were used in the biomimetic formation of HAP. These were predominantly polysaccharides such as agarose [19] and gelatine [20].

The synthesis of enamel-like structures based on the aforementioned approaches requires several days. In recent years, the biomimetic treatment of early caries lesions through the application of various types of bioactive materials with protein analogues has received considerable attention such as PAA-bioactive glass [32] and Glu-apatite nanoparticles [44]. PAA-bioactive glass was applied as slurry to avoid any damage to the enamel lesion structure. Bioactive glass is a ceramic material consisting of a similar amount of sodium, calcium, phosphate, and silicate ions, which are elements naturally found in the human body. The release of ions requires contact with artificial saliva. These bioactive materials dissolve in saliva to liberate ions that diffuse into the enamel subsurface lesion. PAA was used as an analogue to sequester calcium and phosphate ions released from bioactive glass to form amorphous calcium phosphate nano-precursors. The PAA-bioactive glass had the potential to deliver calcium and phosphate ions to subsurface enamel lesions. In an aqueous environment, the protonation of the Si-O groups formed rich silanols on the etched enamel surface. It generated an electronegative surface that provided heterogeneous nucleation sites for the formation of HAP crystals [48]. Amino acids are the basic building blocks of proteins. There is 15-20% Glu in the amino acid composition of amelogenin, and the roles of Glu in the biomimetic mineralisation of enamel have been investigated. Glu has two carboxylate groups that could be preferentially adsorbed onto the (001) face of apatite to induce the oriented crystallisation along the *c*-axis. Several studies demonstrated that apatite nano-particles have the potential to remineralise the initial enamel caries lesions *in vitro*. Nanostructured apatite crystals exhibit high levels of biomimetic properties and surface activity. Apatite nano-particles can act as a calcium and phosphate reservoir that helps to maintain a topical state of super-saturation of these ions on an enamel surface. As a building block, apatite nano-particles could adsorb onto the enamel surface and assemble into oriented HAP crystals under the control of Glu. This resulted in the regeneration of the enamel-like structure under physiological conditions [44].

In biology, the biomineralisation process is an organic matrix particle-mediated non-classical crystallisation pathway involving a mesoscopic transformation process. The presence of transient ACP nanoprecursors has been found during enamel biomineralisation [49]. The formation and stabilisation of amorphous nanoprecursors is an important step in the biomimetic mineralisation of enamel. CPP obtained from milk is an analogue of the proteins involved in the biomineralisation of teeth. CPP contains clusters of phosphorylated serine residues and can, thus, stabilise calcium and phosphate ions through the formation of amorphous nano-complexes [22–24]. CPP-ACP provided a reservoir of calcium and phosphate ions to maintain a state of super-saturation on the enamel surface, resulting in a significant remineralising effect on enamel subsurface lesions [50]. Based on the influence of phosphorylated proteins in enamel biomineralisation, nano-complexes of phosphorylated chitosan and ACP were synthesised to remineralise early enamel caries [25]. These results indicated that the application of natural protein analogues might be an effective strategy to stabilise the amorphous nanoprecursors.

Fluoride was used to prevent and arrest caries by inhibiting bacterial metabolism, inhibiting demineralisation, and enhancing remineralisation. Fluoride ions entered into the directive lattice of HAP crystals and replaced hydroxide ions or adsorbed onto the crystal surface to attract calcium and phosphate ions, resulting in the formation of fluoridated HAP [51]. Fluoridated HAP contains approximately 30,000 ppm fluoride and has a very low solubility in acid. In this review, fluoride was added in different remineralising mediums in 15 studies. The concentration of the fluoride used varied greatly between the studies. Fluoride had a significant effect on the morphology of calcium phosphate crystals. The morphology and orientation of the formed crystals were tuneable by the fluoride concentration. Iijima *et al.* discovered the conversion of ribbon-like octacalcium phosphate into needle-like HAP in a membrane system in the presence of 2 ppm fluoride [52]. Fan *et al.* reported that a critical fluoride at 5 ppm in remineralisation solutions with a pH of 6.74 and 5 mM Ca^{2+} was required to induce the transition from plate-like octacalcium phosphate to octacalcium phosphate rod-like HAP that regenerated on etched enamel surface [29]. The critical concentration of fluoride changed when the pH and the concentrations of Calcium and amelogenin were changed. Rod-like fluoridated HAP was reported in a hydrogel model with a high concentration of fluoride [19]. However, despite the predominant beneficial effect of fluoride on enamel remineralisation, the dental fluorosis induced by excessive intake of fluoride could not be ignored, especially with the application of these products to children.

Apparent progress has been made to understand the process of HAP biomineralisation of enamel, which has the potential to remineralise defective enamel lesions. A number of biomimetic mineralisation methods were reported and published in the literature. However, the overall mechanism of remineralisation *in vivo* has not yet been clarified. Further research is necessary to assess in detail whether the biomimetic mineralisation approaches for enamel caries therapy will be suitable and applicable in daily dental practice. Furthermore, an easy-to-apply, fast-growing system for enamel-like structure regeneration is necessary for clinical use.

4. Methods

Three electronic databases (PubMed, ScienceDirect, and ISI Web of Science) were systematically searched to identify publications before 1 January 2015. The keywords used were: ((biomimetic AND (mineralization OR mineralisation OR regeneration OR remineralization OR remineralisation)) OR biomineralization OR biomineralisation) AND enamel. No publication year or language limit was used. The titles and abstracts of identified articles were independently reviewed by two authors. Duplicated articles were removed. Clinical trials, reviews, non-English articles, animal teeth, non-tooth substrates, and irrelevant studies were excluded. The articles were further reviewed with full-texts if the articles met the inclusion criteria (biomimetic mineralisation methods for demineralised human enamel *in vitro*). Manual screening was conducted on the reference lists of all included articles to identify relevant articles. Disagreements about the inclusion or exclusion of a study were discussed with a third author for a final decision. Data extraction was conducted by one author from full-texts and evaluated by another one. Similar information mentioned in the included articles was summarised (such as protein and protein analogues).

5. Conclusions

Many studies with different methods reported success in the biomimetic mineralisation of enamel, including the use of protein and protein analogues, bioactive materials or components, an agarose hydrogel model, a glycerine-enriched gelatine technique, and ethylenediaminetetraacetic acid.

Acknowledgments

This work was supported by the General Research Fund (HKU 760413M).

Author Contributions

Chris Ying Cao and May Lei Mei searched the databases and prepared the manuscript. Quan-li Li regulated the data extraction. Edward Chin Man Lo and Chun Hung Chu revised the manuscript. All authors read and approved the final manuscript.

Conflicts of Interest

The authors declare no conflict of interest.

References

1. Chen, H.F.; Tang, Z.Y.; Liu, J.; Sun, K.; Chang, S.R.; Peters, M.C.; Mansfield, J.F.; Czajka-Jakubowska, A.; Clarkson, B.H. Acellular synthesis of a human enamel-like microstructure. *Adv. Mater.* **2006**, *18*, 1846–1851.
2. Veis, A. Materials science. A window on biomineralization. *Science* **2005**, *307*, 1419–1420.
3. Cui, F.Z.; Ge, J. New observations of the hierarchical structure of human enamel, from nanoscale to microscale. *J. Tissue Eng. Regen. Med.* **2007**, *1*, 185–191.
4. Fincham, A.G.; Moradian-Oldak, J.; Simmer, J.P. The structural biology of the developing dental enamel matrix. *J. Struct. Biol.* **1999**, *126*, 270–299.
5. Aoba, T. Solubility properties of human tooth mineral and pathogenesis of dental caries. *Oral Dis.* **2004**, *10*, 249–257.
6. Featherstone, J.D. Dental caries: A dynamic disease process. *Aust. Dent. J.* **2008**, *53*, 286–291.
7. Yin, Y.J.; Yun, S.; Fang, J.S.; Chen, H.F. Chemical regeneration of human tooth enamel under near-physiological conditions. *Chem. Commun.* **2009**, 5892–5894.
8. Fan, Y.; Sun, Z.; Moradian-Oldak, J. Effect of fluoride on the morphology of calcium phosphate crystals grown on acid-etched human enamel. *Caries Res.* **2009**, *43*, 132–136.
9. Fowler, C.E.; Li, M.; Mann, S.; Margolis, H.C. Influence of surfactant assembly on the formation of calcium phosphate materials—A model for dental enamel formation. *J. Mater. Chem.* **2005**, *15*, 3317–3325.
10. Ye, W.; Wang, X.X. Ribbon-like and rod-like hydroxyapatite crystals deposited on titanium surface with electrochemical method. *Mater. Lett.* **2007**, *61*, 4062–4065.
11. Yamagishi, K.; Onuma, K.; Suzuki, T.; Okada, F.; Tagami, J.; Otsuki, M.; Senawangse, P. Materials chemistry: A synthetic enamel for rapid tooth repair. *Nature* **2005**, *433*, 819.

12. Chen, H.F.; Clarkson, B.H.; Sun, K.; Mansfield, J.F. Self-assembly of synthetic hydroxyapatite nanorods into an enamel prism-like structure. *J. Colloid. Interface Sci.* **2005**, *288*, 97–103.
13. Moradian-Oldak, J. Protein-mediated enamel mineralization. *Front. Biosci.* **2012**, *17*, 1996–2023.
14. Ruan, Q.; Moradian-Oldak, J. Amelogenin and enamel biomimetics. *J. Mater. Chem. B* **2015**, *3*, 3112–3129.
15. Margolis, H.C.; Beniash, E.; Fowler, C.E. Role of macromolecular assembly of enamel matrix proteins in enamel formation. *J. Dent. Res.* **2006**, *85*, 775–793.
16. Cao, Y.; Liu, W.; Ning, T.; Mei, M.L.; Li, Q.L.; Lo, E.C.; Chu, C.H. A novel oligopeptide simulating dentine matrix protein 1 for biomimetic mineralization of dentine. *Clin. Oral Investig.* **2014**, *18*, 873–881.
17. Zhou, Y.Z.; Cao, Y.; Liu, W.; Chu, C.H.; Li, Q.L. Polydopamine-induced tooth remineralization. *ACS Appl. Mater. Interfaces* **2012**, *4*, 6901–6910.
18. Li, X.; Pan, D.M.; Lin, S.; Zhuang, Z.Y.; Lin, Z. Facile *in vitro* hydroxyapatite remineralization of human enamel with remarkable hardness. *CrystEngComm.* **2013**, *15*, 4351–4356.
19. Cao, Y.; Mei, M.L.; Li, Q.L.; Lo, E.C.; Chu, C.H. Agarose hydrogel biomimetic mineralization model for the regeneration of enamel prismlike tissue. *ACS Appl. Mater. Interfaces* **2014**, *6*, 410–420.
20. Busch, S. Regeneration of human tooth enamel. *Angew. Chem. Int. Edit.* **2004**, *43*, 1428–1431.
21. Xie, R.Q.; Feng, Z.D.; Li, S.W.; Xu, B.B. EDTA-assisted self-assembly of fluoride-substituted hydroxyapatite coating on enamel substrate. *Cryst. Growth. Des.* **2011**, *11*, 5206–5214.
22. Reynolds, E.C. Remineralization of enamel subsurface lesions by casein phosphopeptide-stabilized calcium phosphate solutions. *J. Dent. Res.* **1997**, *76*, 1587–1595.
23. Kumar, V.L.N.; Itthagarun, A.; King, N.M. The effect of casein phosphopeptide-amorphous calcium phosphate on remineralization of artificial caries-like lesions: An *in vitro* study. *Aust. Dent. J.* **2008**, *53*, 34–40.
24. Jayarajan, J.; Janardhanam, P.; Jayakumar, P.; Deepika. Efficacy of CPP-ACP and CPP-ACPF on enamel remineralization—An *in vitro* study using scanning electron microscope and diagnodent. *Indian J. Dent. Res.* **2011**, *22*, 77–82.
25. Zhang, X.; Li, Y.; Sun, X.; Kishen, A.; Deng, X.; Yang, X.; Wang, H.; Cong, C.; Wang, Y.; Wu, M. Biomimetic remineralization of demineralized enamel with nano-complexes of phosphorylated chitosan and amorphous calcium phosphate. *J. Mater. Sci. Mater. Med.* **2014**, *25*, 2619–2628.
26. Fan, Y.; Wen, Z.T.; Liao, S.; Lallier, T.; Hagan, J.L.; Twomley, J.T.; Zhang, J.F.; Sun, Z.; Xu, X. Novel amelogenin-releasing hydrogel for remineralization of enamel artificial caries. *J. Bioact. Compat. Polym.* **2012**, *27*, 585–603.
27. Chung, H.Y.; Huang, K.C. Effects of peptide concentration on remineralization of eroded enamel. *J. Mech. Behav. Biomed.* **2013**, *28*, 213–221.
28. Fan, Y.; Sun, Z.; Moradian-Oldak, J. Controlled remineralization of enamel in the presence of amelogenin and fluoride. *Biomaterials* **2009**, *30*, 478–483.
29. Fan, Y.; Nelson, J.R.; Alvarez, J.R.; Hagan, J.; Berrier, A.; Xu, X. Amelogenin-assisted *ex vivo* remineralization of human enamel: Effects of supersaturation degree and fluoride concentration. *Acta Biomater.* **2011**, *7*, 2293–2302.

30. Chen, L.; Liang, K.N.; Li, J.S.; Wu, D.; Zhou, X.D.; Li, J.Y. Regeneration of biomimetic hydroxyapatite on etched human enamel by anionic PAMAM template *in vitro*. *Arch. Oral Biol.* **2013**, *58*, 975–980.
31. Kirkham, J.; Firth, A.; Vernals, D.; Boden, N.; Robinson, C.; Shore, R.C.; Brookes, S.J.; Aggeli, A. Self-assembling peptide scaffolds promote enamel remineralization. *J. Dent. Res.* **2007**, *86*, 426–430.
32. Milly, H.; Festy, F.; Watson, T.F.; Thompson, I.; Banerjee, A. Enamel white spot lesions can remineralise using bio-active glass and polyacrylic acid-modified bio-active glass powders. *J. Dent.* **2014**, *42*, 158–166.
33. Ruan, Q.; Zhang, Y.; Yang, X.; Nutt, S.; Moradian-Oldak, J. An amelogenin-chitosan matrix promotes assembly of an enamel-like layer with a dense interface. *Acta Biomater.* **2013**, *9*, 7289–7297.
34. Ruan, Q.; Siddiqah, N.; Li, X.; Nutt, S.; Moradian-Oldak, J. Amelogenin-chitosan matrix for human enamel regrowth: Effects of viscosity and supersaturation degree. *Connect. Tissue Res.* **2014**, *55*, 150–154.
35. Cao, Y.; Mei, M.L.; Li, Q.L.; Lo, E.C.M.; Chu, C.H. Enamel prism-like tissue regeneration using enamel matrix derivative. *J. Dent.* **2014**, *42*, 1535–1542.
36. Wu, D.; Yang, J.; Li, J.; Chen, L.; Tang, B.; Chen, X.; Wu, W.; Li, J. Hydroxyapatite-anchored dendrimer for *in situ* remineralization of human tooth enamel. *Biomaterials* **2013**, *34*, 5036–5047.
37. Chen, M.; Yang, J.; Li, J.; Liang, K.; He, L.; Lin, Z.; Chen, X.; Ren, X.; Li, J. Modulated regeneration of acid-etched human tooth enamel by a functionalized dendrimer that is an analog of amelogenin. *Acta Biomater.* **2014**, *10*, 4437–4446.
38. Li, Q.L.; Ning, T.Y.; Cao, Y.; Zhang, W.B.; Mei, M.L.; Chu, C.H. A novel self-assembled oligopeptide amphiphile for biomimetic mineralization of enamel. *BMC Biotechnol.* **2014**, *14*, 32.
39. Hsu, C.C.; Chung, H.Y.; Yang, J.M.; Shi, W.; Wu, B. Influences of ionic concentration on nanomechanical behaviors for remineralized enamel. *J. Mech. Behav. Biomed. Mater.* **2011**, *4*, 1982–1989.
40. Chung, H.Y.; Li, C.C.; Hsu, C.C. Characterization of the effects of 3DSS peptide on remineralized enamel in artificial saliva. *J. Mech. Behav. Biomed. Mater.* **2012**, *6*, 74–79.
41. Chung, H.Y.; Li, C.C. Microstructure and nanomechanical properties of enamel remineralized with asparagine-serine-serine peptide. *Mate. Sci. Eng. C. Mater. Biol. Appl.* **2013**, *33*, 969–973.
42. Hsu, C.C.; Chung, H.Y.; Yang, J.M.; Shi, W.; Wu, B. Influence of 8DSS peptide on nano-mechanical behavior of human enamel. *J. Dent. Res.* **2011**, *90*, 88–92.
43. Chung, H.Y.; Li, C.C. Asparagine-serine-serine peptide regulates enamel remineralization *in vitro*. *J. Mater. Res.* **2013**, *28*, 2890–2896.
44. Li, L.; Mao, C.; Wang, J.; Xu, X.; Pan, H.; Deng, Y.; Gu, X.; Tang, R. Bio-inspired enamel repair via Glu-directed assembly of apatite nanoparticles: An approach to biomaterials with optimal characteristics. *Adv. Mater.* **2011**, *23*, 4695–4701.
45. Ravindranath, R.M.; Moradian-Oldak, J.; Fincham, A.G. Tyrosyl motif in amelogenins binds N-acetyl-D-glucosamine. *J. Biol. Chem.* **1999**, *274*, 2464–2471.
46. Iijima, M.; Moradian-Oldak, J. Control of apatite crystal growth in a fluoride containing amelogenin-rich matrix. *Biomaterials* **2005**, *26*, 1595–1603.

47. Chen, H.; Banaszak Holl, M.; Orr, B.G.; Majoros, I.; Clarkson, B.H. Interaction of dendrimers (artificial proteins) with biological hydroxyapatite crystals. *J. Dent. Res.* **2003**, *82*, 443–448.
48. Dong, Z.H.; Chang, J.A.; Zhou, Y.; Lin, K.L. *In vitro* remineralization of human dental enamel by bioactive glasses. *J. Mater. Sci.* **2011**, *46*, 1591–1596.
49. Beniash, E.; Metzler, R.A.; Lam, R.S.; Gilbert, P.U. Transient amorphous calcium phosphate in forming enamel. *J. Struct. Biol.* **2009**, *166*, 133–143.
50. Cao, Y.; Mei, M.L.; Xu, J.; Lo, E.C.; Li, Q.; Chu, C.H. Biomimetic mineralisation of phosphorylated dentine by CPP-ACP. *J. Dent.* **2013**, *41*, 818–825.
51. Featherstone, J.D. The science and practice of caries prevention. *J. Am. Dent. Assoc.* **2000**, *131*, 887–899.
52. Iijima, M.; Tohda, H.; Suzuki, H.; Yanagisawa, T.; Moriwaki, Y. Effects of F- on apatite-octacalcium phosphate intergrowth and crystal morphology in a model system of tooth enamel formation. *Calcif. Tissue Int.* **1992**, *50*, 357–361.

13

Reinforcement Strategies for Load-Bearing Calcium Phosphate Biocements

Martha Geffers, Jürgen Groll and Uwe Gbureck *

Department for Functional Materials in Medicine and Dentistry, University Hospital Würzburg, Pleicherwall 2, D-97070 Würzburg, Germany; E-Mails: martha.geffers@fmz.uni-wuerzburg.de (M.G.); juergen.groll@fmz.uni-wuerzburg.de (J.G.)

* Author to whom correspondence should be addressed; E-Mail: uwe.gbureck@fmz.uni-wuerzburg.de;

Academic Editor: Amir A. Zadpoor

Abstract: Calcium phosphate biocements based on calcium phosphate chemistry are well-established biomaterials for the repair of non-load bearing bone defects due to the brittle nature and low flexural strength of such cements. This article features reinforcement strategies of biocements based on various intrinsic or extrinsic material modifications to improve their strength and toughness. Altering particle size distribution in conjunction with using liquefiers reduces the amount of cement liquid necessary for cement paste preparation. This in turn decreases cement porosity and increases the mechanical performance, but does not change the brittle nature of the cements. The use of fibers may lead to a reinforcement of the matrix with a toughness increase of up to two orders of magnitude, but restricts at the same time cement injection for minimal invasive application techniques. A novel promising approach is the concept of dual-setting cements, in which a second hydrogel phase is simultaneously formed during setting, leading to more ductile cement–hydrogel composites with largely unaffected application properties.

Keywords: calcium phosphate cements; porosity; fiber reinforcement; dual setting; mechanical properties

---

1. Introduction

Self-setting cements based on calcium phosphate chemistry combine the advantages of the high biocompatibility of calcium phosphates with the free mouldability of cements and the mechanical stability of ceramic implants [1,2]. Such calcium phosphate cements (CPC) are usually based on freshly prepared mixtures of crystalline or amorphous calcium orthophosphate, calcium hydroxide or calcium carbonate powders with an aqueous solution, which undergo setting in a continuous dissolution–precipitation reaction. Although various mixtures of calcium and phosphate sources can serve as raw materials, there are in principle only two cement types as products of the setting reaction: At neutral or basic pH the calcium phosphate cement sets to nanocrystalline hydroxyapatite (HA, with a variable stoichiometric composition between $Ca_9(PO_4)_5HPO_4OH–Ca_{10}(PO_4)_6(OH)_2$), while at low pH < 4.2, orthophosphate ions are protonated and the secondary phosphates brushite ($CaHPO_4 \cdot 2H_2O$, DCPD) and monetite ($CaHPO_4$, DCPA) are the least soluble calcium phosphates [3,4] and hence precipitated during setting of acidic cement pastes until an end pH of close to 5 [1,2,5–7]. Detailed reviews about CPCs reflecting their synthesis, setting reaction, rheological properties or biological performance can be found in literature [2,8,9]. CPC are resorbed *in vivo* and replaced by new bone tissue [10,11], whereas the speed of degradation depends on the final composition of the cement matrix. Hydroxyapatite forming cements degrade only slowly within years since the surrounding extracellular fluid ($[Ca^{2+}]$ ~ 2.5 mmoL/L, $[HPO_4^{2-}]$ ~ 1 mmoL/L [12]) is supersaturated regarding HA (solubility of hydroxyapatite ~ 0.2–0.3 mg/L) [2]. HA forming cements degrade solely by osteoclastic bone remodeling, which is limited to surface degradation since cells cannot penetrate the microporous cement structure. Osteoclastic cells resorb the cement by providing a local acidic environment increasing the solubility of the mineral [13–16]. In contrast, cements forming brushite or monetite have a higher solubility (calculated solubility in water for monetite: 41–48 mg/L, brushite: 85–88 mg/L [17]) and many studies have demonstrated the bone remodelling capacity of such cements in various animal models within a time period of 8–52 weeks [18–21]. A passive resorption of such cements by simple chemical dissolution is a topic of contention in the literature, whereas some authors postulate that the extracellular liquid is in equilibrium with brushite [22], while others have calculated a thermodynamic instability of brushite in simulated body fluid [23]. The latter is supported by the fact that brushite forming cements are indeed dissolved *in vivo* even in the absence of osteoclastic cells (e.g., after intramuscular implantation) [24]. Worth noting is that for brushite forming cements a phase transformation into lower soluble minerals like octacalcium phosphate, hydroxyapatite or whitlockite can occur *in vivo* by a dissolution–reprecipitation reaction, which slows down biodegradation [25,26].

Calcium phosphate bone cements have been shown to provide compressive strength of up to 80 MPa measured under application near conditions without a precompaction of the cement paste leading to lower porosity/higher strength, since this is not applicable under *in vivo* conditions [27]. Set CPC can be considered as porous ceramic materials with an inherent brittleness and comparatively low flexural strength compared to natural hard tissues such as bone or teeth. A comprehensive characterization of the elastic and failure properties for both hydroxyapatite and brushite forming CPC by Charrière *et al.* [28] indicated brushite cements to be suitable as bone fillers, while hydroxyapatite cements were attributed to having the potential to be a structural biomaterial. The low fracture toughness restricts the use of CPC to non-load-bearing defects [29]. Typical applications are the treatment of maxillofacial defects or deformities [1] or the repair of craniofacial defects [30]. An extension of the application of calcium phosphate cements to load-bearing defects, e.g., in vertebroplasty or kyphoplasty [31–33], would require less brittle cements with an increased fracture toughness. This is of high interest since the application of commonly used polymeric cements have strong drawbacks near the spinal cord due to their strong exothermic setting reaction and cytotoxic monomer release [34–36]. Common approaches to reduce brittleness of CPC and to improve their mechanical performance for load-bearing applications cover the modification of the cement liquid with polymeric additives such as collagen [37–40], the addition of fibres to the cement matrix [41,42] or the use of dual-setting cements in which a dissolved monomer is simultaneously cross-linked during cement setting [43–45] (Figure 1). This article aims to feature the most significant reinforcement strategies for calcium phosphate cements based on either intrinsic (porosity) or extrinsic (fiber addition, dual setting cement) material modifications.

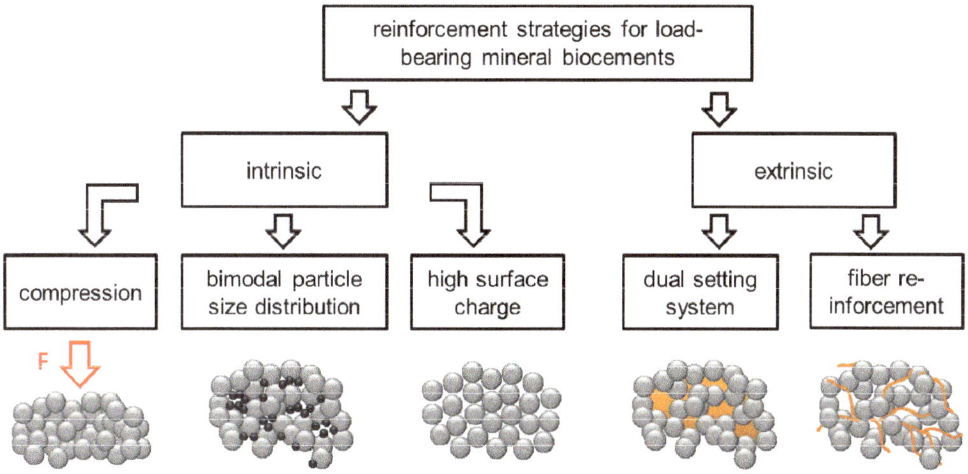

Figure 1. Strategies to reinforce mineral biocement for load-bearing applications.

2. Porosity Reduction for Strength Improvement of CPC

Calcium phosphate biocements set by a dissolution–precipitation reaction, during which the cement raw material continuously dissolves to form a supersaturated solution with regard to the setting product. The latter is precipitated from the aqueous cement phase and forms an entangled cementitious crystal matrix. The mechanical strength of a cement matrix is a direct result of this crystal entanglement and several factors determine the final strength of the matrix, such as degree of conversion, setting product or porosity. The latter is likely the most important factor and it is known from literature that porosity reduction in cements from 50% to 31% by compression can increase compressive strength by nearly an order of magnitude [46]. Porosity in biocements predominately originates from the presence of unreacted cement liquid after setting located in the voids between the entangled crystal matrix. Since any excess of water used for paste mixing, which is not consumed during the setting reaction creates porosity, the main influencing parameter on the total cement porosity is the powder to liquid ratio (PLR) used for cement processing. Pore sizes in CPC typically have a diameter range spanning from a few nanometers to several micrometers [47,48] and are occupying about 22–55 vol% of cements without further paste manipulation (e.g., compaction, porogen addition) [49,50]. Generally, pores in hydroxyapatite cements are smaller than in brushite cements (due to smaller crystal size of HA), whereas the total porosity is mostly smaller for brushite cement. The latter is a result of an increased water consumption during brushite cement setting.

Porosity considerably lowers the strength and stiffness (Young's modulus) of the cements matrix with an inverse exponential relationship between cement porosity and compressive strength:

$$CS = CS_0 \exp^{-2KP} \quad (1)$$

where CS is the compressive strength at a given porosity; CS_0 is the maximum theoretical strength of the material; K is a constant; and P is porosity [46]. Porosity is usually measured by helium pycnometry [51], mercury intrusion porosimetry (MIP) [52] or it is calculated based on the phase composition of the set cements and their densities [53]. Due to the disadvantages of these methods (destructive, long analysis times, toxicity of mercury, misleading results due to amorphous phases), Unosson *et al.* [54] have investigated a method which is based on the assumption that the evaporated water from a dried cement sample equals to the volume of pores within the cement. Since the accuracy of this method depends on a quantitative drying of samples without affecting the phase composition, the authors evaluated several drying conditions (vacuum, elevated temperature) for cement samples and compared the results with porosity determined by the above mentioned methods. Since the measured porosity was found to vary between the different methods, the authors recommended using more than one method to determine cement porosity, whereas the water evaporation method (24 h in vacuum) proved to be fast, easy and precise in estimating the porosity of CPCs.

Porosity reduction by decreasing the amount of cement liquid used for mixing is a key parameter to increase the intrinsic strength of any biocement matrix. This, however, is limited, since every cement powder requires a formulation specific minimum amount of water ("plastic limit") for surface wetting of all cement particles and for filling the space between the particles [55]. A correlation between the powder to liquid ratio used for forming a cement paste and the resulting porosity/compressive strength is displayed in Figure 2 for both HA and brushite forming cements. An effective method to reduce cement porosity is based on both creating a bimodal size distribution of cement raw materials and the creation of a high surface charge (zeta-potential) of the particles. A bimodal size distribution is thought to fill space in cement pastes normally occupied by water. The possibility to reduce porosity has been demonstrated for both hydroxyapatite [27] and brushite [49,50] forming biocements. In addition, a high surface charge (zeta-potential) will help to disperse agglomerates of fine sized particles by reducing attractive interparticulate forces. The zeta-potential can be influenced by using multiple charged ions as additives to the cement liquid, e.g., tatrates or citrates [56], which adsorb at the particle surface and increase the zeta-potential to values of ~−40 to −50 mV. Applying these two principles to a matrix of α-tricalcium phosphate (monomodal size distribution with d_{50} ~9.8 μm) by using 13–33 wt% fine sized $CaHPO_4$ filler (d_{50} ~ 1.16 μm) and 0.5 M trisodium citrate solution increase the plastic limit of the cements from 3.5 to 5.0 g/mL. At the same time, porosity was decreased from 37% to 25% and a strength improvement from 50 to 79 MPa could be found [27]. Another study by Engstrand *et al.* [49] investigated the effect of β-TCP filler particles on the mechanical properties of a brushite forming cement (β-TCP-MCPM system). The results showed that the addition of low amounts of a filler (up to 10%) in combination with 0.8 M citric acid solution can effectively increase the powder to liquid ratio and hence decrease porosity from ~30% to ~23%. This strongly affects compressive strength of the cements with an increase from ~23 MPa (no filler and citric acid) to ~42 MPa. Space in cement pastes may also be filled by using hard agglomerates similar to civil engineering Portland cements as shown by Gu *et al.* [57]. In this study, the dispersion of 20% high-strength β-tricalcium phosphate granules with a size of 200–450 µm in the cement showed an increase of the compressive strength by 70%, while maintaining the rheological properties (injectability through 2.2 mm needle by applying a 5 kg weight on the syringe plunger) of the cement paste.

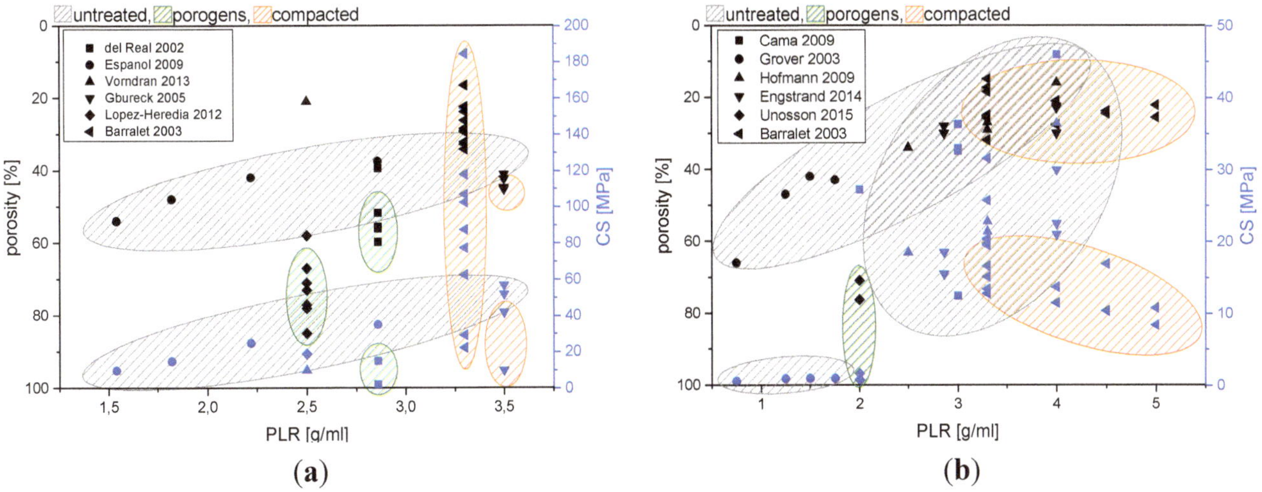

Figure 2. Correlation between powder to liquid ratio and porosity/compressive strength for (**a**) hydroxyapatite and (**b**) brushite cement from different studies. Cements were either set without compacting manipulation (untreated), processed by pre-compaction or porogens were added to create artificial macroporosity. Data were obtained from: (**a**) del Real 2002 [58], Espanol 2009 [47], Vorndran 2013 [59], Gbureck 2005 [27], Lopez-Heredia 2012 [60], Barralet 2003 [61], (**b**) Cama 2009 [62], Grover 2003 [51], Hofmann 2009 [50], Engstrand 2014 [49], Unosson 2015 [63], Barralet 2003 [64].

Caution must be exercised when comparing the obtained strength values from different studies, since many parameters during cement sample preparation and testing can affect the results. Unlike polymeric polymethylmethacrylate (PMMA) based bone cements [65], testing of calcium phosphate bone cement is not regulated, and our own experiences show that strength of set cement can vary by several times depending on the sample preparation and testing conditions. Generally, strength of dried samples is superior to that of (application near) wet specimen, mainly because water acts as a lubricant between the entangled crystals of the precipitated matrix. In addition, sample preparation may cause changes of cement porosity, e.g., by precompacting the paste in a mold. This ejects liquid from the paste (through the narrow gap between mold and plunger) leading to a lower porosity and hence a higher strength compared to uncompacted samples [61,66,67].

3. Fiber Reinforcement of CPC

Similar to reinforcement approaches of sintered hydroxyapatite ceramics [68], the addition of fibers to CPC is one of the most successful reinforcement technique [41,69]. The mechanical behavior of such fiber reinforced calcium phosphate cements (FRCPC) is a result of the complex interaction between all of the composite constituents. Contributions to the macroscopic behavior come from strength and stiffness of both fiber and cementitious matrix, matrix toughness, mechanical interaction between fibers and matrix as well as supplementary effects of polymeric additives or aggregates [69].

Fiber reinforcement studies have been performed with many different types of fibres (degradable *vs.* non-degradable, see Table 1 showing a strong increase of the mechanical strength depending on several parameters such as (1) matrix composition and strength, (2) fibre volume fraction, orientation, aspect ratio and tensile modulus as well as (3) the interface properties between matrix and fibres [69]. In addition to an increase of the bending strength from approx. 10–15 MPa for pure CPC to a maximum strength of 45 MPa (polyglactin fibers)—60 MPa (carbon fibers), especially the work of fracture for fiber reinforced cement composites usually increases by at least one order of magnitude (Table 1).

As illustrated in Figure 3, there is not only a complex interaction of factors, but in clinical application the properties of the fiber–cement composites are also time dependent since both the cement matrix and the fibers may degrade during tissue regeneration.

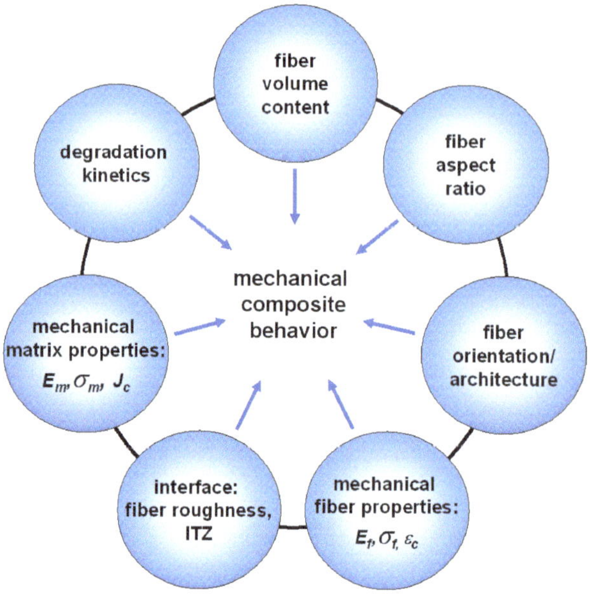

Figure 3. Interaction of material parameters which influence the time dependent mechanical behavior of the FRCPC composite. Reprinted with permission from [69].

Generally, the load-bearing capacity of fibers increases with their Young's modulus, whereas the maximum tensile stress within the fiber is determined by the fiber's modulus and the matrix strain [70]:

$$\sigma_f^{max} = E_f \varepsilon_m \tag{2}$$

When the composite is loaded, differences between Young's moduli of fiber and matrix lead to additional strain near the interface, mainly in the softer material [71]. The diameter of the fibers directly influences the total interface area between fibers and matrix for a given fiber volume fraction and affects both homogeneity and processability of the fiber–cement mixtures. Most biomedical composites are reinforced by discontinuous fibers. Their length and diameter are of great relevance, since substantial load has to be transferred from the matrix to the fiber via the interface for a reinforcing effect. Load is predominantly transferred by shear stresses at the lateral surface of the fibers rather than via the end faces of the fibers. Reinforcement effects are only observed, if the fiber length exceeds a critical value l_c, which can be calculated based on the assumption that the fiber is loaded up to the fracture strength.

$$l_c = \frac{d\sigma_{f,B}}{2\tau_i} \tag{3}$$

where σ_{fB} and τ_i denote fracture strength of the fiber and shear stress at the interface and d is the diameter of the fiber. Optimum fiber volume content has been addressed by many researchers. Civil engineering concretes typically are reinforced with <5 vol% of steel, glass, natural or synthetic polymer fibers [72]. In many studies on medical FRCPC, the fiber content is one order of magnitude higher than in fiber reinforced cements for civil engineering. This is attributed to a frequently observed trend in FRCPC research [73–75] that strength and ductility of the composites increased with fiber content. Moderate load transfer due to non-optimized interface strength and low modulus of the fibers require such high fiber volume fraction. Furthermore, fiber costs are not such a limiting factor, at least in the research stage.

Table 1. Examples for the reinforcement of calcium phosphate cements with either degradable or non-degradable fibres. (3 p.b.: Three point bending, 4 p. b.: Four point bending. # UD: Unidirectional fibers. TTCP: Tetracalcium phosphate. HA_w: Hydroxyapatite whiskers) [69].

Composition Fiber/Additive/Matrix	Fiber Volume Fraction	Strength [MPa]	Work of Fracture [kJ/m²]	Test Method	Ref.
		DEGRADABLE FIBRES			
HA matrix (TTCP + DCPA (+ Na_2HPO_4 – solution))	-	10–15	0.032–0.05	3 p. b.	[76,77]
Polyglactin 910/-/HA (TTCP + DCPA)	25 vol%	17.5–25	2.6–3.6	3 p. b.	[76]
Polyglactin 910/-/HA (TTCP + DCPA)	Mesh multilayer	8.5–24.5	0.75–3.1	3 p. b.	[78]
Polyglactin 910/chitosan lactate/HA (TTCP + DCPA)	45 vol%	41	11	3 p. b.	[74]
Polyglactin 910/chitosan lactate/HA (TTCP + DCPA)	Mesh multilayer	43	9.8	3 p. b.	[79]
Polyglactin 910/(poly(caprolactone))/ brushite (β-TCP + H_3PO_4)	24 vol% random short 6–25 long fibers UD #	7.5–20	n.a.	4 p. b.	[80]
		NON-DEGRADABLE FIBRES			
Carbon/-/HA (TTCP + DCPA)	2–10 vol%	32–60	3.5–6.5	3 p. b.	[72]
CNT/-/HA (α-TCP + HA)	0.2–1.0 wt%	8.2–10.5	n.a.	3 p. b.	[81]
Aramid/-/macroporous HA (TTCP + DCPA + Na_2HPO_4)	6 vol%	7.5–13.5	0.8–6.5	3 p. b.	[75]
HA_w/-/HA (TTCP + DCPA)	10–40 vol%	5.4–7.4	57–102	4 p. b.	[82]

Biodegradable polylactic-co-glycolic acid (PLGA) is one of the most frequently used reinforcement fiber materials for CPC. For a high fiber volume, considerable increase in bending strength has been reported, e.g., from 2.7 MPa (unreinforced CPC) to 17.7 MPa for CPC with 45 vol% polyglactin fibers [74]. This strengthening effect can be further enhanced to 40.5 MPa by incorporation of chitosan lactate into the matrix. The synergistic strengthening of the CPC by chitosan and fibers together is stronger than from either suture fibers or chitosan alone [74], which was explained by both a much stronger cement matrix after chitosan incorporation supporting the suture fibers to better resist cracking as well as an improved suture-matrix bonding [72,74]. Generally, strength increases with length to diameter aspect ratio of fibers, whereas occurrence of fiber aggregation leading to inhomogenities in fiber distribution represents the practical upper limit for the aspect ratio. Xu and co-workers [72] systematically varied the length of carbon fibers in HA cement and found a continuous increase of strength between 3 and 75 mm fiber length (aspect ratio of 1000 and 9000), which was followed by a strength decrease for 200 mm long fibers (aspect ratio 25,000). While the use of such long fibers strongly alters the workability of cement pastes and impedes a minimal invasive application by injection, cement pastes filled with short fibers have been demonstrated to maintain their injection properties up to a fiber length of 1 mm and a fiber volume of 7.5% [83].

Cements may also be modified by using fiber meshes instead of single fibers, especially in cases where biomechanical stresses will primarily be oriented linearly or biaxially to the cement implant. Meshes provide a strength enhancement (in linear or biaxial direction) beyond that of randomly directed fibers and have the advantage that even thin bony structures (e.g., malar, orbital bones) or extensive cranial deficiencies can be reconstructed [30,78,84]. Von Gonten [30] could demonstrate that such a polyglactin mesh–CPC composites have a similar work of fracture to PMMA cements up to seven days' immersion in a buffered electrolyte, which was considered to have potential for structural repair of bone defects.

Most of the studies about FRCPC deal with both non-degradable fibers and with a poorly soluble hydroxyapatite cement matrix (see Table 1 and references [41,69]). This will initially result in long term stable cement composites with only minor changes of mechanical properties. However, even the slow matrix degradation by osteoclastic cells will dissect fibers in a longer time frame, which will be encapsulated in newly formed tissue with the possibility of foreign body reactions. Especially, approaches using technical fiber types (e.g., carbon fibers) or even carbon nanotubes are questionable regarding this point due to their low biocompatibility. The use of degradable fibers in FRCPC may solve this problem and the *in vivo* behavior of such FRCPC has been proven in various studies and is part of a recent review article by Krüger *et al.* [69]. However, at the same time, the use of degradable fibers will result in a time dependent loss of the reinforcement effect due to dissolution of the fibers in an aqueous environment. This effect of fiber degradation on the composite strength was simulated for polyglactin/PLGA fiber material by immersion of reinforced hydroxyapatite cement in a simulated physiological solution [73,74]. These studies confirmed a strength decrease of the reinforced composite after 4–6 weeks' immersion [73], which could be compensated by a simultaneous chitosan infiltration of the cement matrix [74]. As a solution to the above mentioned problems of either a loss of mechanical properties during fiber degradation or the release of non-degraded fibers, the different degradation kinetics of fibers and cement matrix need to be adjusted. An approach is the use of more degradable cements based on the formation of dicalcium phosphate dehydrate (brushite) in conjunction with PLGA fibers [82]. Other promising works are dealing with degradable magnesium phosphate cements, which are reinforced with magnesium metal wires [85]. Especially, the latter provides strong reinforcement effects with a maximum bending strength of the composites of 139 MPa.

4. Dual Setting Cements

While the addition of non-reactive polymers (e.g., collagen, chitosan, hyaluronic acid, cellulose derivates) [37–39,86,87] is commonly used for improving cement cohesion or biological performance, it is only of small benefit for the mechanical cement performance. A reduction of the brittleness of CPC and an increase of strength can be achieved by using polymeric compounds which can be cross-linked by binding calcium ions due to a high density of either carboxylic acid or organic phosphate moieties in the polymer chain, e.g., polyacrylic acid [88–92], polymethyl vinyl ether maleic acid [89,93], poly[bis(carboxylatophenoxy)phosphazene] [94] or poly(vinyl phosphonate) [95]. Such polymer modified cements set both by the aforementioned dissolution–precipitation mechanism as well as by deprotonating the organic acid following the formation of intra- or inter-chained bonding Ca^{2+}–Acid chelates [94] with a highly reactive cement component (mostly tetracalcium phosphate) from the cement powder. Processing of such polymer–cement composites is either possible by reacting an aqueous solution at ambient conditions with the cement powder or by reacting dry cement/polymer mixtures at elevated temperature/pressure in a solid state reaction.

An alternative approach is the use of reactive monomer systems, which are dissolved in the cement liquid and simultaneously react during cement setting by a gelation–polymerisation process. This forms within several minutes a hydrogel matrix with embedded cement particles, which are subsequently converted into the setting product by a continuous dissolution–precipitation reaction. The result is finally an interconnecting hydrogel matrix within the porous cement structure as shown in Figure 4. The advantages of this strategy are the possibility of a high polymer loading of the cement (and hence a large strength and toughness increase) as well as practically unchanged rheological properties of the fresh cement paste. Both are related to the fact that the dissolved monomers are commonly small, water miscible liquids with low viscosity such that even high monomer concentrations are not strongly altering the initial cement viscosity.

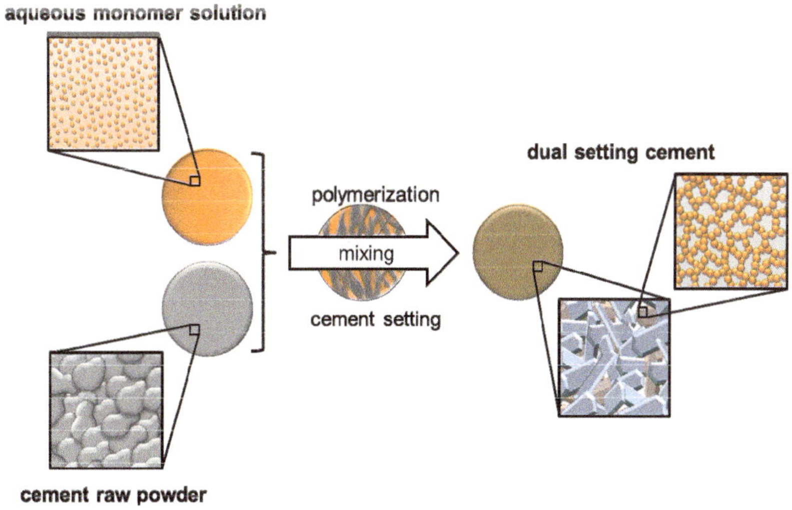

Figure 4. Hardening mechanism of dual-setting cements with the formation of interconnected matrices of hydrogel and precipitated cement crystals.

An early study regarding this concept was using mixtures of triethyleneglycol-dimethacrylate (TEGDMA), Bisphenol-A-dimethacrylate (Bis-GMA), hydroxyethylmethacrylate (HEMA) and 10%

water as cement liquid. After adding this liquid to an equimolar mixture of TTCP and DCPA, polymerization was initiated by benzoylperoxide (coating on cement particles) and di-(N,N)-2-hydroxyethyl-p-toluidine (added to the cement liquid) [96]. Although this study revealed high diametral tensile strengths of up to 26 MPa for such composites, no hydroxyapatite formation of the cement was observed even after 30 d storage in water. This was attributed to the low water content of the cement liquid as well as to an adsorption of the hydrogel on TTCP/DCPA cement particles. This problem was overcome by Dos Santos *et al.* [43,44,97], who modified the cement liquid of an α-tricalcium phosphate cement by the addition of 5%–20% acrylamide and 1% ammonium polyacrylate. While the latter was used to increase initial cement viscosity and to reduce cement wash out in an aqueous environment, the acrylamide was chemically polymerised during cement setting by the use of 0.25% of N,N,N',N'-tetramethylethylenediamide (TEMED) and 0.01% ammonium persulfate. This modification doubled the compressive strength of the set cement from 25–50 MPa while the tensile strength was increased from 9 MPa to <21 MPa. At the same time, the high water content of the cement liquid enabled setting of α-TCP cement particles to calcium deficient hydroxyapatite within seven days. A follow up study by the same authors extended the approach to a fiber reinforced–double setting cement matrix [43] by using 1–4 wt% of 4–10 mm long carbon, nylon and polypropylene fibers. The addition of the fibers was found to reduce the compressive strength of the cement, which was attributed to an increase of porosity. However, this was compensated by strong increase of the cement toughness and tensile strength, which increased from 17–28 MPa.

A major concern about this matrix is the toxicity of non-reacted acrylamide monomer. To overcome this problem, Christel *el al.* [98] investigated the modification of alpha-tricalcium phosphate cement (α-TCP) with 30%–70% of less-toxic 2-hydroxyethylmethacrylate (HEMA), which also resulted in mechanically stable polymer-ceramic composites with interpenetrating organic and inorganic networks. Four-point bending strength was found to increase from 9 MPa to more than 14 MPa when using 50% HEMA, and the bending modulus decreased from 18 GPa to approx. 4 GPa. In addition, cement composites with ≥50% HEMA showed strongly reduced brittle fracture behaviour with an increase of the work of fracture by more than an order of magnitude. While bending of pure ceramic samples was possible only to a maximum of 0.07 mm, samples with 50% or more HEMA monomer had a higher flexibility and bending was possible for 0.4–1.5 mm until fracture. At the same time, the authors could prove that important cement characteristics such as compressive strength or injectability were not significantly altered by using HEMA modification. Another study by Wang *et al.* [45] used methacrylate modified dextran as monomer in a cement matrix of tetracalcium phosphate/dicalcium phosphate anhydrous in a weight ratio of 10:1—1:3 (CPC: Meth.-dextran). The results showed an increase of the compressive strength from 24–83 MPa for a polymer content of 16.7%, as well as improvement of the fracture energy by nearly two orders of magnitude from 0.084–8.35 $kJ \cdot m^{-2}$.

Apart from using organic monomers to form a second network in cements, it is also possible to apply the concept of dual setting cements to pure inorganic materials. Silica addition to CPC is a common approach to modify bioactivity, cement paste cohesion and mechanical cement properties [99,100]. However, most studies either used non-reactive silica fillers in cements [101–103] or they added non-reactive calcium phosphate particles to an *in situ* forming silica matrix prepared by sol-gel processing [104–106]. In contrast, Geffers *et al.* [48] modified a brushite forming cement paste with a second inorganic silica based precursor, which was obtained by pre-hydrolysing tetraethyl orthosilicate (TEOS) under acidic conditions. The addition of the cement powder (mixture of β-tricalcium phosphate and monocalcium phosphate) provoked an increase of the pH of the silica precursor such that cement setting by a dissolution–precipitation process, and the condensation reaction of the hydrolysed TEOS occurred simultaneously. This resulted in an interpenetrating phase composite material in which the macro pores of the cement (pore sizes in µm range) were infiltrated by the micro porous silica gel (pore sizes in nm range), leading to a higher density and a compressive strength approximately 5–10 times higher than the CPC reference.

5. Conclusions and Outlook

This article features reinforcement strategies of biocements to improve their strength and toughness for an application at load-bearing defect sides. While porosity reduction is based on the optimization of an intrinsic cement property leading to higher strength, the addition of fibers or the creation of dual setting cement matrixes are extrinsic approaches not only improving strength but also toughness of the matrix. Surprisingly, most studies devoted to the mechanical properties of calcium phosphate biocements only deal with one of the presented strategies. Here, the simultaneous application of the different methods will definitely bring further improvements such that those optimized cements can likely be applied for load bearing defects. Desired mechanical properties would be likely similar to those of polymeric PMMA cements (bending strength ≥50 MPa, bending modulus ≥1800 MPa) compressive strength ≥70 MPa according to ISO 5833:2002 [107]), whereas few studies have already reached or even exceeded one of these parameters [27,85]. However, practically all strength values for CPC in literature were obtained by test methods under static conditions and there are only few reports dealing with the fatigue properties of calcium phosphate cements in load-bearing defect models [108,109]. Hence, testing of cement strength under cyclic loading is one of the most important parameters which needs to be addressed in future research. In addition, since most of the studies on mechanically reinforced biocements were performed with only slowly degradable hydroxyapatite cement matrices and poorly or even non-degradable additives (fibers, polymers), the major challenge for the future is a transfer of the presented concepts to fully degradable materials. This is demanding since degradable cements based on the formation of brushite have harsh setting conditions (low pH, heat release, fast crystallization) and consume a considerable amount of water during setting. Especially, the latter may interfere with the formation of a second hydrogel phase, since the formation of hydrogel and hydrated cement setting product will compete for the available water in the cement liquid.

Acknowledgments

The authors would like to acknowledge financial support from the Deutsche Forschungsgemeinschaft (DFG GB1/20-1 and DFG GR3232/3-1).

Author Contributions

Martha Geffers is a PhD student and has performed the literature search for this review article and has written major parts of the manuscript. Uwe Gbureck and Jürgen Groll are her PhD supervisors and have supported with writing and correcting the manuscript.

Conflicts of Interest

The authors declare no conflict of interest.

References

1. Bohner, M.; Gbureck, U.; Barralet, J.E. Technological issues for the development of more efficient calcium phosphate bone cements: A critical assessment. *Biomaterials* **2005**, *26*, 6423–6429.
2. Dorozhkin, S.V. Calcium orthophosphate cements for biomedical application. *J. Mater. Sci.* **2008**, *43*, 3028–3057.
3. Chow, L.C. Calcium phosphate cements. *Monogr. Oral Sci.* **2001**, *18*, 148–163.
4. Nancollas, G.H.; Zawacki, S.J. Calcium phosphate mineralization. *Connect. Tissue Res.* **1989**, *21*, 239–244.
5. Tamimi, F.; Sheikh, Z.; Barralet, J. Dicalcium phosphate cements: Brushite and monetite. *Acta Biomater.* **2012**, *8*, 474–487.
6. Sahin, E.; Ciftcioglu, M. Monetite promoting effect of NaCl on brushite cement setting kinetics. *J. Mater. Chem. B* **2013**, *1*, 2943–2950.
7. Bohner, M.; vanLanduyt, P.; Merkle, H.P.; Lemaitre, J. Composition effects on the pH of a hydraulic calcium phosphate cement. *J. Mater. Sci. Mater. Med.* **1997**, *8*, 675–681.
8. Bohner, M. Design of ceramic-based cements and putties for bone graft substitution. *Eur. Cells Mater.* **2010**, *20*, 1–12.
9. Zhang, J.T.; Liu, W.Z.; Schnitzler, V.; Tancret, F.; Bouler, J.M. Calcium phosphate cements for bone substitution: Chemistry, handling and mechanical properties. *Acta Biomater.* **2014**, *10*, 1035–1049.
10. Constantz, B.R.; Barr, B.M.; Ison, I.C.; Fulmer, M.T.; Baker, J.; McKinney, L.A.; Goodman, S.B.; Gunasekaren, S.; Delaney, D.C.; Ross, J.; Poser, R.D. Histological, chemical, and crystallographic analysis of four calcium phosphate cements in different rabbit osseous sites. *J. Biomed. Mater. Res.* **1998**, *43*, 451–461.
11. Apelt, D.; Theiss, F.; El-Warrak, A.O.; Zlinszky, K.; Bettschart-Wolfisberger, R.; Bohner, M.; Matter, S.; Auer, J.A.; von Rechenberg, B. *In vivo* behavior of three different injectable hydraulic calcium phosphate cements. *Biomaterials* **2004**, *25*, 1439–1451.
12. Tas, A.C. The use of physiological solutions or media in calcium phosphate synthesis and processing. *Acta Biomater.* **2014**, *10*, 1771–1792.
13. Grossardt, C.; Ewald, A.; Grover, L.M.; Barralet, J.E.; Gbureck, U. Passive and active *in vitro* resorption of calcium and magnesium phosphate cements by osteoclastic cells. *Tissue Eng. A* **2010**, *16*, 3687–3695.
14. Detsch, R.; Mayr, H.; Ziegler, G. Formation of osteoclast-like cells on HA and TCP ceramics. *Acta Biomater.* **2008**, *4*, 139–148.
15. Wenisch, S.; Stahl, J.P.; Horas, U.; Heiss, C.; Kilian, O.; Trinkaus, K.; Hild, A.; Schnettler, R. *In vivo* mechanisms of hydroxyapatite ceramic degradation by osteoclasts: Fine structural microscopy. *J. Biomed. Mater. Res. A* **2003**, *67*, 713–718.

16. Grover, L.M.; Gbureck, U.; Wright, A.J.; Tremayne, M.; Barralet, J.E. Biologically mediated resorption of brushite cement in vitro. *Biomaterials* **2006**, *27*, 2178–2185.
17. Holzapfel, B.M.; Reichert, J.C.; Schantz, J.T.; Gbureck, U.; Rackwitz, L.; Noth, U.; Jakob, F.; Rudert, M.; Groll, J.; Hutmacher, D.W. How smart do biomaterials need to be? A translational science and clinical point of view. *Adv. Drug Deliv. Rev.* **2013**, *65*, 581–603.
18. Theiss, F.; Apelt, D.; Brand, B.A.; Kutter, A.; Zlinszky, K.; Bohner, M.; Matter, S.; Frei, C.; Auer, J.A.; von Rechenberg, B. Biocompatibility and resorption of a brushite calcium phosphate cement. *Biomaterials* **2005**, *26*, 4383–4394.
19. Tamimi, F.; Torres, J.; Lopez-Cabarcos, E.; Bassett, D.C.; Habibovic, P.; Luceron, E.; Barralet, J.E. Minimally invasive maxillofacial vertical bone augmentation using brushite based cements. *Biomaterials* **2009**, *30*, 208–216.
20. Penel, G.; Leroy, N.; van Landuyt, P.; Flautre, B.; Hardouin, P.; Lemaitre, J.; Leroy, G. Raman microspectrometry studies of brushite cement: *In vivo* evolution in a sheep model. *Bone* **1999**, *25*, 81S–84S.
21. Kuemmerle, J.M.; Oberle, A.; Oechslin, C.; Bohner, M.; Frei, C.; Boecken, I.; von Rechenberg, B. Assessment of the suitability of a new brushite calcium phosphate cement for cranioplasty—An experimental study in sheep. *J. Cranio Maxillofac. Surg.* **2005**, *33*, 37–44.
22. Bohner, M.; Lemaitre, J. Can bioactivity be tested in vitro with SBF solution? *Biomaterials* **2009**, *30*, 2175–2179.
23. Lu, X.; Leng, Y. Theoretical analysis of calcium phosphate precipitation in simulated body fluid. *Biomaterials* **2005**, *26*, 1097–1108.
24. Klammert, U.; Ignatius, A.; Wolfram, U.; Reuther, T.; Gbureck, U. *In vivo* degradation of low temperature calcium and magnesium phosphate ceramics in a heterotopic model. *Acta Biomater.* **2011**, *7*, 3469–3475.
25. Kanter, B.; Geffers, M.; Ignatius, A.; Gbureck, U. Control of *in vivo* mineral bone cement degradation. *Acta Biomater.* **2014**, *10*, 3279–3287.
26. Bohner, M.; Theiss, F.; Apelt, D.; Hirsiger, W.; Houriet, R.; Rizzoli, G.; Gnos, E.; Frei, C.; Auer, J.A.; von Rechenberg, B. Compositional changes of a dicalcium phosphate dihydrate cement after implantation in sheep. *Biomaterials* **2003**, *24*, 3463–3474.
27. Gbureck, U.; Spatz, K.; Thull, R.; Barralet, J.E. Rheological enhancement of mechanically activated alpha-tricalcium phosphate cements. *J. Biomed. Mater. Res. B Appl. Biomater.* **2005**, *73*, 1–6.
28. Charriere, E.; Terrazzoni, S.; Pittet, C.; Mordasini, P.; Dutoit, M.; Lemaitre, J.; Zysset, P. Mechanical characterization of brushite and hydroxyapatite cements. *Biomaterials* **2001**, *22*, 2937–2945.
29. Dorozhkin, S.V. Calcium orthophosphate-based biocomposites and hybrid biomaterials. *J. Mater. Sci.* **2009**, *44*, 2343–2387.
30. Von Gonten, A.S.; Kelly, J.R.; Antonucci, J.M. Load-bearing behavior of a simulated craniofacial structure fabricated from a hydroxyapatite cement and bioresorbable fiber-mesh. *J. Mater. Sci. Mater. Med.* **2000**, *11*, 95–100.
31. Blattert, T.R.; Jestaedt, L.; Weckbach, A. Suitability of a calcium phosphate cement in osteoporotic vertebral body fracture augmentation a controlled, randomized, clinical trial of balloon kyphoplasty comparing calcium phosphate versus polymethylmethacrylate. *Spine* **2009**, *34*, 108–114.

32. Maestretti, G.; Cremer, C.; Otten, P.; Jakob, R.P. Prospective study of standalone balloon kyphoplasty with calcium phosphate cement augmentation in traumatic fractures. *Eur. Spine J.* **2007**, *16*, 601–610.
33. Tarsuslugil, S.M.; O'Hara, R.M.; Dunne, N.J.; Buchanan, F.J.; Orr, J.F.; Barton, D.C.; Wilcox, R.K. Development of calcium phosphate cement for the augmentation of traumatically fractured porcine specimens using vertebroplasty. *J. Biomech.* **2013**, *46*, 711–715.
34. Heini, P.F. Vertebroplastie: Ein Update. *Orthopäde* **2010**, *39*, 658–664.
35. Grafe, I.A.; Baier, M.; Noldge, G.; Weiss, C.; da Fonseca, K.; Hillmeier, J.; Libicher, M.; Rudofsky, G.; Metzner, C.; Nawroth, P.; *et al.* Calcium-phosphate and polymethylmethacrylate cement in long-term outcome after kyphoplasty of painful osteoporotic vertebral fractures. *Spine* **2008**, *33*, 1284–1290.
36. Kiyasu, K.; Takemasa, R.; Ikeuchi, M.; Tani, T. Differential blood contamination levels and powder–liquid ratios can affect the compressive strength of calcium phosphate cement (CPC): A study using a transpedicular vertebroplasty model. *Eur. Spine J.* **2013**, *22*, 1643–1649.
37. Moreau, J.L.; Weir, M.D.; Xu, H.H.K. Self-setting collagen-calcium phosphate bone cement: Mechanical and cellular properties. *J. Biomed. Mater. Res. A* **2009**, *91*, 605–613.
38. Schneiders, W.; Reinstorf, A.; Biewener, A.; Serra, A.; Grass, R.; Kinscher, M.; Heineck, J.; Rehberg, S.; Zwipp, H.; Rammelt, S. *In Vivo* Effects of modification of hydroxyapatite/collagen composites with and without chondroitin sulphate on bone remodeling in the sheep tibia. *J. Orthop. Res.* **2009**, *27*, 15–21.
39. Tamimi, F.; Kumarasami, B.; Doillon, C.; Gbureck, U.; le Nihouannen, D.; Cabarcos, E.L.; Barralet, J.E. Brushite-collagen composites for bone regeneration. *Acta Biomater.* **2008**, *4*, 1315–1321.
40. O'Hara, R.M.; Orr, J.F.; Buchanan, F.J.; Wilcox, R.K.; Barton, D.C.; Dunne, N.J. Development of a bovine collagen-apatitic calcium phosphate cement for potential fracture treatment through vertebroplasty. *Acta Biomater.* **2012**, *8*, 4043–4052.
41. Canal, C.; Ginebra, M.P. Fibre-reinforced calcium phosphate cements: A review. *J. Mechan. Behav. Biomed. Mater.* **2011**, *4*, 1658–1671.
42. Dos Santos, L.A.; de Oliveira, L.C.; Rigo, E.C.D.; Carrodeguas, R.G.; Boschi, A.O.; de Arruda, A.C.F. Fiber reinforced calcium phosphate cement. *Artif. Org.* **2000**, *24*, 212–216.
43. Dos Santos, L.A.; Carrodeguas, R.G.; Boschi, A.O.; de Arruda, A.C.F. Fiber-enriched double-setting calcium phosphate bone cement. *J. Biomed. Mater. Res. A* **2003**, *65*, 244–250.
44. Dos Santos, L.A.; Carrodeguas, R.G.; Boschi, A.O.; de Arruda, A.C.F. Dual-setting calcium phosphate cement modified with ammonium polyacrylate. *Artif. Org.* **2003**, *27*, 412–418.
45. Wang, J.; Liu, C.S.; Liu, Y.F.; Zhang, S. Double-network interpenetrating bone cement via in situ hybridization protocol. *Adv. Funct. Mater.* **2010**, *20*, 3997–4011.
46. Barralet, J.E.; Gaunt, T.; Wright, A.J.; Gibson, I.R.; Knowles, J.C. Effect of porosity reduction by compaction on compressive strength and microstructure of calcium phosphate cement. *J. Biomed. Mater. Res.* **2002**, *63*, 1–9.
47. Espanol, M.; Perez, R.A.; Montufar, E.B.; Marichal, C.; Sacco, A.; Ginebra, M.P. Intrinsic porosity of calcium phosphate cements and its significance for drug delivery and tissue engineering applications. *Acta Biomater.* **2009**, *5*, 2752–2762.

48. Geffers, M.; Barralet, J.E.; Groll, J.; Gbureck, U. Dual-setting brushite-silica gel cements. *Acta Biomater.* **2015**, *11*, 467–476.
49. Engstrand, J.; Persson, C.; Engqvist, H. The effect of composition on mechanical properties of brushite cements. *J. Mech. Behav. Biomed. Mater.* **2014**, *29*, 81–90.
50. Hofmann, M.P.; Mohammed, A.R.; Perrie, Y.; Gbureck, U.; Barralet, J.E. High-strength resorbable brushite bone cement with controlled drug-releasing capabilities. *Acta Biomater.* **2009**, *5*, 43–49.
51. Grover, L.M.; Knowles, J.C.; Fleming, G.J.P.; Barralet, J.E. *In vitro* ageing of brushite calcium phosphate cement. *Biomaterials* **2003**, *24*, 4133–4141.
52. Ginebra, M.P.; Delgado, J.A.; Harr, I.; Almirall, A.; del Valle, S.; Planell, J.A. Factors affecting the structure and properties of an injectable self-setting calcium phosphate foam. *J. Biomed. Mater. Res. A* **2007**, *80*, 351–361.
53. Burguera, E.F.; Guitian, F.; Chow, L.C. A water setting tetracalcium phosphate-dicalcium phosphate dihydrate cement. *J. Biomed. Mater. Res. A* **2004**, *71*, 275–282.
54. Unosson, J.E.; Persson, C.; Engqvist, H. An evaluation of methods to determine the porosity of calcium phosphate cements. *J. Biomed. Mater. Res. B Appl. Biomater.* **2015**, *103*, 62–71.
55. Bohner, M.; Baroud, G. Injectability of calcium phosphate pastes. *Biomaterials* **2005**, *26*, 1553–1563.
56. Barralet, J.E.; Tremayne, M.; Lilley, K.J.; Gbureck, U. Modification of calcium phosphate cement with alpha-hydroxy acids and their salts. *Chem. Mater.* **2005**, *17*, 1313–1319.
57. Gu, T.; Shi, H.; Ye, J. Reinforcement of calcium phosphate cement by incorporating with high-strength ß-tricalcium phosphate aggregates. *J. Biomed. Mater. Res. B Appl. Biomater.* **2012**, *100*, 350–359.
58. Del Real, R.P.; Wolke, J.G.C.; Vallet-Regi, M.; Jansen, J.A. A new method to produce macropores in calcium phosphate cements. *Biomaterials* **2002**, *23*, 3673–3680.
59. Vorndran, E.; Geffers, M.; Ewald, A.; Lemm, M.; Nies, B.; Gbureck, U. Ready-to-use injectable calcium phosphate bone cement paste as drug carrier. *Acta Biomater.* **2013**, *9*, 9558–9567.
60. Lopez-Heredia, M.A.; Sariibrahimoglu, K.; Yang, W.; Bohner, M.; Yamashita, D.; Kunstar, A.; van Apeldoorn, A.A.; Bronkhorst, E.M.; Lanao, R.P.F.; Leeuwenburgh, S.C.G.; *et al.* Influence of the pore generator on the evolution of the mechanical properties and the porosity and interconnectivity of a calcium phosphate cement. *Acta Biomater.* **2012**, *8*, 404–414.
61. Barralet, J.E.; Hofmann, M.; Grover, L.M.; Gbureck, U. High-strength apatitic cement by modification with alpha-hydroxy acid salts. *Adv. Mater.* **2003**, *15*, 2091–2094.
62. Cama, G.; Barberis, F.; Botter, R.; Cirillo, P.; Capurro, M.; Quarto, R.; Scaglione, S.; Finocchio, E.; Mussi, V.; Valbusa, U. Preparation and properties of macroporous brushite bone cements. *Acta Biomater.* **2009**, *5*, 2161–2168.
63. Unosson, J.E.; Montufar, E.B.; Engqvist, H.; Ginebra, M.P.; Persson, C. Brushite foams—The effect of Tween 80 and Pluronic F-127 on foam porosity and mechanical properties. *J. Biomed. Res. B Appl. Biomater.* **2015**, doi:10.1002/jbm.b.33355.
64. Barralet, J.E.; Grover, L.M.; Gbureck, U. Ionic modification of calcium phosphate cement viscosity. Part II: hypodermic injection and strength improvement of brushite cement. *Biomaterials* **2004**, *25*, 2197–2203.

65. Gravius, S.; Wirtz, D.C.; Marx, R.; Maus, U.; Andereya, S.; Mueller-Rath, R.; Mumme, T. Mechanical *in vitro* testing of fifteen commercial bone cements based on polymethylmethacrylate. *Z. Orthop. Unfall.* **2007**, *145*, 579–585.
66. Chow, L.C.; Hirayama, S.; Takagi, S.; Parry, E. Diametral tensile strength and compressive strength of a calcium phosphate cement: Effect of applied pressure. *J. Biomed. Mater. Res.* **2000**, *53*, 511–517.
67. Ishikawa, K.; Asaoka, K. Estimation of ideal mechanical strength and critical porosity of calcium-phosphate cement. *J. Biomed. Mater. Res.* **1995**, *29*, 1537–1543.
68. Dewith, G.; Corbijn, A.J. Metal fiber reinforced hydroxy-apatite ceramics. *J. Mater. Sci.* **1989**, *24*, 3411–3415.
69. Krueger, R.; Groll, J. Fiber reinforced calcium phosphate cements—On the way to degradable load bearing bone substitutes? *Biomaterials* **2012**, *33*, 5887–5900.
70. Rösler, J.; Harders, H.; Bäker, M. *Mechanisches Verhalten der Werkstoffe*, 2nd ed.; Teubner Verlag: Wiesbaden, Germany, 2008.
71. Callister, W.D.; Rethwisch, D.G. *Materials Science and Engineering: An Introduction*, 8th ed.; John Wiley & Sons, Inc.: Hoboken, NJ, USA, 2010.
72. Brandt, A.M. *Cement-Based Composites—Materials, Mechanical Properties and Performance*, 2nd ed.; Taylor & Francis: Abingdon, UK, 2009.
73. Xu, H.H.K.; Eichmiller, F.C.; Barndt, P.R. Effects of fiber length and volume fraction on the reinforcement of calcium phosphate cement. *J. Mater. Sci. Mater. Med.* **2001**, *12*, 57–65.
74. Xu, H.H.K.; Eichmiller, F.C.; Giuseppetti, A.A. Reinforcement of a self-setting calcium phosphate cement with different fibers. *J. Biomed. Mater. Res.* **2000**, *52*, 107–114.
75. Zhang, Y.; Xu, H.H.K. Effects of synergistic reinforcement and absorbable fiber strength on hydroxyapatite bone cement. *J. Biomed. Mater. Res. A* **2005**, *75*, 832–840.
76. Xu, H.H.K.; Quinn, J.B.; Takagi, S.; Chow, L.C.; Eichmiller, F.C. Strong and macroporous calcium phosphate cement: Effects of porosity and fiber reinforcement on mechanical properties. *J. Biomed. Mater. Res.* **2001**, *57*, 457–466.
77. Xu, H.H.K.; Quinn, J.B. Calcium phosphate cement containing resorbable fibers for short-term reinforcement and macroporosity. *Biomaterials* **2002**, *23*, 193–202.
78. Xu, H.H.K.; Simon, C.G. Self-hardening calcium phosphate cement-mesh composite: Reinforcement, macropores, and cell response. *J. Biomed. Mater. Res.* A **2004**, *69*, 267–278.
79. Xu, H.H.K.; Quinn, J.B.; Takagi, S.; Chow, L.C. Synergistic reinforcement of in situ hardening calcium phosphate composite scaffold for bone tissue engineering. *Biomaterials* **2004**, *25*, 1029–1037.
80. Gorst, N.J.S.; Perrie, Y.; Gbureck, U.; Hutton, A.L.; Hofmann, M.P.; Grover, L.M.; Barralet, J.E. Effects of fibre reinforcement on the mechanical properties of brushite cement. *Acta Biomater.* **2006**, *2*, 95–102.
81. Zhao, P.; Sun, K.; Zhao, T.; Ren, X. Effect of CNTs on property of calcium phosphate cement. *Key Eng. Mater.* **2007**, *336–338*, 1606–1608.
82. Muller, F.A.; Gbureck, U.; Kasuga, T.; Mizutani, Y.; Barralet, J.E.; Lohbauer, U. Whisker-reinforced calcium phosphate cements. *J. Am. Ceram. Soc.* **2007**, *90*, 3694–3697.

83. Maenz, S.; Kunisch, E.; Muehlstaedt, M.; Boehm, A.; Kopsch, V.; Bossert, J.; Kinne, R.W.; Jandt, K.D. Enhanced mechanical properties of a novel, injectable, fiber-reinforced brushite cement. *J. Mech. Behav. Biomed. Mater.* **2014**, *39*, 328–338.
84. Weir, M.D.; Xu, H.H.K.; Simon, C.G. Strong calcium phosphate cement-chitosan-mesh construct containing cell-encapsulating hydrogel beads for bone tissue engineering. *J. Biomed. Mater. Res. A* **2006**, *77*, 487–496.
85. Krueger, R.; Seitz, J.-M.; Ewald, A.; Bach, F.-W.; Groll, J. Strong and tough magnesium wire reinforced phosphate cement composites for load-bearing bone replacement. *J. Mech. Behav. Biomed. Mater.* **2013**, *20*, 36–44.
86. Khairoun, I.; Driessens, F.C.M.; Boltong, M.G.; Planell, J.A.; Wenz, R. Addition of cohesion promoters to calcium phosphate cements. *Biomaterials* **1999**, *20*, 393–398.
87. Alkhraisat, M.H.; Rueda, C.; Marino, F.T.; Torres, J.; Jerez, L.B.; Gbureck, U.; Cabarcos, E.L. The effect of hyaluronic acid on brushite cement cohesion. *Acta Biomater.* **2009**, *5*, 3150–3156.
88. Chen, W.-C.; Ju, C.-P.; Wang, J.-C.; Hung, C.-C.; Lin, J.-H.C. Brittle and ductile adjustable cement derived from calcium phosphate cement/polyacrylic acid composites. *Dent. Mater.* **2008**, *24*, 1616–1622.
89. Watson, K.E.; Tenhuisen, K.S.; Brown, P.W. The formation of hydroxyapatite-calcium polyacrylate composites. *J. Mater. Sci. Mater. Med.* **1999**, *10*, 205–213.
90. Khashaba, R.M.; Moussa, M.; Koch, C.; Jurgensen, A.R.; Missimer, D.M.; Rutherford, R.L.; Chutkan, N.B.; Borke, J.L. Preparation, physical-chemical characterization, and cytocompatibility of polymeric calcium phosphate cements. *Int. J. Biomater.* **2011**, *2011*, 467641.
91. Majekodunmi, A.O.; Deb, S.; Nicholson, J.W. Effect of molecular weight and concentration of poly(acrylic acid) on the formation of a polymeric calcium phosphate cement. *J. Mater. Sci. Mater. Med.* **2003**, *14*, 747–752.
92. Majekodunmi, A.O.; Deb, S. Poly(acrylic acid) modified calcium phosphate cements: The effect of the composition of the cement powder and of the molecular weight and concentration of the polymeric acid. *J. Mater. Sci. Mater. Med.* **2007**, *18*, 1883–1888.
93. Matsuya, Y.; Antonucci, J.M.; Matsuya, S.; Takagi, S.; Chow, L.C. Polymeric calcium phosphate cements derived from poly(methyl vinyl ether-maleic acid). *Dent. Mater.* **1996**, *12*, 2–7.
94. Greish, Y.E.; Brown, P.W.; Bender, J.D.; Allcock, H.R.; Lakshmi, S.; Laurencin, C.T. Hydroxyapatite-polyphosphazane composites prepared at low temperatures. *J. Am. Ceram. Soc.* **2007**, *90*, 2728–2734.
95. Greish, Y.E.; Brown, P.W. Chemically formed HAp-Ca poly(vinyl phosphonate) composites. *Biomaterials* **2001**, *22*, 807–816.
96. Sugawara, A.; Antonucci, J.M.; Takagi, S.; Chow, L.C.; Ohashi, M. Formation of hydroxyapatite in hydrogels from tetracalcium phosphate/dicalcium phosphate mixtures. *J. Nihon Univ. Sch. Dent.* **1989**, *31*, 372–381.
97. Rigo, E.C.S.; dos Santos, L.A.; Vercik, L.C.O.; Carrodeguas, R.G.; Boschi, A.O. alpha-tricalcium phosphate- and tetracalcium phosphate/dicalcium phosphate-based dual setting cements. *Lat. Am. Appl. Res.* **2007**, *37*, 267–274.
98. Christel, T.; Kuhlmann, M.; Vorndran, E.; Groll, J.; Gbureck, U. Dual setting alpha-tricalcium phosphate cements. *J. Mater. Sci. Mater. Med.* **2013**, *24*, 573–581.

99. Zhou, H.; Luchini, T.J.F.; Agarwal, A.K.; Goel, V.K.; Bhaduri, S.B. Development of monetite-nanosilica bone cement: A preliminary study. *J. Biomed. Mater. Res. B Appl. Biomater.* **2014**, *102*, 1620–1626.
100. Ahn, G.; Lee, J.Y.; Seol, D.-W.; Pyo, S.G.; Lee, D. The effect of calcium phosphate cement-silica composite materials on proliferation and differentiation of pre-osteoblast cells. *Mater. Lett.* **2013**, *109*, 302–305.
101. Van den Vreken, N.M.F.; de Canck, E.; Ide, M.; Lamote, K.; van der Voort, P.; Verbeeck, R.M.H. Calcium phosphate cements modified with pore expanded SBA-15 materials. *J. Mater. Chem.* **2012**, *22*, 14502–14509.
102. Hesaraki, S.; Alizadeh, M.; Borhan, S.; Pourbaghi-Masouleh, M. Polymerizable nanoparticulate silica-reinforced calcium phosphate bone cement. *J. Biomed. Mater. Res. B Appl. Biomater.* **2012**, *100*, 1627–1635.
103. Hamdan Alkhraisat, M.; Rueda, C.; Blanco Jerez, L.; Marino, F.T.; Torres, J.; Gbureck, U.; Lopez Cabarcos, E. Effect of silica gel on the cohesion, properties and biological performance of brushite cement. *Acta Biomater.* **2010**, *6*, 257–265.
104. Andersson, J.; Areva, S.; Spliethoff, B.; Linden, M. Sol-gel synthesis of a multifunctional, hierarchically porous silica/apatite composite. *Biomaterials* **2005**, *26*, 6827–6835.
105. Heinemann, S.; Heinemann, C.; Bernhardt, R.; Reinstorf, A.; Nies, B.; Meyer, M.; Worch, H.; Hanke, T. Bioactive silica-collagen composite xerogels modified by calcium phosphate phases with adjustable mechanical properties for bone replacement. *Acta Biomater.* **2009**, *5*, 1979–1990.
106. Sousa, A.; Souza, K.C.; Sousa, E.M.B. Mesoporous silica/apatite nanocomposite: Special synthesis route to control local drug delivery. *Acta Biomater.* **2008**, *4*, 671–679.
107. The International Organization for Standardization (ISO). *International Standard ISO 5833 Implants for Surgery—Acrylic Resin Cements*; ISO: Geneva, Switzerland, 2002.
108. Wilke, H.-J.; Mehnert, U.; Claes, L.E.; Bierschneider, M.M.; Jaksche, H.; Boszczyk, B.M. Biomechanical evaluation of vertebroplasty and kyphoplasty with polymethyl methacrylate or calcium phosphate cement under cyclic loading. *Spine* **2006**, *31*, 2934–2941.
109. Lewis, G.; Schwardt, J.D.; Slater, T.A.; Janna, S. Evaluation of a synthetic vertebral body augmentation model for rapid and reliable cyclic compression life testing of materials for balloon kyphoplasty. *J. Biomed. Mater. Res. B Appl. Biomater.* **2008**, *87*, 179–188.

14

Experimental Evidence of Mechanical Isotropy in Porcine Lung Parenchyma

Benjamin Weed [1], Sourav Patnaik [1], Mary Rougeau-Browning [2], Bryn Brazile [1], Jun Liao [1], Raj Prabhu [1] and Lakiesha N. Williams [1,*]

[1] Department of Agricultural and Biological Engineering, Mississippi State University, Starkville, MS 39762, USA; E-Mails: benjaminweed@gmail.com (B.W.); sourav.s.patnaik@gmail.com (S.P.); blb285@gmail.com (B.B.); jliao@abe.msstate.edu (J.L.); rprabhu@abe.msstate.edu (R.P.)

[2] Department of Genetics, University of Georgia, Athens, GA 30602, USA; E-Mail: mar3088@uga.edu

* Author to whom correspondence should be addressed; E-Mail: lwilliams@abe.msstate.edu;

Academic Editor: Amir A. Zadpoor

Abstract: Pulmonary injuries are a major source of morbidity and mortality associated with trauma. Trauma includes injuries associated with accidents and falls as well as blast injuries caused by explosives. The prevalence and mortality of these injuries has made research of pulmonary injury a major priority. Lungs have a complex structure, with multiple types of tissues necessary to allow successful respiration. The soft, porous parenchyma is the component of the lung which contains the alveoli responsible for gas exchange. Parenchyma is also the portion which is most susceptible to traumatic injury. Finite element simulations are an important tool for studying traumatic injury to the human body. These simulations rely on material properties to accurately recreate real world mechanical behaviors. Previous studies have explored the mechanical properties of lung tissues, specifically parenchyma. These studies have assumed material isotropy but, to our knowledge, no study has thoroughly tested and quantified this assumption. This study presents a novel methodology for assessing isotropy in a tissue, and applies these methods to porcine lung parenchyma. Briefly, lung parenchyma samples were dissected so as to be aligned with one of the three anatomical planes, sagittal, frontal, and transverse, and then subjected to compressive mechanical testing. Stress-strain curves from these tests were statistically compared by a novel method for differences in stresses and strains at percentages of the curve. Histological samples aligned with the anatomical planes were also examined by qualitative and quantitative methods to determine any differences in the microstructural morphology. Our study showed significant evidence to support the hypothesis that lung parenchyma behaves isotropically.

Keywords: pulmonary trauma; blast lung; lung parenchyma; mechanical isotropy

---

1. Introduction

Pulmonary injury, including pulmonary contusion and pulmonary laceration, is a serious source of morbidity and mortality following blunt or blast trauma [1–7]. Common causes for civilians include motor vehicle accidents, falls, assaults, and sports injuries [8–10]. Additionally, explosions and other sources of blast trauma can cause severe pulmonary injuries, and are commonly seen in military or civilian victims of such events [1–3,11,12]. Blast lung injury is the primary cause of death in those who initially survive an explosion [12]. The prevalence of these types of injuries has made pulmonary injury a crucial area of trauma research.

Lungs have a complex structure, which can be summarized as a network of stiff airway tubes, bronchi and bronchioles, embedded in a soft and porous tissue, lung parenchyma. The lungs are enclosed by the thin pleural membrane and immersed in pleural fluid. The parenchyma contains the gas-exchanging alveoli which are very soft and highly susceptible to damage. Damage to the parenchyma leads to bleeding, edema, and collapse of the microstructure, which prevents gas exchange [13]. In serious cases this ultimately leads to the inability of the lungs to adequately oxygenate the blood and even medically supplemented oxygen may not be sufficient to prevent death [4]. The degree of lung damage a trauma victim has sustained is difficult to diagnose in a clinical setting. Physical experiments which could create a predictive index for these injuries involving human cadavers or live animals are costly and logistically difficult to perform. Finite element methods are a promising option for performing controlled experiments of different forms and severities of traumatic events.

Finite element methods are commonly used in research for simulating different mechanisms of injury [4,7,11,14,15]. These simulations require constitutive relationships to represent the mechanical properties of the materials. Previous studies have explored the mechanical responses of lung parenchyma under different loading conditions [16–18]. These studies have all assumed that lung parenchyma behaves isotropically, but to our knowledge, no study has verified this assumption. Moreover, reports that have addressed isotropy in other materials have used the comparison of values derived from stress-strain curves such as modulus, extensibility, or stress and strain at failure [19–22]. These methods of comparisons offer insight into the isotropy in a material, but do not compare the entire stress-strain relationships of different loading directions along quantifiable histological data, which may be useful for certain materials or loading multiaxial conditions.

In this study we evaluated lung tissue in compression with three experimental groups comprising specimens aligned with the three anatomical planes: sagittal, frontal, and coronal. We subjected these groups to uniaxial compression, with each group corresponding to one of the anatomical planes. Stress-strain curves of each group were generated from the test data, and the groups were compared to determine the differences among loading directions in the stress and strain values at 10% intervals of the curve. These differences were evaluated for statistical significance between direction groups. Furthermore, histological investigations (Movat, Massons) and analysis of the histological micrographs were used to determine if any morphological differences were apparent at the microstructure level.

2. Materials and Methods

2.1. Tissue Procurement and Preparation

Porcine lungs were obtained from a local abattoir, in accordance with Mississippi State University Institutional Animal Care and Use Committee regulations. All specimens were obtained from market ready pigs of approximately 12 months of age weighing 225 pounds. Lungs were obtained immediately following sacrifice, placed in sealed plastic bags in a cooler on ice and transferred within thirty minutes to the laboratories in the Department of Agricultural and Biological Engineering at Mississippi State University for mechanical testing (Figure 1a). To alleviate the effects of post mortem tissue changes, all mechanical tests were performed within twelve h of sacrifice. Lungs were dissected of the pleural membranes and a section of the lung, approximately 60 mm by 100 mm rectangular sections were isolated from the center area (Figure 1b). Twenty mm cylindrical test specimens were dissected from the larger section of lung. These cylindrical samples were 10 mm thick (Figure 2c). The cylindrical samples were prepared such that they were aligned with one of the three anatomical planes; sagittal, frontal, and coronal. The three alignments were used as our three comparison groups.

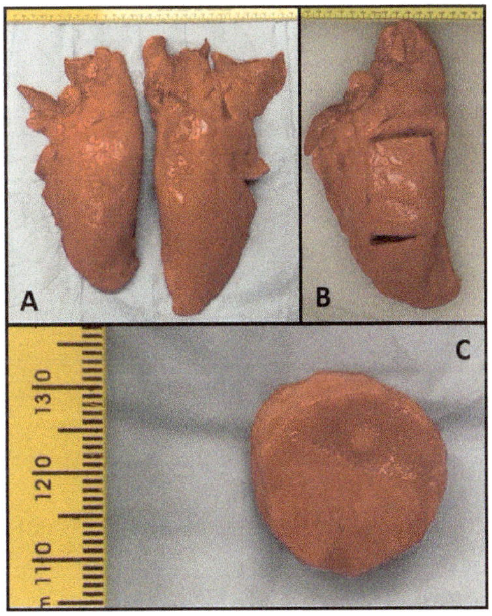

Figure 1. (**A**) Freshly obtained pig lung. (**B**) 60 mm × 100 mm rectangular sections were removed. (**C**) 10 mm thick cylindrical samples were prepared.

2.2. Histological Preparation and Light Microscopy

Three specimens, one from each of the aforementioned anatomical planes, were prepared for histological examination. Specimens were fixed using 10% neutral buffered formalin, paraffin embedded, sectioned to 5 μm thickness, and mounted to glass slides. One set of slides was stained with

Movat's Pentachrome to observe the composition of structural proteins, and a second set of slides was stained with Masson's Trichrome to observe the blood vessels and fibrous tissue. Prepared specimens were analyzed via light microscopy using a Leica DM2500 light microscope (Leica Microsystems, Wetzlar, Germany) at 100× magnification to observe any qualitative differences in microstructure.

Specimens were quantitatively analyzed for differences in microstructure using ImageJ (National Institute of Health, Bethesda, MD, USA). For each orientation, a series of twenty five overlapping light microscope images were obtained. These micrographs were compiled using the MosaicJ plugin (Ecole Polytechnique Federale De Lausanne, Lausanne, Switzerland) for ImageJ. Briefly, overlapping micrographs were positioned within the MosaicJ window and the software was used to create a single cohesive image based on overlapping portions of separate micrographs. This allowed for larger regions of the slide-mounted sample to be imaged without omitting portions of larger alveoli. The cohesive images were then cropped to create the largest rectangle available. Figure 2 illustrates how individual micrographs contributed to the composites, and how the uneven edges were cropped. The final composites for each direction were thresholded, inverted, and analyzed using the "Analyze Particles" feature in ImageJ. In this tissue the most relevant, and most readily observed by this method, feature is the alveolar spaces (white) of the lung tissue. "Analyze Particles" was configured to filter out particles with an area of less than 2000 pixels, approximately 2839 square microns, to eliminate extremely small particles or artifacts. Edge particles were also filtered. Roundness, Circularity, Solidity, and Aspect Ratio of the Fit Ellipse were plotted as histograms and analyzed to discern the 1st, 2nd and 3rd quartiles of the histograms to compare any differences among the alignment planes. These parameters describe particle shape and provide insight into particle elongation and tortuosity. Formulas for these parameters are shown in Appendix.

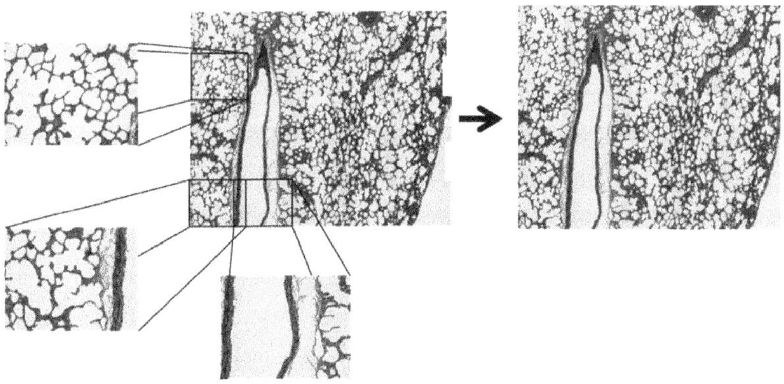

Figure 2. Composite micrograph composed of individual micrographs.

2.3. Compression Testing

Specimens were subjected to uniaxial compressive loading using the Mach1 Micromechanical Testing System® (Biomomentum, Laval, QC, Canada). A flat, disc-shaped metal platen was used for compression testing. The sample was attached with a small amount of high-viscosity cyanoacrylate ester adhesive (Permabond LLC, Pottstown, PA, USA) to prevent slipping of the sample from below the compression head, without introducing confinement effects that would compromise the uniform stress state assumption. Samples were submerged in Phosphate buffer saline (PBS) to simulate physiological conditions and prevent samples from drying during the test. Figure 3 shows the testing apparatus. The samples were pre-loaded to two grams-force, and pre-conditioned by cyclic loading to ten percent strain for ten cycles. We chose to precondition because lung is under constant cyclic loading during normal physiological conditions, and the preconditioning mimics this cycling. Specimens were then loaded to ten kilograms-force at a rate of ten percent strain (engineering) per second, with the loading velocity adjusted accordingly for each specimen. Load and displacement data were collected by the Mach1 software for further analysis.

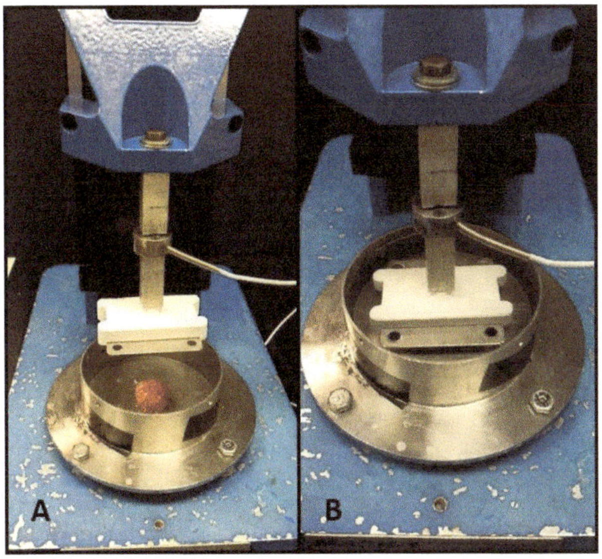

Figure 3. Mach1 Micromechanical Testing System®. (**A**) Sample in dish with compression platen removed; (**B**) Compression platen lowered onto sample in testing dish.

3. Data Analysis

3.1. True Stress versus True Strain Conversions

Data from our mechanical tests were processed in Microsoft Excel (Microsoft, Redmond, WA, USA) to create true stress-true strain curves. Data from each test were converted to engineering stress and engineering strain using the following Equations (1) and (2) (Appendix). Data were further processed to true stress and true strain using Equations (3) and (4) (Appendix). These true stress-strain conversions assume a uniform stress state and incompressibility. These assumptions are discussed further in the limitations section of this manuscript.

3.2. Interpolation of True Stress-True Strain Curves

We observed the true stress and true strain at equivalent percentages of the stress-strain curve in order to compare tests within each anatomical plane and between anatomical planes. Because the sampling rate during testing yields results that do not coincide with even percentages, an Excel-based interpolation script was used to create an equivalent curve with 10 data points at multiples of 10% (10%, 20%, …, 100%) for each experimental data set. Briefly and for a given set of test data, this interpolation script determined the maximum true stress and true strain values and normalized all true stress and true strain values relative to the maximum, respectively. The distance was then determined between each recorded data point by using the Pythagorean distance formula. For each data point, the accumulated distance was divided by the total distance for the data set to determine the percentage of the total for that data point. Using these newly assigned percentages for each data point, the true stress-true strain data points closest to a 10% strain were then selected. A linear interpolation was done between these two points to create a new data point, which represented the 10% point of the experimentally recorded curves. This process was repeated for each multiple of 10% until 90% was reached. The final experimentally recorded data point was used as the 100% data point. This script was run for each experimental data set to generate a bank of curves that would allow for precise analysis.

3.3. Comparison of Recorded Data among Anatomical Planes

The interpolated data sets from each experimental test were compiled in Excel and the data at increments of 10% strain within each anatomical plane were averaged to create a characteristic curve for each anatomical plane. The average of the characteristic curves were plotted with standard deviation (Figure 4).

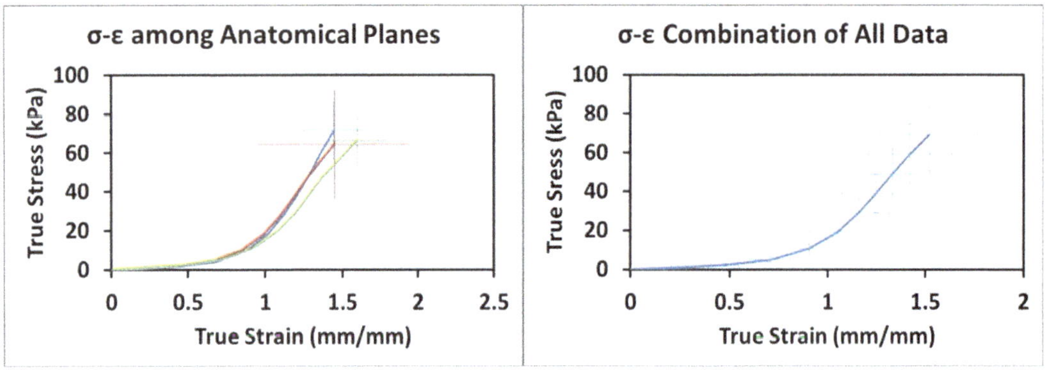

(A) (B)

Figure 4. (**A**) Stress-Strain relationship across anatomical planes (Frontal (blue): $n = 18$, Sagittal (red): $n = 13$, Transverse: $n = 12$). (**B**) Combination of all Stress-Strain data ($n = 43$). Error bars indicate +/− one standard deviation.

Statistical analysis was performed to compare the differences in stress and strain in each of the three anatomical planes. At 10% strain increments, the stress-strain data were tested using a single-factor ANOVA with a 95% confidence interval to determine if there was a significant difference among each of the three anatomical planes. Statistics were performed using in Microsoft Excel.

4. Results

Micrographs for each anatomical plane are presented in Figure 5. Figure 5A-B, C-D, and E-F show Movat's Pentachrome for the sagittal, frontal, and transverse planes, respectively. These micrographs do not indicate any apparent qualitative differences in morphology or alignment of microstructural features among the three anatomical planes. Composite Micrographs created with MosaicJ are shown in Figure 6; Figure 6A is sagittal, Figure 6B is frontal, Figure 6C is transverse. The corresponding histograms for a particle analysis, particles being alveolar spaces, are shown in Figure 7. Histograms for Roundness (Figure 7A), Circularity (Figure 7B), and Solidity (Figure 7C), were prepared with bins from 0 to 1, incremented by 0.05, where the columns indicate the percentage of particles that fell into each bin. The aspect ratio (Figure 7D), was prepared with bins from 1.0 to 3.4, incremented by 0.1, and columns denoting the percentage of particles for each bin. It should be noted that some aspect ratio data was not displayed in this histogram in the interest of more effectively highlighting the majority of the data; 3.4 was chosen as the end of the axis because this was the final value at which more than two particles occurred within a single bin for any of the three alignments. Table 1 shows the 1st quartile, median, and 3rd quartiles for the histograms. Histograms of particle analysis parameters had little difference among the different orientation planes.

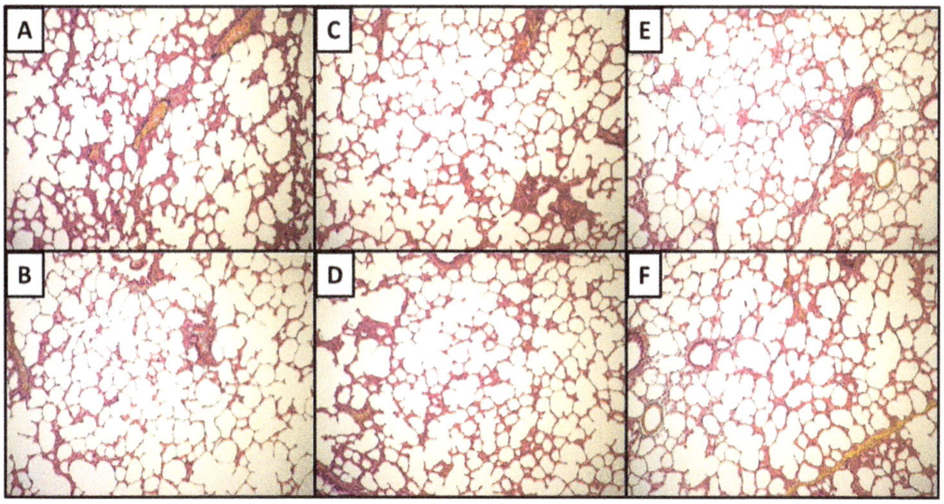

Figure 5. Movat's Pentachrome for the sagittal (**A**,**B**), frontal (**C**,**D**) and transverse (**E**,**F**) planes.

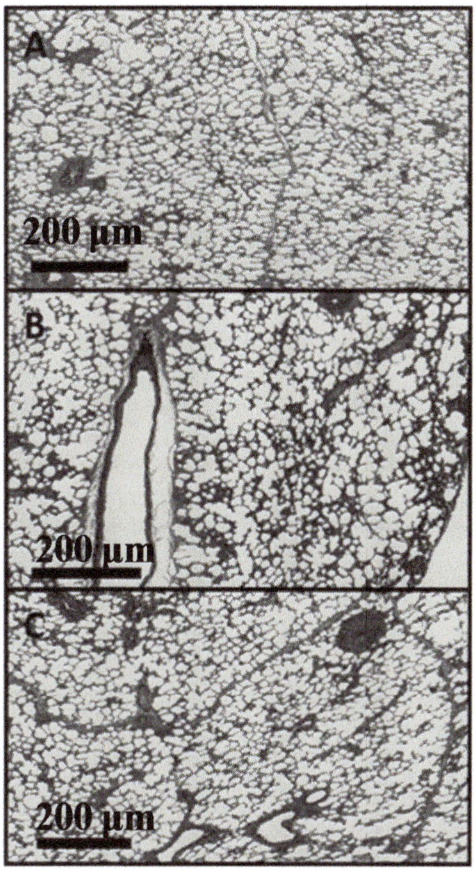

Figure 6. Composite micrographs of (**A**) saggital, (**B**) frontal, (**C**) transverse planes (200 µm scale).

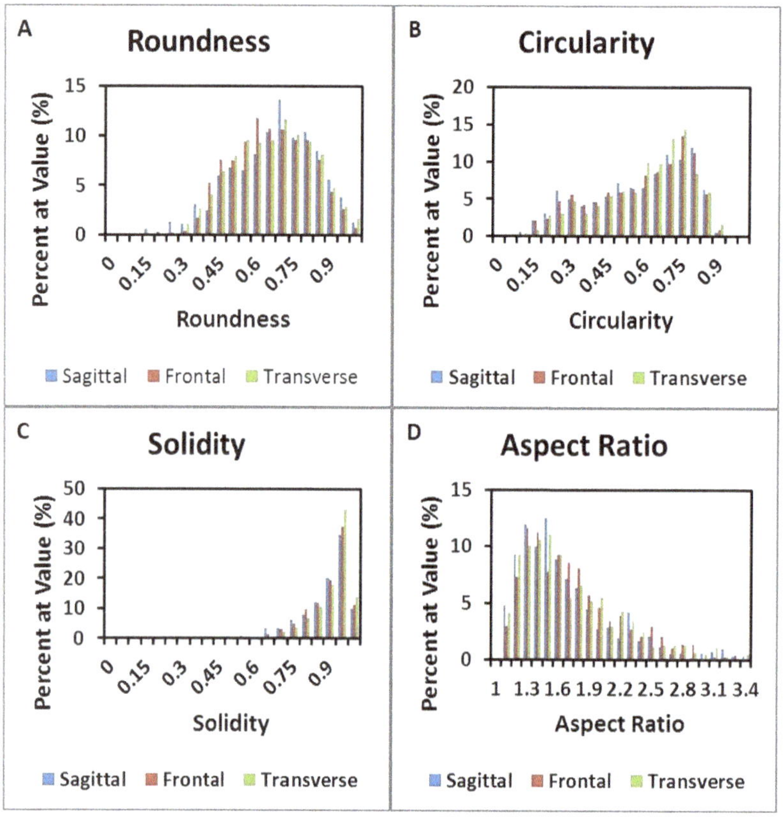

Figure 7. Histograms for (**A**) Roundness, (**B**) Circularity, and (**C**) Solidity. (**D**) shows the aspect ratio of samples.

Table 1. The 1st quartile, median, and 3rd quartiles for the histograms.

Roundness	**Sagittal**	**Frontal**	**Transverse**
1st Quartile	0.53375	0.512	0.512
Median	0.66	0.626	0.647
3rd Quartile	0.77225	0.75	0.761
Circularity	**Sagittal**	**Frontal**	**Transverse**
1st Quartile	0.385	0.411	0.451
Median	0.5855	0.6	0.618
3rd Quartile	0.72475	0.722	0.723
Solidi	**Sagittal**	**Frontal**	**Transverse**
1st Quartile	0.812	0.824	0.853
Median	0.889	0.898	0.912
3rd Quartile	0.93525	0.936	0.94
Aspect Ratio	**Sagittal**	**Frontal**	**Transverse**
1st Quartile	1.29475	1.333	1.314
Median	1.515	1.598	1.546
3rd Quartile	1.87475	1.953	1.955

True stress-true strain curves for each anatomical plane are shown in Figure 4A. The composite curve for the average of the three planes is shown in Figure 4B. The results of the statistical comparisons for each percentage are presented in Tables 2 and 3. There were no significant differences in either stress or strain among the three anatomical planes for any percentage level.

Table 2. Results of statistical comparisons for the samples.

Path	Frontal	Sagittal	Transverse	*F*	*p*-value	Frontal	Sagittal	Transverse	*F*	*P*-value
%	Avg. True σ	Avg. True σ	Avg. True σ			Avg. True ε	Avg. True ε	Avg. True ε		
10	619.39	729.75	957.42	2.6893	0.0793	0.2293	0.2251	0.2477	0.0470	0.9542
20	1,588.29	1,934.04	2,390.74	2.5329	0.0912	0.4580	0.4488	0.4937	0.0468	0.9544
30	3,984.76	4,568.52	5,304.14	1.4654	0.2423	0.6817	0.6642	0.7301	0.0462	0.9549
40	9,563.28	9,819.38	10,977.00	0.3470	0.7087	0.8778	0.8506	0.9318	0.0453	0.9557
50	18,377.42	17,719.62	19,178.35	0.0115	0.9886	1.0198	0.9872	1.0789	0.0448	0.9562
60	28,466.14	26,711.31	28,380.77	0.0223	0.9780	1.1261	1.0896	1.1895	0.0444	0.9566
70	39,023.73	36,054.37	37,928.60	0.0771	0.9259	1.2130	1.1766	1.2817	0.0442	0.9568
80	49,760.69	45,514.93	47,479.43	0.1367	0.8726	1.2907	1.2584	1.3727	0.0443	0.9567
90	60,601.53	54,822.69	56,676.20	0.2193	0.8040	1.3625	1.3489	1.4820	0.0453	0.9558
100	71,323.68	64,005.78	65,778.25	0.2841	0.7541	1.4406	1.4472	1.5977	0.0461	0.9550

Table 3. Standard deviations for statistical comparisons of the samples. Dev.: Deviations.

Part %	Frontal		Sagittal		Transverse	
	Strain Dev.	Stress Dev.	Strain Dev.	Stress Dev.	Strain Dev.	Stress Dev.
10	0.030915423	127.2781798	0.075739326	240.5005433	0.027138761	331.5140315
20	0.061629963	347.8942716	0.150734704	648.4193448	0.054122752	817.186667
30	0.090535682	829.7268617	0.221484316	1556.705447	0.078778275	1624.193247
40	0.110923627	1890.534598	0.280026781	3438.735718	0.097867071	2602.47132
50	0.123904198	3539.070152	0.32436637	6711.154692	0.113630437	3650.112918
60	0.137258653	5585.462839	0.359870173	10716.82231	0.127062902	5247.527794
70	0.150673246	7795.522934	0.392218749	14986.77743	0.139112415	7176.190338
80	0.164835636	10072.46936	0.424828203	19386.53237	0.151445696	9218.119181
90	0.180556847	12424.48858	0.46166626	23807.36975	0.168660868	11333.30934
100	0.208370407	14937.09112	0.504503056	28290.23423	0.198543701	13672.60307

5. Discussion

True stress-true strain curves for each anatomical plane appear to be similar, and the error bars indicate the variation among specimens is much greater than the variation between the averaged curves for each anatomical plane. It is well known that biological tissues exhibit a large degree of inter-specimen variation, and the degree to which the plane curves vary appears to be within one standard deviation for each other curve, respectively. This observation, while qualitative, strongly supports our hypothesis of isotropy. Statistical comparison of the stress and strain values for each multiple of 10% indicates no significant differences between any of the three planes or the average of the three planes. This quantitative observation also supports our hypothesis of isotropy.

The interpolation method presented was formulated with the interest of comparing curves effectively. The method for determining the percentages of the curve was chosen as a means of giving equivalent weight to both stress and strain. We believed this approach to be preferable to basing percentages on only one parameter because the characteristic curves for biological materials, which are often described as hyperelastic, are commonly observed to have two distinct regions of tissue behavior. The first region is described as the toe-in region and is dominated by the realignment of tissue structures to resist loading. The second region is described as the linear region and is dominated by significant resistance to loading as the tissue builds towards failure. A comparison based on strain percentages would capture more of the toe-in region and less of the linear region, essentially hiding a large amount of tissue behavior in the higher strain percentages. Likewise, a percentage comparison based on stress would capture more of the linear region and less of the toe-in, hiding a large amount of tissue behavior in the lower stress values. The combined weight of the two parameters allows the entire stress-strain path to be considered equally.

Quantitative analysis of micrographs did not indicate apparent differences in the morphology of porcine lungs (Figure 7 and Table 1). The quantitative particle analysis, which showed the morphological parameters had very similar values and distributions among the orientation planes, complimented the qualitative observations. Overall, the histological examinations supported the hypothesis that porcine lung is an isotropic material. Analysis of microstructural features is an important part of understanding the hierarchical organization of biological tissues. The change in organization, shape, and size of some features under deformation lends insight to possible physiological changes in the biological system. Microstructural changes during loading can be used to understand the sub structural evolution of lung, or other biological tissues, under loading and thus aid in modeling complex tissue behaviors. The data presented here strongly supports the widely held belief that lung parenchyma is an isotropic material. Despite this belief, we are not aware of a thorough analysis of this assumption. Beyond the implications for lung tissue, we believe the methods presented here represent a strong system for assessing isotropy in other tissues. This is valuable for future research as biological tissues may be isotropic or anisotropic, and a confident understanding of a given tissue's behavior is crucial for effective constitutive modeling. Additionally, verification of isotropy allows the analysis of more complex material properties such as viscoelasticity, strain rate dependency, and stress state dependency without considering the complexity of anisotropy. The ultimate result is a more efficient path towards a complete understanding of the material properties of lung, and other biological tissues of interest to the research community.

6. Conclusions

This study presents a novel method for testing mechanical isotropy in biological tissues. This method was used to demonstrate the isotropic nature of porcine lung parenchyma. This was a widely used assumption in the body of literature, but had not been explicitly analyzed or proven. This is significant for soft-tissue biomechanics in general, as it presents a means for testing isotropy in other tissues. It is significant for lung biomechanics research because it allows for future research to be conducted without concern for overlooking anisotropy in more complex material properties. Future studies will address important behaviors such as viscoelasticity, strain rate dependency, stress-state dependency, and microstructural evolution of lung parenchyma.

7. Limitations

The conversion formulas for true stress and true strain rely on assumptions of uniform stress state and incompressibility. Our mechanical testing setup is intended to provide the most uniform stress state possible, but there may be some uneven distributions that limit this aspect of the assumption.

Moreover, the assumption of incompressibility is inexact, as the lung contains solid and gas components, which have a high probability of being compressible. Considering that our test procedure does not actively track the changing cross-sectional area it is necessary to use some form of conversion to achieve true stress, as this method was used in our previous publication [23].

The interpolation method described uses a linear interpolation process, which slightly overestimates the interpolated values given that our curves are concave upward. Future studies will explore the incorporation of Newton-Raphson methodology to alleviate this limitation.

Acknowledgments

This material is based upon work supported by the National Nuclear Security Administration, Department of Energy, under award number (DE-FC26-06NT42755). The authors would also like to thank Sansing Meat Service, Maben, MS, USA, for their generous provision of porcine lung tissue.

Author Contributions

Benjamin Weed was a graduate student who led and directed the mechanical and microstructural testing and analysis of this study. Sourav Patnaik and Bryn Brazile are graduate students who worked with Benjamin on obtaining the tissues, as well as equipment design and calibration for mechanical testing. Mary Rougeau-Browning was an undergraduate student working in the laboratory under the mentorship of the corresponding author. She helped in obtaining tissues, tissue dissection, and mechanical testing. Dr. Jun Liao and Dr. Raj Prabhu assisted with data analysis. Dr. Lakiesha N. Williams assisted with the study design and was a lead in writing the manuscript. All authors contributed to an aspect of writing the manuscript.

Appendix

Formulas for Stress-Strain Calculations

1. $\sigma_{engineering} = \frac{force}{area_{undeformed}}$
2. $\varepsilon_{engineering} = \frac{length_{undeformed} - displacement}{length_{undeformed}}$
3. $\sigma_{true} = \sigma_{engineering} * \left(1 + \varepsilon_{engineering}\right)$
4. $\varepsilon_{true} = \ln(1 + \varepsilon_{engineering})$

Formulas for Shape Descriptors Used in Particle Analysis

1. $Roudness = 4 \times \left(\frac{Area}{\pi \times (Major\ Axis\ of\ Fit\ Ellipse)^2}\right)$
2. $Circularity = 4\pi \times \left(\frac{Area}{Perimeter^2}\right)$
3. $Solidity = \frac{Area}{Convex\ Area}$
4. $Aspect\ Ratio = \frac{Major\ Axis\ of\ Fit\ Ellipse}{Minor\ Axis\ of\ Fit\ Ellipse}$

Conflicts of Interest

The authors declare no conflict of interest.

References

1. Champion, H.R.; Holcomb, J.B.; Young, L.A. Injuries from explosions: Physics, biophysics, pathology, and required research focus. *J. Trauma Acute Care Surg.* **2009**, *66*, 1468–1477.
2. Wolf, S.J.; Bebarta, V.S.; Bonnett, C.J.; Pons, P.T.; Cantrill, S.V. Blast injuries. *Lancet* **2009**, *374*, 405–415.
3. Mayorga, M.A. The pathology of primary blast overpressure injury. *Toxicology* **1997**, *121*, 17–28.
4. Stuhmiller, J.H. Biological response to blast overpressure: A summary of modeling. *Toxicology* **1997**, *121*, 91–103.
5. Stuhmiller, J.H.; Ho, K.H.; Vander Vorst, M.J.; Dodd, K.T.; Fitzpatrick, T.; Mayorga, M. A model of blast overpressure injury to the lung. *J. Biomech.* **1996**, *29*, 227–234.
6. Cooper, G.J.; Townend, D.J.; Cater, S.R.; Pearce, B.P. The role of stress waves in thoracic visceral injury from blast loading: Modification of stress transmission by foams and high-density materials. *J. Biomech.* **1991**, *24*, 273–285.
7. Stuhmiller, J.H.; Chuong, C.J.; Phillips, Y.Y.; Dodd, K.T. Computer modeling of thoracic response to blast. *J. Trauma Acute Care Surg.* **1988**, *28*, 132–139.
8. Dehghan, N.; de Mestral, C.; McKee, M.D.; Schemitsch, E.H.; Nathens, A. Flail chest injuries: A review of outcomes and treatment practices from the national trauma data bank. *J. Trauma Acute Care Surg.* **2014**, *76*, 462–468.
9. Pauzé, D.R.; Pauzé, D.K. Emergency management of blunt chest trauma in children: An evidence-based approach. *Pediatr. Emerg. Med. Pract.* **2013**, *10*, 1–22.
10. Idriz, S.; Abbas, A.; Sadigh, S.; Padley, S. Pulmonary laceration secondary to a traumatic soccer injury: A case report and review of the literature. *Am. J. Emerg. Med.* **2013**, *31*, 1625.
11. D'Yachenko, A.I.; Manyuhina, O.V. Modeling of weak blast wave propagation in the lung. *J. Biomech.* **2006**, *39*, 2113–2122.
12. Born, C.T. Blast trauma: The fourth weapon of mass destruction. *Scand. J. Surg.* **2005**, *94*, 279–285.
13. Sasser, S.M.; Sattin, R.W.; Hunt, R.C.; Krohmer, J. Blast lung injury. *Prehosp. Emerg. Care* **2006**, *10*, 165–172.
14. Gayzik, F.S.; Hoth, J.J.; Stitzel, J. Finite element–based injury metrics for pulmonary contusion via concurrent model optimization. *Biomech. Model. Mechanobiol.* **2011**, *10*, 505–520.
15. Vawter, D.L. A finite element model for macroscopic deformation of the lung. *J. Biomech. Eng.* **1980**, *102*, 1–7.
16. Suki, B.; Bates, J.H.T. Lung tissue mechanics as an emergent phenomenon. *J. Appl. Physiol.* **2011**, *110*, 1111–1118.
17. Gao, J.; Huang, W.; Yen, R.T. Mechanical properties of human lung parenchyma. *Biomed. Sci. Instrum.* **2006**, *42*, 172–180.

18. Freed, A.D.; Einstein, D.R. An implicit elastic theory for lung parenchyma. *Int. J. Eng. Sci.* **2013**, *62*, 31–47.
19. Feng, Y.; Okamoto, R.J.; Namani, R.; Genin, G.M.; Bayly, P.V. Measurements of mechanical anisotropy in brain tissue and implications for transversely isotropic material models of white matter. *J. Mech. Behav. Biomed. Mater.* **2013**, *23*, 117–132.
20. Lillie, M.A.; Shadwick, R.E.; Gosline, J.M. Mechanical anisotropy of inflated elastic tissue from the pig aorta. *J. Biomech.* **2010**, *43*, 2070–2078.
21. Sacks, M.S.; Sun, W. Multiaxial mechanical behavior of biological materials. *Annu. Rev. Biomed. Eng.* **2003**, *5*, 251–284.
22. Sacks, M. Biaxial mechanical evaluation of planar biological materials. *J. Elast.* **2000**, *61*, 199–246.
23. Weed, B.C.; Borazjani, A.; Patnaik, S.S.; Prabhu, R.; Horsteemyer, M.F.; Ryan, P.L.; Franz, T.; Williams, L.N.; Liao, J. Stress state and strain rate dependence of human placenta. *Ann. Biomed. Eng.* **2012**, *40*, 2255–2265.

Impact of Packing and Processing Technique on Mechanical Properties of Acrylic Denture Base Materials

Touraj Nejatian [1,2], Farshid Sefat [3,4,*] and Tony Johnson [2]

1 Dental Materials Unit, School of Dental Sciences, University of Newcastle, Newcastle upon Tyne NE2 4BW, UK; E-Mail: Touraj.Nejatian@newcastle.ac.uk
2 School of Clinical Dentistry, University of Sheffield, Sheffield, South Yorkshire S10 2TA, UK; E-Mail: a.johnson@sheffield.ac.uk
3 Department of Biomedical Engineering, School of Engineering, King Faisal University, Al-Hofuf, Al-Ahsa 31982, Saudi Arabia
4 Tissue Engineering Group, Department of Chemistry, Chemical Biology and Biomedical Engineering, Stevens Institute of Technology, Hoboken, NJ 07030, USA

* Author to whom correspondence should be addressed; E-Mail: fsefat@kfu.edu.sa;

Academic Editors: Maryam Tabrizian and Ihtesham ur Rehman

Abstract: The fracture resistance of polymethylmethacrylate (PMMA) as the most popular denture base material is not satisfactory. Different factors can be involved in denture fracture. Among them, flexural fatigue and impact are the most common failure mechanisms of an acrylic denture base. It has been shown that there is a correlation between the static strength and fatigue life of composite resins. Therefore, the transverse strength of the denture base materials can be an important indicator of their service life. In order to improve the fracture resistance of PMMA, extensive studies have been carried out; however, only a few promising results were achieved, which are limited to some mechanical properties of PMMA at the cost of other properties. This study aimed at optimizing the packing and processing condition of heat-cured PMMA as a denture base resin in order to improve its biaxial flexural strength (BFS). The results showed that the plain type of resin with a powder/monomer ratio of 2.5:1 or less, packed conventionally and cured in a water bath for 2 h at 95 °C provides the highest BFS. Also, it was found that the performance of the dry heat processor is inconsistent with the number of flasks being loaded.

Keywords: polymethylmethacrylate (PMMA); acrylic denture; mechanical properties

1. Introduction

Polymethyl methacrylate (PMMA) has been the most popular material for construction of dentures since the 1930s due to many advantages, including good aesthetics, accurate fit, stability in the oral environment, easy laboratory and clinical manipulation, and inexpensive equipment. However, the fracture resistance of PMMA is not satisfactory [1,2]. According to a survey, two-third of dentures had broken within three years of their provision [3]. Darbar *et al.* 1994, reported that 33% of the repairs carried out by three laboratories were due to debonded/detached teeth; 29% were due to midline fractures which were most common in the upper dentures; and the rest (38%) were other types of fracture. Denture fracture is a multifactorial phenomenon, and even strengthening measures could not efficiently prevent denture fracture [4].

Flexural fatigue and impact fracture have been implicated as a mechanism of denture fracture. Therefore, the transverse strength of the denture base materials can be an important indicator of their performance [5,6]. To reduce the denture fracture incidence, four different aspects can be considered based on the etiology of the fracture: (1) retaining the mechanical characteristics through corrective surgery of anatomic abnormalities such as high frenum and palatal torus, improving denture fit and balanced occlusion; (2) optimizing chemical structure through modifying packing and processing techniques [7]; (3) improving adhesion between acrylic teeth and the denture base resin [8]; (4) altering the composition either chemically by, for example, changing brittle polymers to a high impact polymer through the addition of rubber particles [9]; or making physical alterations that in this case incorporate materials into PMMA such as fibers, metal inserts, and particles [10]. An additional method is using alternative materials such as polyamides, epoxy resin, polystyrene vinyl acrylic, polycarbonate, polyurethane and nylon [11–13].

Despite a great deal of effort, the fracture PMMA denture base material is still a matter of concern for both dentists and patients [14]. Few promising results, limited to some mechanical properties of PMMA, have often been achieved at the cost of other properties [10,15]. This study was designed to evaluate the effect of processing methods, packing techniques, and the type of resin on the fracture resistance of PMMA as a denture base material. The study design is based on a multi-factorial experiment to evaluate the influence of five factors; resin type, powder-toliquid ratio, packing technique, processing method and curing time on flexural strength of the resin. In addition, the performance of acrylic processors, dry heat oven and water bath, were assessed and compared. The effect of extruder angle on mechanical properties of PMMA packed using an injection technique was also studied. In this study, a biaxial flexural strength (BFS) test was used to measure bending strength of PMMA. BFS and four-point bend tests have been employed to study PMMA and resin composites [16,17] as they are known to be better indicators of pure bending strength in comparison with the three-point bend test due to less shear stress generated in the specimen [16,18–20]. In the case of the BFS test, the specimen is supported near to its periphery and, as the stress rises, far from the edge that makes the test less sensitive to edge defects, which inevitably occur during the preparation. Therefore, it provides less scatter in the data [18,21]. Also, easy sample preparation and its size match to the clinical application make the BFS test a reproducible test to measure the flexural strength of these materials [18,22].

2. Results and Discussion

2.1. Effect of the Resin Type (Factor A)

The mean BFS of the samples, made from plain PMMA (149 MPa, SD = 30), were higher than the veined type (133 MPa, SD = 32), and the difference was statistically significant ($n = 300$, $p < 0.05$) (Figure 1).

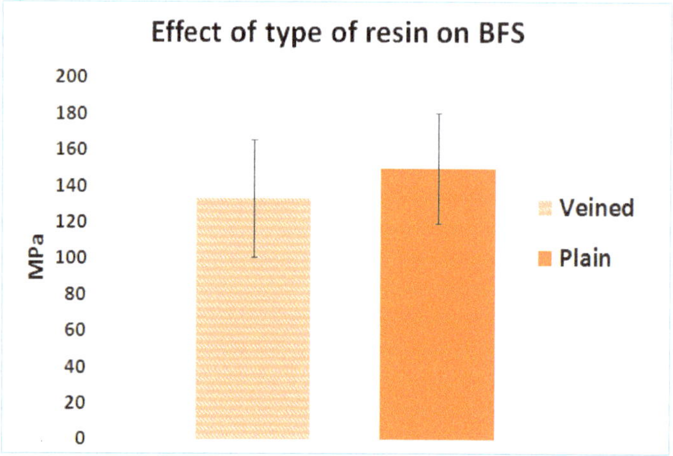

Figure 1. Effect of resin type on Biaxial Flexural Strength.

2.2. Effect of Packing Method on Biaxial Flexural Strength (Factor B)

Figure 2A,B represent the data for the effect of the packing method on the BFS of PMMA. The mean BFS of conventionally packed plain (159 MPa, SD = 22), and veined PMMA (145 MPa, SD = 32), discs was greater than the injection-packed ones (139 MPa, SD = 34 and 123 MPa, SD = 31 respectively), and the differences were statistically significant ($n = 150$, $p < 0.05$).

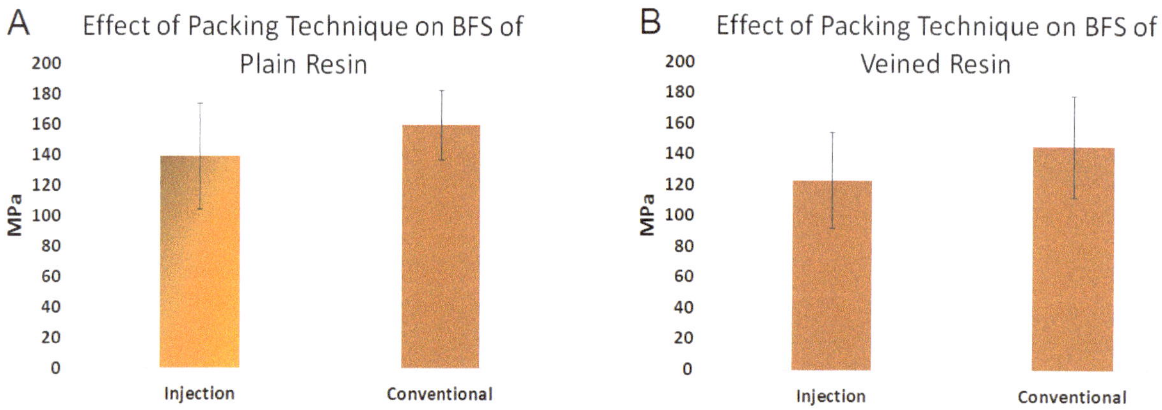

Figure 2. Effect of packing method on BFS of (**A**) plain polymethylmethacrylate (PMMA); and (**B**) veined PMMA for conventional and injection.

2.3. Effect of Powder-to-Liquid Ratio on BFS (Factor C)

The data for the effect of the powder-to-liquid ratio on the BFS are illustrated in Figure 3A,B. Increasing the powder/liquid ratio of plain resin from 1.5/l to 3.5/l had no effect on the BFS of the resin (161 MPa, SD = 18 and 161 MPa, SD = 17 and 169 MPa, SD = 27 and159 MPa, SD = 25 and 159 MPa, SD = 24, respectively), ($n = 30$, $p > 0.05$). Similarly, changing powder/liquid ratio of veined resin from 1.5/l to 2.5/l had no significant effect on the BFS of the resin (150 MPa, SD = 26 and 156 MPa, SD = 30 and 158 MPa, SD = 41, respectively) ($p > 0.05$). Raising the powder ratio to 3 and 3.5 with respect to the monomer reduced the mean strength (135 MPa, SD = 26 and 124 MPa, SD = 25, respectively) compared to the 2.5/l ratio. The difference was statistically significant ($n = 30$, $p < 0.05$).

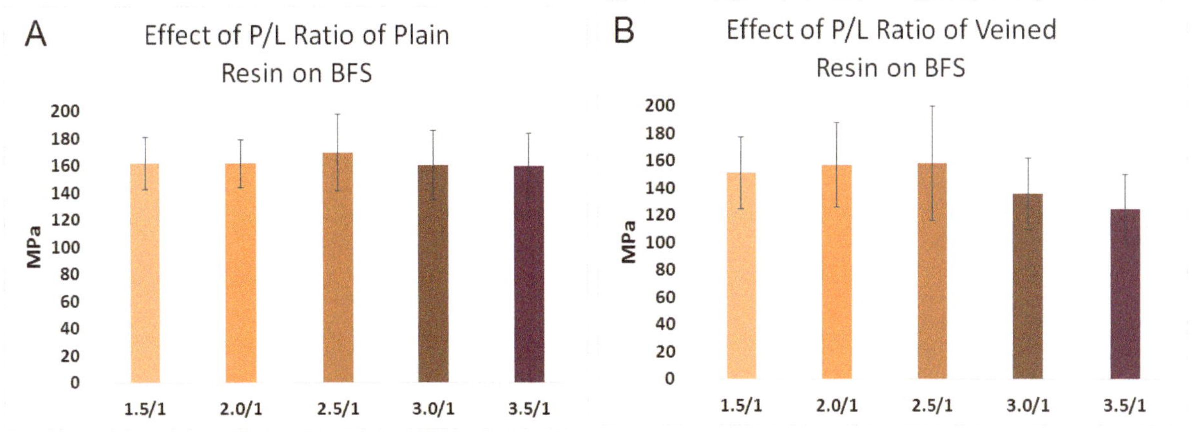

Figure 3. Effect of power/liquid ratio on BFS of (**A**) plain PMMA; and (**B**) veined PMMA for both plain and veined PMMA.

2.4. Effect of Curing Time in Water Bath (Factor E)

The data for the effect of the curing time in the water bath on the BFS are presented in Figure 4A,B. Changing the curing time in the water bath from 8 h at 75 °C and 2 h at 95 °C to 4 h at 75 °C and 2 h at 95 °C and then 2 h at 95 °C did not have any effect on the mean BFS of the plain (170 MPa, SD = 15 and 164 MPa, SD = 31 and 161 MPa, SD = 27, respectively) and veined PMMA (172 MPa, SD = 42 and 159 MPa, SD = 42 and 144 MPa, SD = 39, respectively) ($n = 10$, $p > 0.05$) (Figure 4).

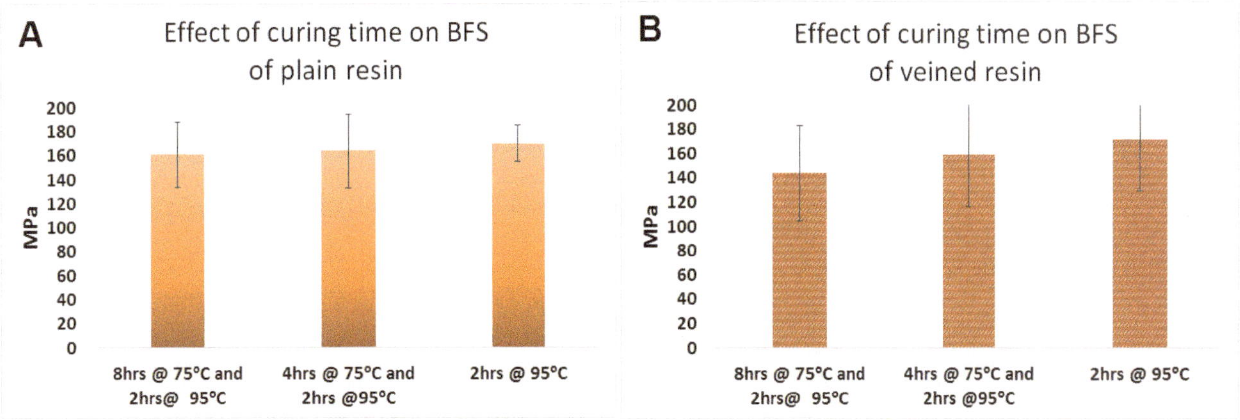

Figure 4. Effect of curing time in water bath on BFS of (**A**) plain and (**B**) veined PMMA.

2.5. Effect of Extruder Angle on Mechanical Properties of PMMA

2.5.1. Effect of Extruder Angle on BFS of PMMA

The data for the effect of extruder angle on the BFS of PMMA are shown in Figure 5A. Altering the extruder angle from 90° to 160° did not improve the mean BFS of veined type of the resin (156 MPa, SD = 17 and 143 MPa, SD = 24, respectively) ($n = 10$, $p > 0.05$).

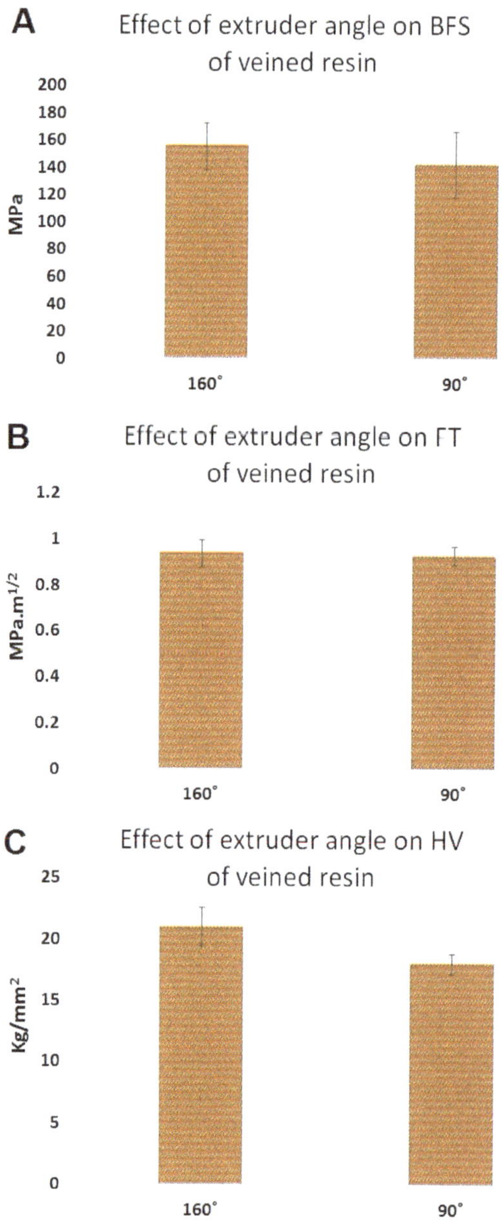

Figure 5. Effect of extruder angle on (**A**) BFS; (**B**) fracture toughness (FT); and (**C**) HV of veined PMMA. VIm W (veined type of resin packed using modified extruder, 160°, cured in the water bath), VIW (veined type of resin packed using standard extruder, 90°, cured in the water bath).

2.5.2. Effect of Extruder Angle on Fracture Toughness of PMMA

The data for the effect of extruder angle on fracture toughness of PMMA are shown in Figure 4B; altering the extruder angle from 90° to 160° did not improve the mean fracture toughness of veined type of the resin (0.94 MPa·m$^{1/2}$, SD = 0.06 and 0.93 MPa·m$^{1/2}$, SD = 0.04, respectively) ($n = 30$, $p > 0.05$).

2.5.3. Effect of Extruder Angle on Hardness of PMMA

The data for the effect of extruder angle on Vicker's hardness, HV, of PMMA shown in Figure 4C. Altering the extruder angle from 90° to 160° improved the mean Vicker's hardness of veined type of the resin, but the difference was statistically significant (21 kg/mm^2, SD = 1.6 and 18.4 kg/mm^2, SD = 0.8, respectively) ($n = 20$, $p < 0.05$).

2.6. Analysis of Temperature within Curing Baths; Inside and Outside Curing Resin

The overall data for the effect of the resin type on the BFS are presented in Figure 6A.

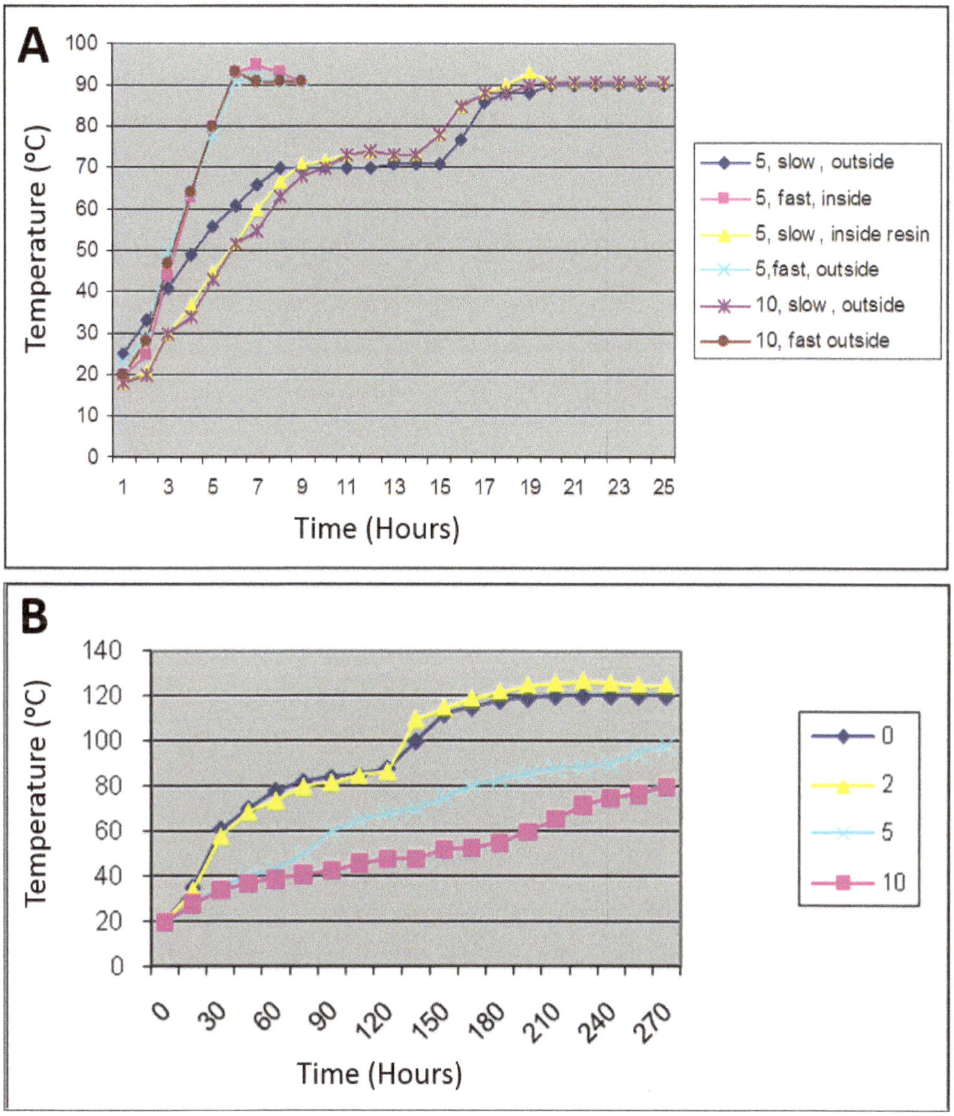

Figure 6. *Cont.*

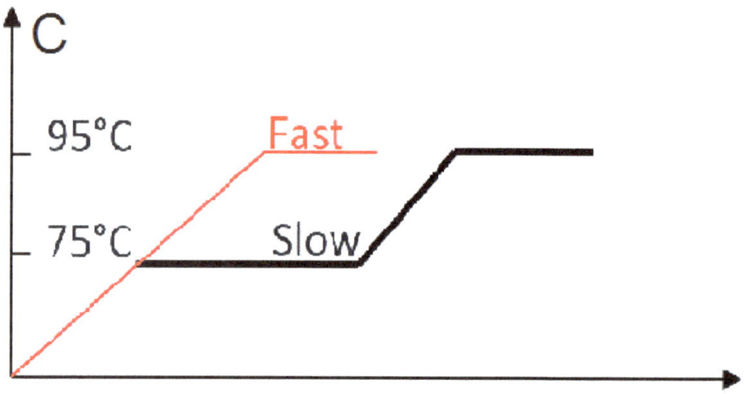

Figure 6. Pattern of temperature in (**A**) water bath loaded with 5 and 10 flasks and cure under slow (4 h 75 °C + 2 h 95 °C) and fast (2 h 95 °C) processing methods was measured; (**B**) Pattern of temperature inside the curing resin while water bath loaded with 5 flasks was measured; (**C**) Pattern of temperature in dry heat oven loaded with 0, 2, 5 and 10 flasks.

Analysis of temperature climb and hold within the curing chambers showed a consistent performance for the water bath irrespective of the number of flasks being cured. It was also found that the pattern of the temperature change inside the resin during curing is identical to that seen outside the resin even with different numbers of flasks being loaded (Figure 6A). In comparison with the standard setting (Figure 6C), the dry heat processor showed inconsistent results with flask numbers having a major effect on the rate of climb and holding temperature. In other words, the more flasks that were loaded, the slower the temperature elevated. When the oven is empty or loaded with up to two flasks, the chamber reached 120 °C, which is beyond the optimum curing temperature of 95 °C. Conversely, when loaded with 10 flasks, the oven could only reach a temperature of 80 °C, which is less than the required curing temperature. The accepted result was achieved when the oven was loaded with five flasks. (Figure 6B).

The standard setting of two curing methods in the dry heat processor: (1) the fast curing method in which temperature starts increasing steadily from 0 to 95 °C and then remains constant for a period of time (the time is adjustable); (2) The slow curing method in which temperature increases from 0 to 75 °C and remains constant for a period of time (the time is adjustable) then climbs up again and remains constant at 95 °C for a period of time (the time is adjustable).

2.7. Discussion

This study was essentially designed to evaluate the effect of resin type and its processing and packing variables on the BFS of heat-cure denture base PMMA. This was to optimize the condition to improve the BFS of PMMA, which may help increase the fracture resistance of denture base. Evaluating the temperature within the curing chambers during processing revealed that the number of flasks loaded in a dry heat oven seriously affects the performance of the processor. However, there was no difference between thermal changes inside and outside the curing resin, within the curing chamber. No evidence was found to study and compare the performance of processing. Due to the inconsistency of temperature and rate of climb within the dry heat processor, this curing method was not used for the main body of the work.

Plain resin showed higher BFS results than the veined resin, which may be caused by weak bonding between the fibers and matrix. In this case, fibers may act as inclusions that weaken the resin. This might possibly be investigated by scanning electron microscopy. When considering the packing method, the conventionally packed plain and veined resins showed higher BFS results than the injection-packed ones, which may be because of disruption in the polymer chain formation due to the stretch caused by the turbulence effect that occurs inside the standard extruder syringe during the injection (Figure 11A). This can result in short and less cross-linked polymer chains, which is related to low strength of the resin. By altering the extruder angle from 90° to 160°, the mechanical properties of the resin were not improved. However, the standard deviation was decreased, possibly due to uniform flow and orientation of the fibers in the direction of injection and a reduced turbulence effect during the injection (Figure 7). Hardness value (HV) is directly related to elastic modulus and the resistance of material against plastic deformation, whereas BFS and FT are representative of resistance of the material against fracture. In other words, they are representative of two different types of properties of the material, which are not necessarily correlated. Unlike BFS and FT, improvement of HV was statistically significant which could be due to the concentration of the fibers on the superficial layer of the specimens. According to anecdotal evidence, this may happen because of marginalization of the fiber due to the friction of the resin with the internal surface of the mold during injection. As PMMA shows brittle behavior against impact forces, it is considered as a notch-sensitive material. Poor hardness value makes the material easily scratched by food or negligence, and the resulting micro-cracks and scratches can compromise the fracture resistance of the material. This part of the study was unique in the context of dental materials. To confirm the aforementioned hypotheses, further chemical analyses and electron microscopic studies would need to be carried out. Previous studies on the effects of packing method on flexural strength showed inconsistent results, probably due to different testing methods, sample size, and shape [23,24].

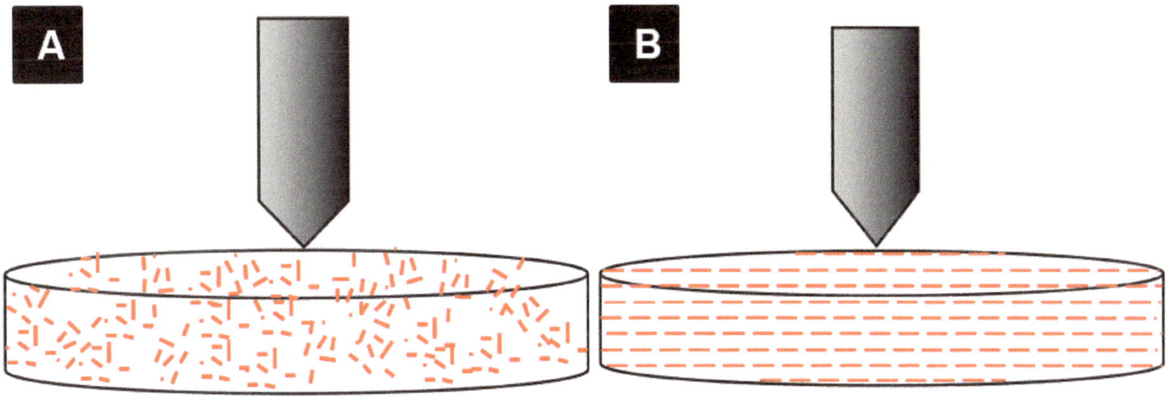

Figure 7. Schematic feature of fibers oriented (**A**) randomly (produced by right angle extruder); (**B**) organized perpendicular to direction of load (produced by 160° extruder).

Although powder/liquid ratio of equivalent to 2:1 (volume) was recommended by the manufacturer, in this study, at least in the case of veined resin, increasing the powder/liquid ratio up to 2.5:1 increased the strength of the resin, but beyond that ratio, the strength of the resin was decreased. Therefore, using powder/liquid of 2.5:1 seems to be recommendable. No specific study has been found to investigate the optimal powder/liquid ratio of the resin; however, Dogan *et al.* [25] stated that excess monomer has a deteriorating effect on mechanical characteristics of the resin.

Reducing the delay time at 75 °C from 8 to 4 and 0 hours has no significant effect on the BFS of the resin, but reducing the curing time to 1 hour at 95 °C resulted in a soft specimen, which was deformed due to the load and could not be tested. Softness of the specimens was probably due to the incomplete setting reaction and retaining excessive amount of residual monomer, which was retained in the material [26]. Therefore, in order to cure the resin, at least 2 h at 95 °C is required.

In previous studies, no evidence was found to show the effect of elimination of the delay time on the strength of PMMA. However, some of them reported that increasing the curing time from 30 to 60 min not only improves the mechanical properties of the resin [25,27], but also decreases the residual monomer in the cured PMMA. This evidence suggests that raising curing time to 2 h at 95 °C is required to achieve an optimum BFS of PMMA. However, extended curing time does not have any extra benefit. In addition, the curing time could not be extended beyond 2 hours due to the limitation in the processor time adjustability. Hayhurst and Johnson (2004) in a similar study investigated the effect of the type of resin, packing, and processing variables on the BFS of PMMA, and they found no significant improvement effect on the BFS of cured PMMA [28]. The sample size of that study, it should be noted, was small and may not have been enough to show some of the differences.

3. Experimental Section

3.1. Materials

Plain and veined types of PMMA as a cadmium-free co-polymer (Oracryl heat cure denture material, Bracon, Etchingham, West Sussex, UK), base plate wax (Cavex, Cavex Holland BV, 2003 RW Haarlem, The Netherlands), plaster coating solution (Cold Mould Seal, Metrodent Ltd., Paddock, UK), yellow soft paraffin (Ecolab Ltd., Leeds, UK) were used in this study.

3.2. Sample Preparation

Samples were prepared in order to evaluate resin type, processing, and packing factors. In the first stage, five processing variables were considered, including: (1) type of the resin (plain and veined shown by "P" and "V"); (2) five different powder-to-liquid ratios by volume (1.5:1, 2:1, 2.5:1, 3:1, 3.5 represented by numbers 1–5); (3) two packing techniques (conventional and injection shown by "C" and "I") (4) two processing methods (dry heat and water bath presented by "D" and "W"); 5) and finally three different curing times including 4 h at 75 °C and 2 h at 95 °C (as a usual curing time), 8 h at 75 °C and 2 h at 95 °C (as an upper extreme) and 2 h at 95 °C as the lowest curing time. Table 1 is a multifactorial test table designed to evaluate the effect of the mentioned factors.

Table 1. Multifactorial test table was designed to evaluate the effect of five variables on the BFS of PMMA.

Multifactorial test to evaluate the effect of five variables on the BFS of PMMA																							
				A																			
				2 (n = 300)										1 (n = 300)									
				B										**B**									
				1 (n = 150)					2 (n = 300)					1 (n = 150)					2 (n = 150)				
				C (n = 30)					**C** (n = 30)					**C** (n = 30)					**C** (n = 30)				
				1	2	3	4	5	1	2	3	4	5	1	2	3	4	5	1	2	3	4	5
D	1	**E**	1																				
			2																				
			3																				
	2	**E**	1																				
			2																				
			3																				

3.3. Factor Codes

Each factor given a code in this study, including types of resin (Factor A: veined 1, plain 2), packing method (Factor B: conventional 1, injection 2), powder/ liquid ratio (Factor C: (Cell No. 1 (1.50:1), Cell No. 2 (2: 1), Cell No.3 (2.50:1), Cell No. 4 (3:1), Cell No. 5 (3.50:1), processing method (Factor D: water bath 1, dry heat 2) and, finally, curing times (Factor E: a) 2 h at 95 °C; b) 4 h at 75 °C and 2 h at 95 °C; 3) 8 h at 75 °C and 2 h at 95 °C). To assess the fracture resistance of the resin, the biaxial flexural strength (BFS) of the specimens was considered. The performance of the curing devices, changing temperature inside the resin during the curing procedure and the effect of changing extruder angle of injection machine, were also assessed.

3.4. Wax Discs' Preparation

Single discs (Figure 8A) were prepared using a rubber mold; a tree-shaped 15-disc rubber mold was designed and made to prepare wax discs (Figure 8B) for the conventional- and injection-packing methods, respectively. Melted wax was poured into the rubber mold, left to cool, and then removed.

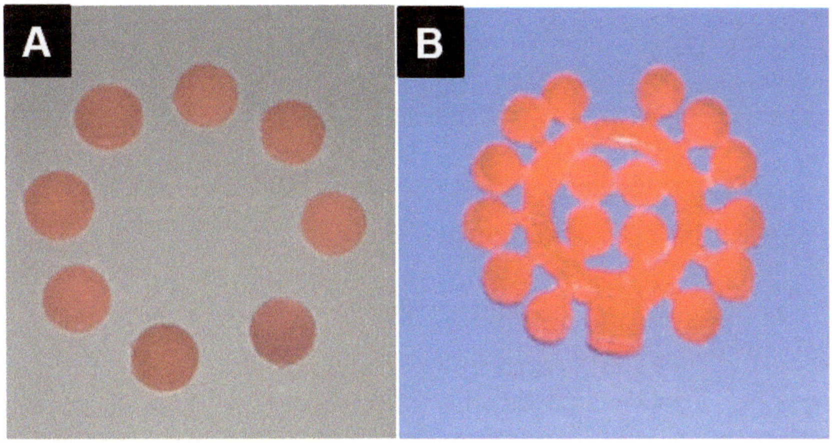

Figure 8. (**A**) Single-wax discs; (**B**) Tree-shaped arranged wax discs.

3.5. Flasking and Packing the Wax Discs

3.5.1. Conventional Technique (Factor B1)

Plaster of Paris was mixed with water and poured inside one half of a two-part brass flask and then 15 prepared wax discs—with a 12 mm diameter and 2 mm thick—were inserted on the surface of the plaster. After initial setting of the plaster, all exposed disc surfaces were cleaned of plaster using water. Once the plaster was set, it was then lubricated with yellow soft paraffin, the flask was then filled with plaster of Paris and turned onto the first half, pushed together, and left to set completely (Figure 9A).

The flask was subsequently placed in the boiling water to remove the wax. After opening, the remaining wax was washed out using boiling water (Figure 9B). The mold surface was dried and painted with plaster coating solution. Two types of denture base acrylic resin powder, PMMA, plain and veined (Factor A), with five powder-to-liquid ratios 1.5:1, 2:1, 2.5:1, 3:1, 3.5:1 (Factor C) were mixed in a rubber bowl and left covered until ready to pack. The mixture was packed in the mold, pressed, and clamped conventionally (Figure 9B).

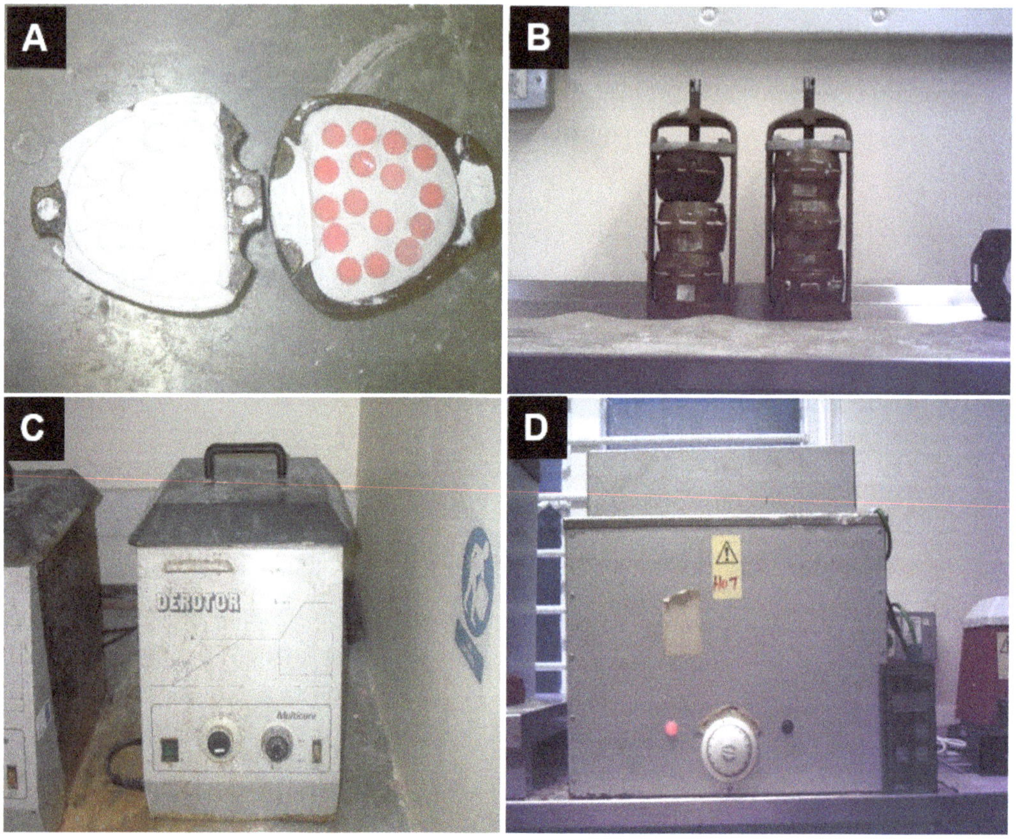

Figure 9. (**A**) Conventional method of flasking discs; (**B**) Conventionally packed and clamped flasks; (**C**) water bath (**D**) dry heat oven.

The clamped flask was put either in the water bath (Derotor water curing bath Quayle Dental Manufacturing CO. Ltd., Worthing, West Sussex, UK) (Figure 9C) or in the dry heat processor (Ditton dry acrylic processor, Chaperlin and Jacobs, Sutton, Surrey) (Figure 9D) (Factor D) and was cured with different curing times: (1) 8 h at 75 °C and 2 h at 95 °C; (2) 4 h at 75 °C and 2 h at 95 °C; (3) 2 h at 95 °C (Factor E).

3.5.2. Injection Technique (Factor B2)

The two halves of the injection flasks were lubricated internally. A plastic investing dummy was located on the opening of the flask (Figure 10A). The rest of the investing was identical to the conventional technique which was described previously. The plastic investing dummy was replaced with the metal injection nozzle and the two halves of the flask screwed together. The acrylic mixture (mixed as described previously for the conventional technique) was placed into a syringe which was installed on the nozzle and fixed in the injection machine (Success Injection System™, Dentsply DeTrey, Dreieich, Germany) (Figure 10B). The resin was injected under an atmosphere pressure of 4 into the flask and the pressure was maintained for 5 min.

After removing the syringe, a pressure-maintaining device was screwed on the nozzle to keep the pressure constant during the processing (Figure 10C). The resin was then processed in both the water and dry heat bath as described previously for the conventional technique.

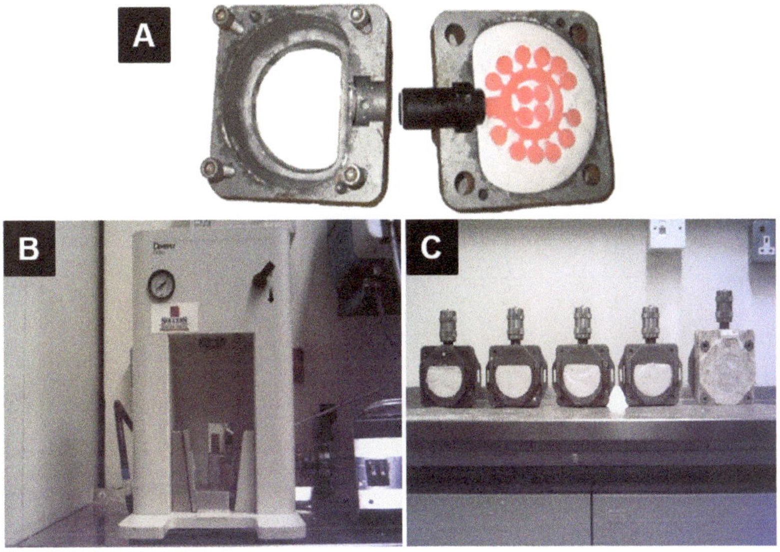

Figure 10. (**A**) Tree-shaped wax pattern invested in the flask with a plastic dummy; (**B**) Success Injection System™ (DENTSPLY International Inc., York, PA, USA); (**C**) Five injection-packed flasks.

3.6. De-Flasking and Preparing the Discs

After processing, the flasks were left to cool at room temperature and then opened up before removing the discs from the plaster. In the case of the injection technique, the discs were cut from the tree-shaped resin template and all excess resin "flash" trimmed using a hand piece and tungsten carbide bur. Finally, the discs were sandpapered with 600 μm silicon carbide sandpaper to achieve a completely flat surface and uniform thickness (±0.2 mm). The finished discs were stored at 37 °C for 48 ± 2 h in tap water before testing to comply with ISO standard (BS EN ISO 1567: 2002).

3.7. Modification of the Injection-Packing Technique

Based on anecdotal evidence, the turbulence effect in the injection-molding system is a common problem in polymer engineering, which has been dealt with by altering the extruder angle. The old prefabricated extruder with a 90° angle was suspected of generating turbulence and stretch of the resin inside the extruder. An extruder with a modified angle of 160° was designed to reduce the probable deteriorating effect of the right angle extruder on the resin. Veined resin was prepared under optimal conditions and packed using the modified extruder. Biaxial flexural strength, Vickers hardness, and micro indentation fracture toughness tests were carried out on the samples, and the results were compared with those achieved using the standard right angle (Figure 11).

Due to the inconsistency of the results produced by the dry heat oven, the performance of the processors was suspected; an assessment of these processors was therefore undertaken. Water bath and dry heat oven were loaded with varying numbers of flasks to see the effect of flask numbers on the temperature and rate of climb. The thermal reaction inside the curing resin was also measured and compared with the internal temperature of curing chambers. This was carried out by drilling a hole in the middle of the flask into which a thermocouple (Jenco Electronics Ltd., Taipei, Taiwan, model 7001H) was placed to be in contact with the curing resin inside the flask (Figure 12A–C).

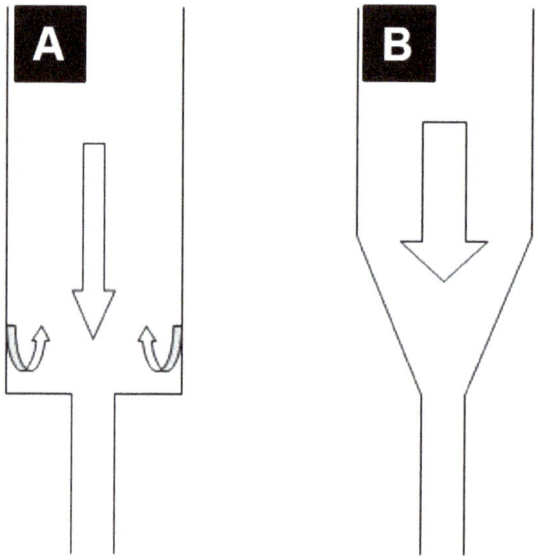

Figure 11. (**A**) Right angle, 90°, extruder; (**B**) modified angle, 160°, extruder.

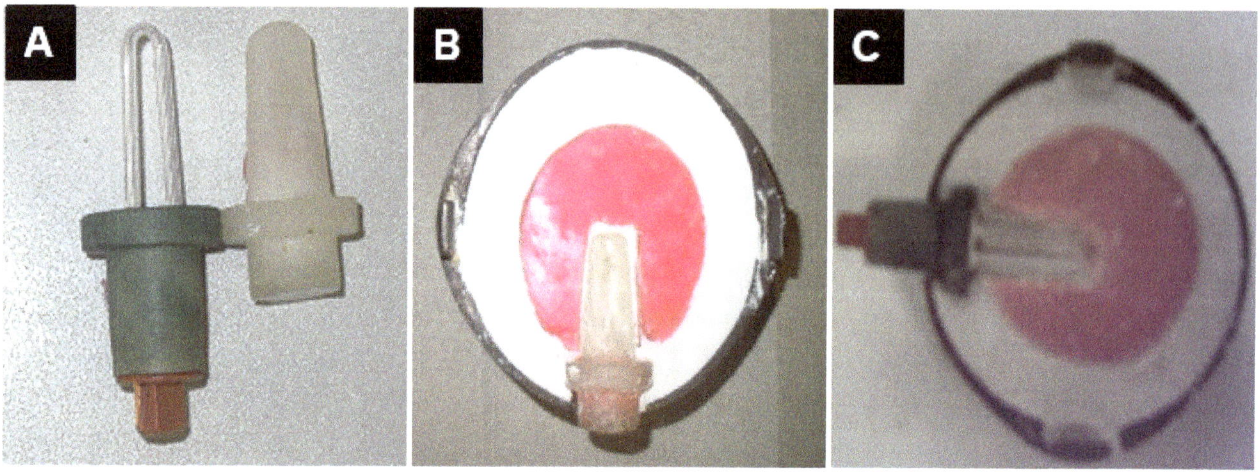

Figure 12. (**A**) Duplicated thermocouple head; (**B**) Investing the duplicated dummy; (**C**) Thermocouple head located inside the resin.

3.8. Evaluation of the Performance of the Processors

During processing, temperature readings were recorded every 15 min. The trend of the climb and hold patterns inside the curing chambers and acrylic resin was recorded using a Microcomputer Thermometer (Figure 13A). To evaluate the temperature within the curing bath, the tip of the thermometer was suspended inside the curing chambers. The temperature was recorded with no flasks in the processors and with varying numbers of flask inside the processors.

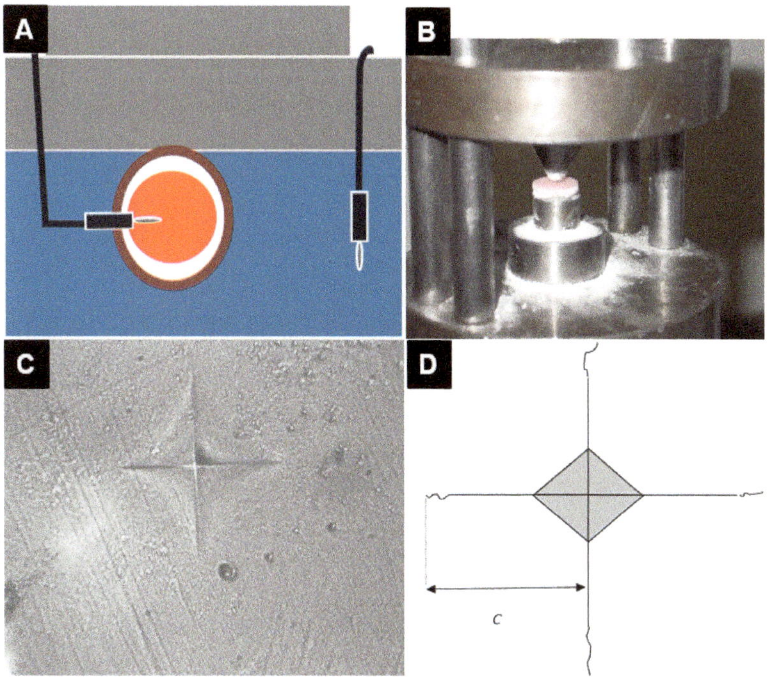

Figure 13. (**A**) Schematic of temperature measurement inside the resin and water bath; (**B**) Biaxial flexural strength (BFS) testing (ball on ring) using a tensile tester; (**C**) Microindentation Vickers hardness tests; (**D**) Crack length, *c*, on microindentation.

3.9. Biaxial Flexural Strength Testing

In this experiment, a sample size of 10 was calculated based on power = 0.85, Sigma = 20, Alpha = 0.05 and Maximum Difference = 30. Ten discs out of fifteen in each batch were selected based on having uniform thickness and no visual defect or porosity. The thickness of each disc was measured at three points with a digital micrometer (Quickmini, Mitutoyo Corp, Kawasaki, Japan) and the average thickness was calculated for each disc (±0.01 mm). A tensile-testing machine (Lloyd 2000R universal testing machine, Lloyd instrument Ltd., Fareham, Hants, UK) was employed to determine BFS of the discs. The discs were centrally placed onto an "O" ring and a ball tip instrument was used to apply load on the central point of the discs at a crosshead speed of 1 mm/min (Figure 13B). The BFS of the specimens was calculated using the following equation (Equation (1)).

$$\sigma_{max} = \frac{P}{h^2}[0.606\log_e(a/h) + 1.13] \quad (1)$$

where σ_{max} is the maximum biaxial flexural strength; *P* is the load to fracture; *a* is the radius of the knife-edged support (O ring) and *h* is the sample thickness [29].

In order to measure the hardness of the composites, two discs were prepared for each composition and sandpapered with 240, 600 and 1200 grades of silicon carbide paper and then polished using 6 and 1 μm polycrystalline diamond abrasives (Buehler) to achieve a well-polished surface and even thickness (±0.1 mm). The discs were stored in water for 24 h at 37 °C. Ten Vickers indentations of 25 gf by microindenter (Mitutoyo hardness testing, Mitutoyo Ltd., Painesville, OH, USA) (Figure 13C) were applied on one of each of the disc surfaces and the hardness of the surfaces was calculated. To determine the microindentation fracture toughness [30], 30 indentations of 500 gf were carried out and the length of the cracks (2c) were measured by a Polyvar camera microscope (Reichert Polyvar Met Microscope, Wien, Austria) (Figure 13D). Using the average length of the radial cracks (*c*) and the following Evans and Charles equation (Equation (2)) [31], the fracture toughness of the composites was calculated:

$$Klc = 0.0752\ P/c^{3/2} \tag{2}$$

3.10. Data Analysis

The data were analyzed using one-way ANOVA and Tukey comparison at the 95% confidence level ($p = 0.05$) (Minitab release 13.1).

4. Conclusions

This research was undertaken for evaluating the effect of type of resin, packing, and processing variables on the BFS of PMMA as a denture base resin. According to the results, the maximum biaxial flexural strength is achieved when the plain type of resin with a powder/liquid ratio of 2.5:1 was packed conventionally and cured in a water bath. The type of resin, packing procedure, and processing variables can have a significant effect on the BFS of PMMA. Based on the data presented, it can be concluded that plain resin is preferred over veined resin; a powder/liquid ratio between 1.5:1 and 2.5:1 can be used; conventional packing is preferred over injection molding; a water bath should be used instead of a dry heat oven; and, finally, a curing time of two hours at 95 °C is the optimum.

Author Contributions

Touraj Nejatian planned the experiment and was responsible for sample preparation and obtaining results; Farshid Sefat was responsible for results development, coordinated all tasks in the paper and finalized all figures and tables for publication; Tony Johnson planned the experiment and supervised the study.

Conflicts of Interest

The authors declare no conflict of interest.

References

1. Jagger, D.C.; Harrison, A.; Jandt, K.D. The reinforcement of dentures. *J. Oral Rehabil.* **1999**, *26*, 185–194.

2. John, J.; Gangadhar, S.A.; Shah, I. Flexural strength of heat-polymerized polymethyl methacrylate denture resin reinforced with glass, aramid, or nylon fibers. *J. Prosthet. Dent.* **2001**, *86*, 424–427.
3. Johnston, E.P.; Smith, D.E. Flexural fatigue of 10 commonly used denture base resins. *J. Prosthet. Dent.* **1981**, *46*, 478–480.
4. Darbar, U.R.; Huggett, R.; Harrison, A. Denture fracture—A survey. *Br. Dent. J.* **1994**, *176*, 342–345.
5. Vallittu, P.K.; Alakuijala, P.; Lassila, V.P.; Lappalainen, V. *In vitro* fatigue fracture of an acrylic resin-based partial denture: An exploratory study. *J. Prosthet. Dent.* **1994**, *72*, 289–295.
6. Vallittu, P.K.; Lassila, V.P.; Lappalainen Niom, R. The effect of notch shape and self-cured acrylic resin repair on the fatigue resistance of an acrylic resin denture base. *J. Oral Rehabil.* **1996**, *23*, 108–113.
7. Ganzarolli, S.M.; de Mello, J.A.; Shinkai, R.S.; del Bel Cury, A.A. Internal adaptation and some physical properties of methacrylate-based denture base resins polymerized by different techniques. *J. Biomed. Mater. Res. Part B Appl. Biomater.* **2007**, *82*, 169–173.
8. Lamb, D.J.; Ellis, B.; van Noort, R. The fracture topography of acrylic dentures broken in service. *Biomaterials* **1985**, *6*, 110–112.
9. Kim, S.H.; Watts, D.C. The effect of reinforcement with woven E-glass fibers on the impact strength of complete dentures fabricated with high-impact acrylic resin. *J. Prosthet. Dent.* **2004**, *91*, 274–280.
10. Nejatian, T.; Johnson, A.; van Noort, R. Reinforcement of denture base resin. *Adv. Sci. Technol.* **2006**, *49*, 124–129.
11. Hedzelek, W.; Gajdus, P. Comparison of mechanical strength of palatal denture bases made from various plastic materials. *Int. J. Prosthodont.* **2006**, *19*, 193–194.
12. Stafford, G.D.; Huggett, R.; MacGregor, A.R.; Graham, J. The use of nylon as a denture-base material. *J. Dent.* **1986**, *14*, 18–22.
13. Yunus, N.; Rashid, A.A.; Azmi, L.L.; Abu-Hassan, M.I. Some flexural properties of a nylon denture base polymer. *J. Oral Rehabil.* **2005**, *32*, 65–71.
14. Radford, D.R.; Juszczyk, A.S.; Clark, R.K.F. The bond between acrylic resin denture teeth and the denture base: Recommendations for best practice. *Br. Dent. J.* **2014**, *216*, 165–167.
15. Praveen, B.; Babaji, H.V.; Prasanna, B.G.; Rajalbandi, S.K.; Shreeharsha, T.V.; Prashant, G.M. Comparison of Impact Strength and Fracture Morphology of Different Heat Cure Denture Acrylic Resins: An *In vitro* Study. *J. Int. Oral Health* **2014**, *6*, 12–16.
16. Kanchanavasita, W.; Anstice, H.M.; Pearson, G.J. Long-term flexural strengths of resin-modified glass-ionomer cements. *Biomaterials* **1998**, *19*, 1703–1713.
17. Higgs, W.A.J.; Lucksanasombool, P.; Higgs, R.J.E.D.; Swain, M.V. Evaluating acrylic and glass-ionomer cement strength using the biaxial flexure test. *Biomaterials* **2001**, *22*, 1583–1590.
18. Wagner, W.C.; Chu, T.M. Biaxial flexural strength and indentation fracture toughness of three new dental core ceramics. *J. Prosthet. Dent.* **1996**, *76*, 140–144.
19. Wen, M.Y.; Mueller, H.J.; Chai, J.; Wozniak, W.T. Comparative mechanical property characterization of 3 all-ceramic core materials. *Int. J. Prosthodont.* **1999**, *12*, 534–541.
20. Yilmaz, H.; Aydin, C.; Gul, B.E. Flexural strength and fracture toughness of dental core ceramics. *J. Prosthet. Dent.* **2007**, *98*, 120–128.

21. Pick, B.; Meira, J.B.C.; Driemeier, L.; Braga, R.R. A critical view on biaxial and short-beam uniaxial flexural strength tests applied to resin composites using Weibull, fractographic and finite element analyses. *Dent. Mater.* **2010**, *26*, 83–90.
22. O'Brien, W.J. *Dental Materials and Their Selection*; Quintessence Pub. Co.: Chicago, IL, USA, 2002.
23. Uzun, G.; Hersek, N. Comparison of the fracture resistance of six denture base acrylic resins. *J. Biomater. Appl.* **2002**, *17*, 19–28.
24. Gharechahi, J.; Asadzadeh, N.; Shahabian, F.; Gharechahi, M. Flexural strength of acrylic resin denture bases processed by two different methods. *J. Dent. Res. Dent. Clin. Dent. Prospect.* **2014**, *8*, 148–152.
25. Doğan, A.; Bek, B.; Çevik, N.N.; Usanmaz, A. The effect of preparation conditions of acrylic denture base materials on the level of residual monomer, mechanical properties and water absorption. *J. Dent.* **1995**, *23*, 313–318.
26. Honorez, P.; Catalan, A.; Angnes, U.; Grimonster, J. The effect of three processing cycles on some physical and chemical properties of a heat-cured acrylic resin. *J. Prosthet. Dent.* **1989**, *61*, 510–517.
27. Wright, D.D.; Lautenschlager, E.P. The effect of processing temperature and time on the structure and fracture characteristics of self-reinforced composite poly(methyl methacrylate). *J. Mater. Sci. Mater. Med.* **1999**, *10*, 503–512.
28. Hayhurst, L.; Johnson, A. The effect of resin type, packing and processing methods on the Biaxial Flexural strength of polymethyl methacrylate denture base materials. *Quintessence J. Dent. Technol.* **2004**, *2*, 414–420.
29. Piddock, V.; Marquis, P.M.; Wilson, H.J. Comparison of the strengths of aluminous porcelain fired on to platinum and palladium foils. *J. Oral Rehabil.* **1986**, *13*, 31–37.
30. Tan, F.; Qiao, X.L.; Chen, J.G.; Wang, H.S. Effects of coupling agents on the properties of epoxy-based electrically conductive adhesives. *Int. J. Adhes. Adhes.* **2005**, *26*, 406–413.
31. Li, A.G.; Kobayashi, A.S.; Bradt, R.C. Indentation fracture toughness of sintered silicon carbide in the palmqvist crack regime. *J. Am. Ceram. Soc.* **1989**, *72*, 904–911.

Permissions

All chapters in this book were first published in Materials, by MDPI; hereby published with permission under the Creative Commons Attribution License or equivalent. Every chapter published in this book has been scrutinized by our experts. Their significance has been extensively debated. The topics covered herein carry significant findings which will fuel the growth of the discipline. They may even be implemented as practical applications or may be referred to as a beginning point for another development.

The contributors of this book come from diverse backgrounds, making this book a truly international effort. This book will bring forth new frontiers with its revolutionizing research information and detailed analysis of the nascent developments around the world.

We would like to thank all the contributing authors for lending their expertise to make the book truly unique. They have played a crucial role in the development of this book. Without their invaluable contributions this book wouldn't have been possible. They have made vital efforts to compile up to date information on the varied aspects of this subject to make this book a valuable addition to the collection of many professionals and students.

This book was conceptualized with the vision of imparting up-to-date information and advanced data in this field. To ensure the same, a matchless editorial board was set up. Every individual on the board went through rigorous rounds of assessment to prove their worth. After which they invested a large part of their time researching and compiling the most relevant data for our readers.

The editorial board has been involved in producing this book since its inception. They have spent rigorous hours researching and exploring the diverse topics which have resulted in the successful publishing of this book. They have passed on their knowledge of decades through this book. To expedite this challenging task, the publisher supported the team at every step. A small team of assistant editors was also appointed to further simplify the editing procedure and attain best results for the readers.

Apart from the editorial board, the designing team has also invested a significant amount of their time in understanding the subject and creating the most relevant covers. They scrutinized every image to scout for the most suitable representation of the subject and create an appropriate cover for the book.

The publishing team has been an ardent support to the editorial, designing and production team. Their endless efforts to recruit the best for this project, has resulted in the accomplishment of this book. They are a veteran in the field of academics and their pool of knowledge is as vast as their experience in printing. Their expertise and guidance has proved useful at every step. Their uncompromising quality standards have made this book an exceptional effort. Their encouragement from time to time has been an inspiration for everyone.

The publisher and the editorial board hope that this book will prove to be a valuable piece of knowledge for researchers, students, practitioners and scholars across the globe.

List of Contributors

Xiyun Zhao
College of Life Science, Northeast Forestry University, Harbin 150040, China

Jie Zhang
College of Life Science, Northeast Forestry University, Harbin 150040, China

Binsong Wang
Key Laboratory of Chemical Engineering Process & Technology for High-Efficiency Conversion,
School of Chemistry and Material Sciences, Heilongjiang University, Harbin 150080, China

Amir Zada
Department of Chemistry, Abdul Wali Khan University Mardan, KPK, Mardan 23200, Pakistan

Muhammad Humayun
Department of Chemistry, Abdul Wali Khan University Mardan, KPK, Mardan 23200, Pakistan

Mira Moussa
Division of Fixed Prosthodontics and Biomaterials, University of Geneva, University Clinics of Dental Medicine, 19, rue Barthélemy-Menn, Geneva 1205, Switzerland

Jean-Pierre Carrel
Department of Maxillofacial and Oral Surgery, Division of Oral and Maxillofacial Pathology (HUG), University Clinics of Dental Medicine, 19, rue Barthélemy-Menn, Geneva 1205, Switzerland

Susanne Scherrer
Division of Fixed Prosthodontics and Biomaterials, University of Geneva, University Clinics of Dental Medicine, 19, rue Barthélemy-Menn, Geneva 1205, Switzerland

Maria Cattani-Lorente
Division of Fixed Prosthodontics and Biomaterials, University of Geneva, University Clinics of Dental Medicine, 19, rue Barthélemy-Menn, Geneva 1205, Switzerland

Anselm Wiskott
Division of Fixed Prosthodontics and Biomaterials, University of Geneva, University Clinics of Dental Medicine, 19, rue Barthélemy-Menn, Geneva 1205, Switzerland

Stéphane Durual
Division of Fixed Prosthodontics and Biomaterials, University of Geneva, University Clinics of Dental Medicine, 19, rue Barthélemy-Menn, Geneva 1205, Switzerland

Eisner Salamanca
School of Dentistry, College of Oral Medicine, Taipei Medical University, 250 Wu-Hsing Street, Taipei 110, Taiwan

Wei-Fang Lee
School of Dental Technology, College of Oral Medicine, Taipei Medical University, 250 Wu-Hsing Street, Taipei 110, Taiwan

Chin-Yi Lin
School of Dentistry, College of Oral Medicine, Taipei Medical University, 250 Wu-Hsing Street, Taipei 110, Taiwan

Haw-Ming Huang
Graduate Institute of Biomedical Materials & Tissue Engineering, College of Oral Medicine,
Taipei Medical University, 250 Wu-Hsing Street, Taipei 110, Taiwan

Che-Tong Lin
School of Dentistry, College of Oral Medicine, Taipei Medical University, 250 Wu-Hsing Street, Taipei 110, Taiwan

Sheng-Wei Feng
School of Dentistry, College of Oral Medicine, Taipei Medical University, 250 Wu-Hsing Street, Taipei 110, Taiwan

Wei-Jen Chang
School of Dentistry, College of Oral Medicine, Taipei Medical University, 250 Wu-Hsing Street, Taipei 110, Taiwan
Dental Department of Taipei Medical University, Shuang-Ho Hospital, Taipei 110, Taiwan

Saad Liaqat
Division of Biomaterials and Tissue Engineering, UCL Eastman Dental Institute, 256 Gray's Inn Road, London WC1X 8LD, UK

Anas Aljabo
Division of Biomaterials and Tissue Engineering, UCL Eastman Dental Institute, 256 Gray's Inn Road, London WC1X 8LD, UK

Muhammad Adnan Khan
Division of Biomaterials and Tissue Engineering, UCL Eastman Dental Institute, 256 Gray's Inn Road, London WC1X 8LD, UK

Hesham Ben Nuba
Division of Biomaterials and Tissue Engineering, UCL Eastman Dental Institute, 256 Gray's Inn Road, London WC1X 8LD, UK

Laurent Bozec
Division of Biomaterials and Tissue Engineering, UCL Eastman Dental Institute, 256 Gray's Inn Road, London WC1X 8LD, UK

Paul Ashley
Department of Paediatric Dentistry, UCL Eastman Dental Institute, 256 Gray's Inn Road, London WC1X 8LD, UK

Anne Young
Division of Biomaterials and Tissue Engineering, UCL Eastman Dental Institute, 256 Gray's Inn Road, London WC1X 8LD, UK

Manfred M. Maurer
Institute of Mechanical Systems, ETH Zurich, Leonhardstrasse 21, Zurich 8092, Switzerland

Barbara Röhrnbauer
Institute of Mechanical Systems, ETH Zurich, Leonhardstrasse 21, Zurich 8092, Switzerland

Andrew Feola
Center for Surgical Technologies, Faculty of Medicine, Universitair Ziekenhuis "Gasthuisberg"
Leuven, Katholieke Universiteit Leuven, Leuven 3000, Belgium

Jan Deprest
Center for Surgical Technologies, Faculty of Medicine, Universitair Ziekenhuis "Gasthuisberg"
Leuven, Katholieke Universiteit Leuven, Leuven 3000, Belgium

Edoardo Mazza
Institute of Mechanical Systems, ETH Zurich, Leonhardstrasse 21, Zurich 8092, Switzerland
Empa—Swiss Federal Laboratories for Materials Science and Technology, Überlandstrasse 129, Dübendorf 8600, Switzerland

Sebastian Wille
Department of Prosthodontics, Propaedeutics and Dental Materials, School of Dentistry, Christian-Albrechts University, Arnold-Heller-Straße 16, Kiel D-24105, Germany

Vincent van Broeck
Chair for Multicomponent Materials, Institute for Materials Science, Faculty of Engineering,
Christian-Albrechts University at Kiel, Kaiserstr. 2, Kiel D-24143, Germany

Thomas Strunskus
Chair for Multicomponent Materials, Institute for Materials Science, Faculty of Engineering,
Christian-Albrechts University at Kiel, Kaiserstr. 2, Kiel D-24143, Germany

Franz Faupel
Chair for Multicomponent Materials, Institute for Materials Science, Faculty of Engineering,
Christian-Albrechts University at Kiel, Kaiserstr. 2, Kiel D-24143, Germany

Matthias Kern
Department of Prosthodontics, Propaedeutics and Dental Materials, School of Dentistry, Christian-Albrechts University, Arnold-Heller-Straße 16, Kiel D-24105, Germany

Muhammad Kaleem
Army Medical College, National University of Sciences and Technology, Islamabad 44000, Pakistan

Abdul S. Khan
Interdisciplinary Research Centre in Biomedical Materials, COMSATS Institute of Information Technology, Lahore 54000, Pakistan
Centre for Oral Growth and Development, Institute of Dentistry, Barts and the London School of Medicine and Dentistry, Queen Mary University of London, London E1 2AD, UK

Ihtesham U. Rehman
Kroto Research Institute, Department of Materials Science and Engineering, University of Sheffield, Sheffield S3 7HQ, UK

Ferranti S. Wong
Centre for Oral Growth and Development, Institute of Dentistry, Barts and the London School of Medicine and Dentistry, Queen Mary University of London, London E1 2AD, UK

Bianca Posocco
Centro di Riferimento Oncologico, Via Franco Gallini 2, I-33081 Aviano (PN), Italy
Department of Engineering and Architecture, University of Trieste, Via Alfonso Valerio, 6/A, I-34127 Trieste, Italy

Eva Dreussi
Centro di Riferimento Oncologico, Via Franco Gallini 2, I-33081 Aviano (PN), Italy
Department of Engineering and Architecture, University of Trieste, Via Alfonso Valerio, 6/A, I-34127 Trieste, Italy

Jacopo de Santa
Centro di Riferimento Oncologico, Via Franco Gallini 2, I-33081 Aviano (PN), Italy
Department of Engineering and Architecture, University of Trieste, Via Alfonso Valerio, 6/A, I-34127 Trieste, Italy

Giuseppe Toffoli
Centro di Riferimento Oncologico, Via Franco Gallini 2, I-33081 Aviano (PN), Italy

Michela Abrami
Department of Life Sciences, Cattinara University Hospital, Trieste University, Strada di Fiume 447, I-34149 Trieste, Italy

Francesco Musiani
Laboratory of Bioinorganic Chemistry, Department of Pharmacy and Biotechnology, University of Bologna, I-40127 Bologna, Italy

Mario Grassi
Department of Engineering and Architecture, University of Trieste, Via Alfonso Valerio, 6/A, I-34127 Trieste, Italy

Rossella Farra
Department of Engineering and Architecture, University of Trieste, Via Alfonso Valerio, 6/A,\ I-34127 Trieste, Italy

Federica Tonon
Department of Engineering and Architecture, University of Trieste, Via Alfonso Valerio, 6/A, I-34127 Trieste, Italy

Gabriele Grassi
Department of Life Sciences, Cattinara University Hospital, Trieste University, Strada di Fiume 447, I-34149 Trieste, Italy

Barbara Dapas
Department of Life Sciences, Cattinara University Hospital, Trieste University, Strada di Fiume 447, I-34149 Trieste, Italy

Marit Øilo
Department of Clinical Dentistry, Faculty of Medicine and Dentistry, University of Bergen, Aarstadveien 19, Bergen NO-5009, Norway

Vidar Bakken
Department of Clinical Science, Faculty of Medicine and Dentistry, University of Bergen, Jonas Lies vei 65, The Laboratory Building, Bergen NO-5021, Norway

Mohamed A. Hussein
Department of Mechanical Engineering, King Fahd University of Petroleum & Minerals, Dhahran 31261, Saudi Arabia
Department of Mechanical Engineering, Kafrelsheikh University, Kafrelsheikh 33516, Egypt

Abdul Samad Mohammed
Department of Mechanical Engineering, King Fahd University of Petroleum & Minerals, Dhahran 31261, Saudi Arabia

Naser Al-Aqeeli
Department of Mechanical Engineering, King Fahd University of Petroleum & Minerals, Dhahran 31261, Saudi Arabia

Giacomo Derchi
Department of Surgical, Medical, Molecular and Critical-Area Pathology, University of Pisa, via Paradisa 2, I-56124 Pisa, Italy
Tuscan Stomatological Institute, via Aurelia 335, Lido di Camaiore, I-55041 Lucca, Italy

Enrico Manca
Tuscan Stomatological Institute, via Aurelia 335, Lido di Camaiore, I-55041 Lucca, Italy

Amirreza Shayganpour
Nanophysics Department, Istituto Italiano di Tecnologia, via Morego 30, I-16163 Genova, Italy
University of Genova, viale Causa 13, I-16145 Genova, Italy

Antonio Barone
Department of Surgical, Medical, Molecular and Critical-Area Pathology, University of Pisa, via Paradisa 2, I-56124 Pisa, Italy

Alberto Diaspro
Nanophysics Department, Istituto Italiano di Tecnologia, via Morego 30, I-16163 Genova, Italy
University of Genova, viale Causa 13, I-16145 Genova, Italy

Marco Salerno
Nanophysics Department, Istituto Italiano di Tecnologia, via Morego 30, I-16163 Genova, Italy

Chris Ying Cao
Faculty of Dentistry, the University of Hong Kong, Hong Kong, China

May Lei Mei
Faculty of Dentistry, the University of Hong Kong, Hong Kong, China

Quan-li Li
Stomatological Hospital and College, Anhui Medical University, Key Lab. of Oral Diseases Research of Anhui Province, Hefei 230032, China

Edward Chin Man Lo
Faculty of Dentistry, the University of Hong Kong, Hong Kong, China

Chun Hung Chu
Faculty of Dentistry, the University of Hong Kong, Hong Kong, China

Martha Geffers
Department for Functional Materials in Medicine and Dentistry, University Hospital Würzburg, Pleicherwall 2, D-97070 Würzburg, Germany

Jürgen Groll
Department for Functional Materials in Medicine and Dentistry, University Hospital Würzburg, Pleicherwall 2, D-97070 Würzburg, Germany

Uwe Gbureck
Department for Functional Materials in Medicine and Dentistry, University Hospital Würzburg, Pleicherwall 2, D-97070 Würzburg, Germany

Benjamin Weed
Department of Agricultural and Biological Engineering, Mississippi State University, Starkville, MS 39762, USA

Sourav Patnaik
Department of Agricultural and Biological Engineering, Mississippi State University, Starkville, MS 39762, USA

Mary Rougeau-Browning
Department of Genetics, University of Georgia, Athens, GA 30602, USA

Bryn Brazile
Department of Agricultural and Biological Engineering, Mississippi State University, Starkville, MS 39762, USA

Jun Liao
Department of Agricultural and Biological Engineering, Mississippi State University, Starkville, MS 39762, USA

Raj Prabhu
Department of Agricultural and Biological Engineering, Mississippi State University, Starkville, MS 39762, USA

Lakiesha N. Williams
Department of Agricultural and Biological Engineering, Mississippi State University, Starkville, MS 39762, USA

Touraj Nejatian
Dental Materials Unit, School of Dental Sciences, University of Newcastle, Newcastle upon Tyne NE2 4BW, UK
School of Clinical Dentistry, University of Sheffield, Sheffield, South Yorkshire S10 2TA, UK

Farshid Sefat
Department of Biomedical Engineering, School of Engineering, King Faisal University, Al-Hofuf, Al-Ahsa 31982, Saudi Arabia
Tissue Engineering Group, Department of Chemistry, Chemical Biology and Biomedical Engineering, Stevens Institute of Technology, Hoboken, NJ 07030, USA

Tony Johnson
School of Clinical Dentistry, University of Sheffield, Sheffield, South Yorkshire S10 2TA, UK